Flexible Automation in Japan

John Hartley

IFS (Publications) Ltd., UK

Springer-Verlag
Berlin Heidelberg New York Tokyo
1984

British Library Cataloguing Publication Data

Hartley, John, 19 - - -
Flexible automation in Japan.
1. Automation – Japan
I. Title
670.42′7′0952 T59.5

ISBN 0-903608-64-2 IFS (Publications) Ltd.
ISBN 3-540-13499-9 Springer-Verlag Berlin Heidelberg New York Tokyo
ISBN 0-387-13499-9 New York Heidelberg Berlin Tokyo

Typesetting by Fotographics (Bedford) Ltd., UK
Printed by Butler and Tanner Ltd., Frome, UK

Foreword

Much has been said and written about Japan's manufacturing prowess. Most of the comment comes from people who are merely visitors to the country and can be best classified as 'observers looking in from the outside'. Other views come from the Japanese themselves in which the double barrier of culture and language filters out much information that would be of real value to Western industrialists.

Neither of these limitations apply to John Hartley, who has been resident in Japan for the past five years. He understands the culture, can speak the language and has extensive contacts at the highest level. Therefore, he is in a unique position to report on the Japanese scene and its activities in advanced manufacturing technology. This he has been doing on a regular basis to IFS magazines: *The Industrial Robot, Assembly Automation, Sensor Review* and *The FMS Magazine.*

Most of the material in this book is from John Hartley's 'pen' and represents his most significant contributions on flexible automation in Japan to these journals over the last three years. It is augmented with a few other articles written by leading authorities on new technology in Japanese manufacturing industry.

A valuable contribution to the book is a compilation of the specifications of all known Japanese robots presently being manufactured . This information is compiled by IFS as a result of its close relationship with Japan, not only through John Hartley but also through its own direct contacts with Japanese robot manufacturers and users. The robot manufacturers are also indexed to provide a valuable cross reference of suppliers and agents of their equipment in the USA, the UK and other European countries.

I believe that this book is unique in its concept and in its contents. It should assist manufacturing executives and managers, production engineers and those engaged in research to more fully understand 'what makes Japan tick', and to assist them in their modernisation programmes and in making their companies more competitive.

BRIAN ROOKS

Editor of *Assembly Automation* and *The Industrial Robot*

Contents

All articles unless indicated otherwise are written by John Hartley

1. The Japanese Scene

Statistics and reality are often in conflict, and nowhere is the contrast more marked than in the world of robotics in Japan. In the 1970s, statistics showed that Japan was using robots at a prodigious pace, yet visitors had difficulty in finding more than a few. Moreover, when they did find robots, the applications were unimaginative.

More recently, though, the Japanese have been obliged to turn to flexible automation because their products, especially in the consumer electronics and automotive industries have such short lives. In one case, on average, the model is changed significantly every six months, so that it is impractical to completely retool for the new product. The answer was *flexible automation.*

In the first article in this book an attempt is made to reconcile the conflict between those early statistics and the reality of how flexible automation has developed while Kanji Yonemoto's article shows just how the use of automated handling equipment including robots has increased sharply over the past decade, and is set to increase at a fast pace in the future. Fully fledged robots are not needed to give flexibility in all areas of manufacture, though, and in many assembly lines simple pick-and-place devices are being used to allow models to be changed at low cost.

Professor Hiroshi Makino, the 'father' of the SCARA robot, the only robot type to actually originate in Japan, is a well-known expert on assembly automation. He developed a number of cam designs many years ago, and also carried out a thorough evaluation of design for assembly over ten years ago.

It was from that work that the concept of SCARA evolved, and since then he has proposed a system for evaluating assemblies for different types of assembly methods – fixed automation, a mixture of hard automation and robots, and robots only. In the interview, Professor Makino explains his philosophy towards automation. He explains that 70–80% of assembly jobs can easily be automated, and that simple robots are ideal for much of this work.

Katuo Aoki was the leading light in the programme that brought automation to Nippondenso. It is true that in that period, Nippondenso was assured of growth as the Japanese auto industry expanded at a fast rate. Nevertheless, in those years, and when the situation changed in the early 1980s, Nippondenso installed some excellent examples of hard automation, and then some interesting examples of flexible automation.

Japan is unusual in that many houses are prefabricated, while many large buildings are produced as structures which are subsequently clad. Thus, its housing industry is a suitable area for robotics, as Professor Hasegawa shows in his forward-looking article.

1.1 An introduction to Japan

Are there really so many robots in Japan?

The Japanese have adopted robots as workmates but will their use continue to expand?

ENGINEERS visiting Japan in search of robots have usually been perplexed. On the one hand, the statistics and filmshows have indicated that there is an enormous number of robots in Japan, and that the Japanese have a tremendous enthusiasm for robots. But on the other hand, visitors have generally been disappointed in the lack of robots they have seen in use in factories.

At the same time, many Japanese production engineers seem to be just as sceptical of robots as do people elsewhere. And they explain that they resist using more robots because they are too slow, too expensive, or too unreliable.

So what is the reality of Japan's enthusiasm for robots? First, it has to be remembered that the Japanese robot population appears to be so large because almost any form of automated handling device is counted as a robot under the Japanese definition of these machines. Secondly, the Japanese media always use these statistics but report on NC or playback robots as if the statistics refer to these types of robots only. Finally, almost any robot with some form of sensor is called an 'intelligent robot'.

Robot fever

However, for all that, the Japanese did begin installing robots in some numbers in the mid-1970s – mainly in car factories where the aim was to reduce arduous manual work. These installations incorporated no new ideas, though they encouraged many small companies to look into robotics.

By the late 1970s the need for greater flexibility was well understood, and companies like Toyota with a conservative background were realising they must use more robots to increase flexibility. This was because following the energy crisis the growth in car sales had become less predictable, with some models needing to be redesigned much earlier than expected. Not surprisingly the car companies set about installing robots in earnest. And the media reckoned that robot fever had set in.

There is no doubt that in many companies this was the case. For example, one of the Japanese delegates to the Robot Symposium in Tokyo in 1981 said that he had been sent by his company president to find applications for robots in his factories producing grinding wheels. Another production engineer, this time in the auto industry, who had visited the Yamaha factories said: 'The engineers told me that they had been instructed by their president to use robots, but they were having difficulty in finding many good applications in assembly.'

Clearly, there are many companies in Japan whose presidents and employees think that they need robots to compete. And in Japan competition is very severe, whether the products are small portable radios or large construction machines.

But people making robots and responsible for the production engineering in large plants take a more sanguine view, Seiuemon Inaba, president of Fanuc, is one of the leading proponents of robots.

In 1982 he opened a factory to produce servo motors containing 101 robots and staffed by only 60 people. Yet, commenting on 'robot fever', he said: 'The actual situation is not as it is played up by the media. There are many problems to be solved before robots can be used widely –technical ones, cost, and their effect on employment. We do not expect a rapid increase in sales.'

Equally keen to refute the impression that robots are the only answer is Minoru Morita, managing director, production engineering, Matsushita Electric. Believe it or not, one of his reasons is the short model life in consumer electronics, which he says is no more than four months with many products!

'We prefer sequence-controlled arms to robots,' he said, 'because they are more reliable and they are also flexible to some extent. We do use NC controlled devices to insert components in printed circuit boards, and these could be called robots. Their individual downtime is about 0.01%, so we are able to operate the line with 99.5% uptime.'

Fig. 1. Body welding was one of the first applications in Japan.

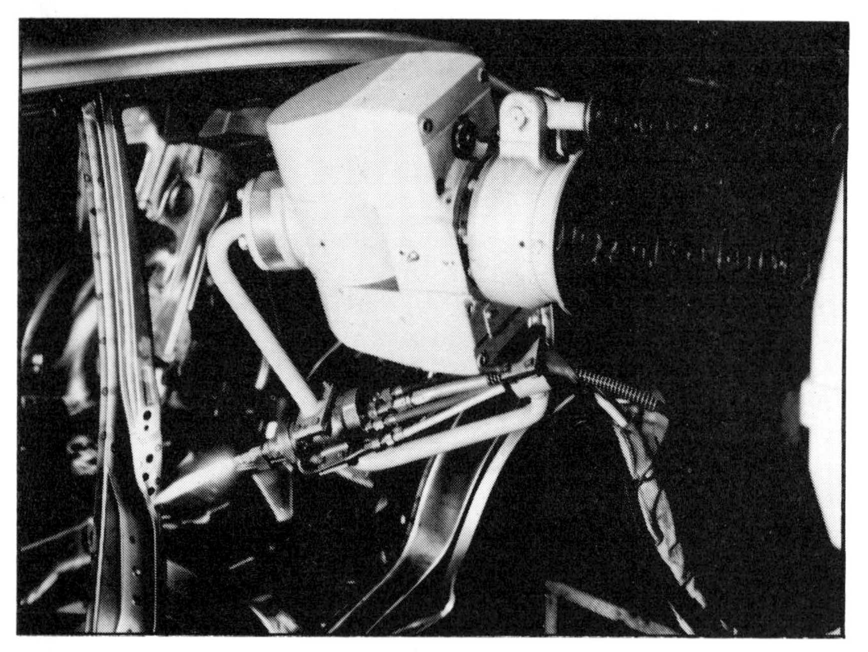

Fig. 2. Gas welding is becoming more common as sensors improve.

Morita doubts whether this would be possible with fully-fledged robots having five or six axes of movement. For this reason he is looking for simpler and cheaper robots to work in his factories and, until these are available, he will concentrate or the use of simple arms.

But it is still aquestion of horses for courses, with Matsushita developing a whole range of robots for welding, inspection and assembly. Fanuc, Mitsubishi Electric and Hitachi are following a similar path.

Although many people may now be trying to play down Japan's 'robot fever', it is certainly true the Japanese have been much more enthusiastic about the use of robots than Europeans or Americans. The Japanese believe European and American trade unionists will not accept robots in factories, arguing that this explains why there are more robots in Japan than elsewhere. But the reality is different.

When the Japanese car companies installed their first robots they gained a distinct advantage in productivity and quality. And many other people assumed therefore that robots were the road to improved productivity.

Many robots were built and installed to work where productivity gains were marginal, For example, Shin Meiwa and Kobe Steel used their own robots to arc weld without an apparent reduction in manning levels. Daido Steel sold Nissan Shatai a pair of robots for handling at presses where conventional automated hands would have done the job equally well at far less cost.

Flexibility and productivity

The point is that in Japan the philosophy is to install the robot first, and then make it productive later.

However, we are moving now into a new era where the combination of the need for flexibility, and high productivity is making robots more attractive. And because competition in Japan is so fierce it follows that many small companies will install robots simply to keep up.

But large companies are increasing their robot use also in areas where the advantages must be marginal.

For example, in the car industry, the spotlight is now on robot painting and robot assembly. Both Nissan and Toyo Kogyo have installed robots for spray painting, but in both cases there are doubts about the economics.

For example, Toyo Kogyo has installed 20 Tokico robots in its new paint plant at Hofu. Four of these are used to apply sealer along welding seams, while three pairs of robots are used to apply pvc coating, underseal and anti-chipping compound. These can be justified easily enough because the work is dirty, and the operator is involved in a lot of bending and crouching – in other words, the operators cannot work efficiently for a full shift.

But there are also 10 robots to apply primer/surfacer and colour coats. With such a number of robots, it would be expected that they might apply most of the paint. But not a bit of it; the bulk of the painting is still done by conventional reciprocators, leaving the robots to do the work that cannot be done by normal automation.

In fact, five people are still needed in each colour coat booth to spray inaccessible areas. Outside Japan, it would be difficult to justify the use of robots in such a situation. Management would normally take the view that if men were needed anyway, it would be better not to use the robots.

At Toyo Kogyo employees are aware that the company must improve productivity to match that of Nissan and Toyota, both of which have greater economies of scale. In addition, of course, it was agreed from the outset that the robots would be used to improve quality, and every effort has been made to ensure that this happens.

In fact, the reject rate on painted bodies in the old factory is 10%; while in the new plant it is only 5%, and the target is 2%. Of course, rectification of reject paintwork is very costly, and its reduction can easily justify the robots.

This is a typical attitude in Japan: first, the robots are installed because productivity must be increased; secondly, targets are set to make sure

Fig. 3. Handling at machine tools is an area for big increases.

Fig. 4. Robots are coming into use for spray painting, despite marginal economies.

that the robots pay.

Raising productivity is seen in Japan as vital, and managers are often prepared to pay for automation when the actual return on investment seems low. Managers are helped by low interest rates and the fact that robots can be leased at low cost. They do not need to obtain such a high return as European managers to justify the investment.

Equally, there is no doubt that the Japanese are far more susceptible to crazes and trends in manufacturing than elsewhere.

In Europe managers often seem keener to show that a certain piece of equipment will not pay than to find a way of making it pay. In Japan, people will often buy one or two robots with little or no idea of payback.

Economics

Even if the robot is not really economic it will be used and investment costs put down to development. And the employees in the shop will be encouraged to come up with ideas to improve the productivity of the robot. If the robot does pay, then more and more robots will be introduced.

Recently, a turning point was reached in robot use at NC machine tools. Until recently few companies were using robots at machine tools, but now 'FMS fever' seems to be developing in Japan, with small companies wanting to buy FMS rather than an NC machine.

Now that Yamazaki and others have shown the advantages of using robots at machine tools a trend has been set, and use will accelerate. In many cases, the machine-with-robot will be installed because that is the trend, rather than because there is mathematical evidence of a big increase in productivity.

Soon a similar situation will develop in assembly, not least because assembly robots are becoming cheaper. The Scara breed is priced at around £10,000-£15,000 while Hitachi is introducing a simple assembly robot priced at about £2,500. By 1984 it is almost certain that many other companies will be selling small simple robots for around £2,500. Production targets will have an aggregate of thousands rather than hundreds of machines a month.

Once a major manufacturer starts to install such robots in volume, then a whole host of companies will follow.

So why did the Fanuc assembly factory not inspire that breakthrough? In the first place, it is recognised that Fanuc can well afford to equip such a factory with robots for assembly to promote sales of robots. Secondly, the product is very specialised and Fanuc's competitors are generally large companies. The use will take off when it is clear that robots are economic in the assembly of products found in thousands of factories.

Official estimates put robot growth in Japan at about 20% a year for the next few years. Most manufacturers seem to agree with those figures although they expect the growth curve will have its stop and go periods.

In the next decade, it seems certain that robot use will spread through a wider range of jobs and into the small companies. Within a few years, we can expect almost all injection moulding machines to be sold equipped with robot arms. Many machine tools will be similarly equipped, but the larger ones will rely on separate robots. Equally, robots will be used far more in car body welding, pushing the robot welding level up from 50-75% to 75-90%.

Assembly is clearly on the threshold of a major expansion in robot usage, although this may take several years to materialise in a significant way. At the same time Japan's Ministry of Inter-

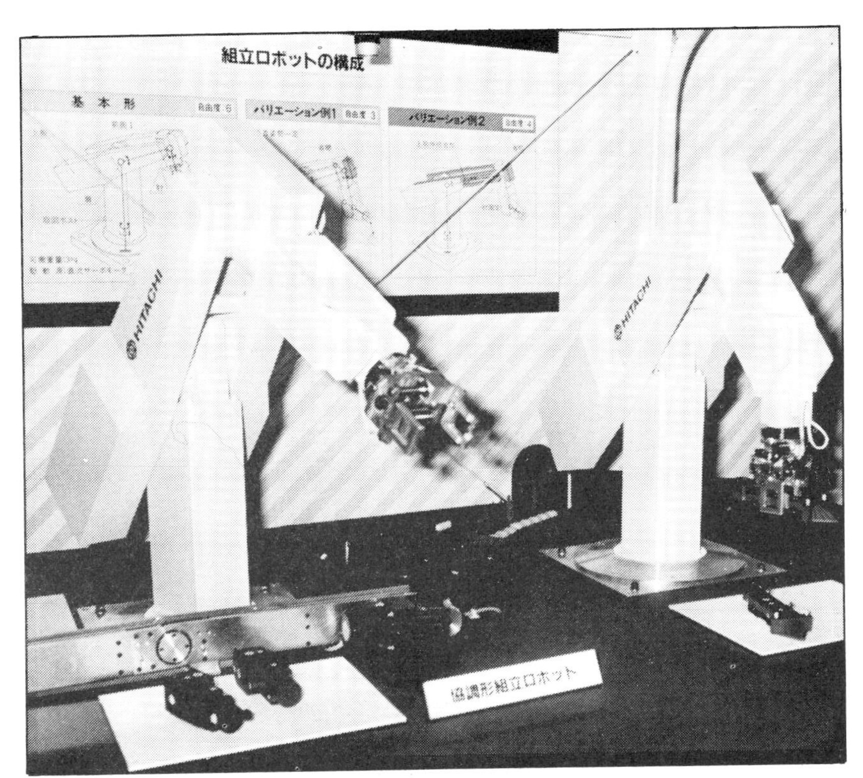

Fig 5. Cheap assembly robots will help open the floodgates to robot assembly.

national Trade and Industry (MITI) is sponsoring an ambitious project to develop an intelligent robot, while even more intelligent(!) robots are expected to emerge from the government-sponsored fifth-generation computer project.

By 1993 industrial robots will have become everyday machines that yield flexibility; but the signs are that the lion's share will be found in Japan. Some very complex robots will be used to supervise unmanned workshops, while robots will be used to inspect nuclear reactors and other dangerous areas as a matter of course.

Unemployment

At the bottom end of the market the low cost of electronics will make it practical for any handling device to be programmable so that it can deal with many different products. Indeed, robots should be doing almost all jobs involving manual effort. The Japanese, who take such a long view on productivity and business gains but such a short view on the effects on society, are certain to be in the forefront as users of these new robots.

Can the Japanese find a way to overcome the problems of employment? The answer is not clear. In the past the Japanese answer always has been to increase GNP to mop up the extra workforce.

Even now the Japanese distribution system and retail outlets employ far more people than necessary by European standards, so there seems to be little scope for a wholesale switch from manufacture to the traditional services.

On the other hand, the emerging information industry, from computer software to data banks and multichannel television will generate major business which will certainly change the balance of society in such a way that worries about robots taking over from humans are likely to be groundless – so long as industry remains competitive, with good products and high productivity.

Whatever happens, it is difficult to see anything stopping the Japanese from pursuing their goals of higher productivity through the use of robots –wherever that may take them.

Why the robot will replace the Japanese navvie

Professor Yukio Hasegawa, Waseda University, Japan

Waves of automation have been penetrating various industrial processes, ranging from the simple to the complex. Included in these is the industrial robot which has been helping to accelerate the diffusion of automation and bring breakthroughs into many difficult fields.

THE MOST successful fields of robot application today are mainly in mass and medium scale manufacturing operations such as those found in building automobiles, electric appliances, electronics and so forth.

On the other hand even today many industrial operations remain in a poor state as a result of not receiving any of the benefits of innovative robot technology. For example, productivity in the construction industry is behind that found in advanced industries; indeed there exists a big gap between them.

For several years the author and a group of colleagues have been participating in a construction industry robot research project. Last year the group reformed under the name of WASCOR (Waseda Construction Robot) research project group. Recently the Japanese government also has begun provisional activities to study robot reasearch in this field.

This article introduces a methodology for the research into such robots, examines problems in the construction industry, and looks at conceptional design of construction industry robots. Also shown are examples of robot development. Some of the problems still to be solved for the successful introduction of robots are discussed as well.

First let us look at some of the problems of introducing robots in the

Fig. 1. Manual constructing operations.

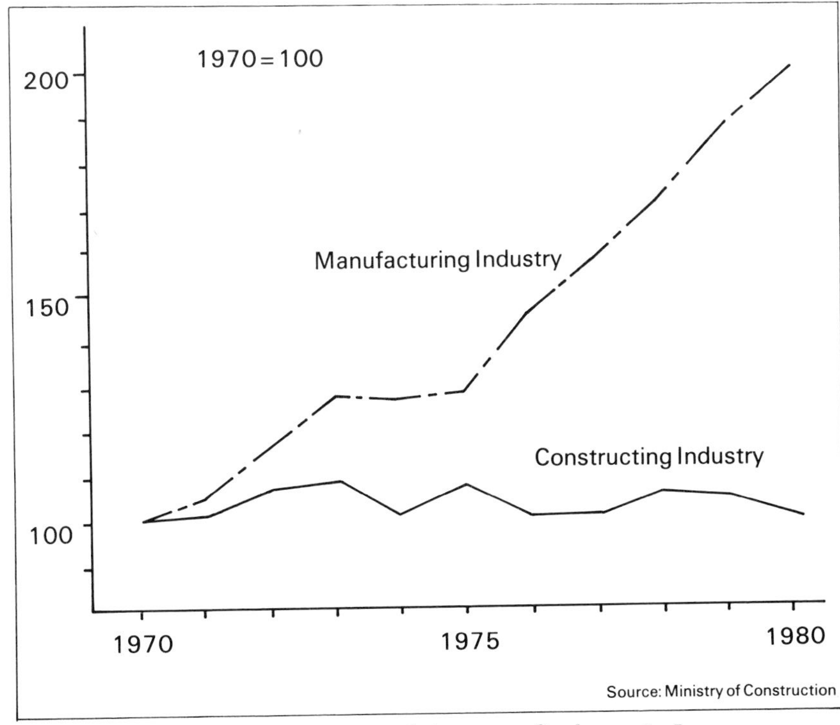

Fig. 2. Productivity increment trends in types of industry in Japan.

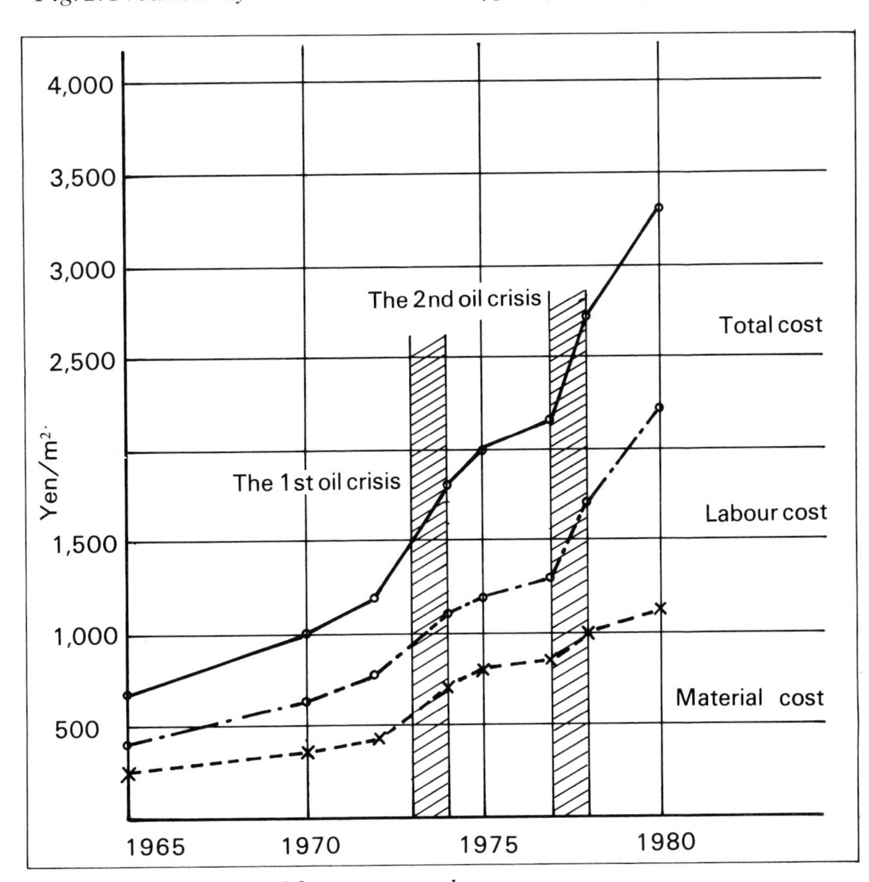

Fig. 3. Recent plywood form cost trends.

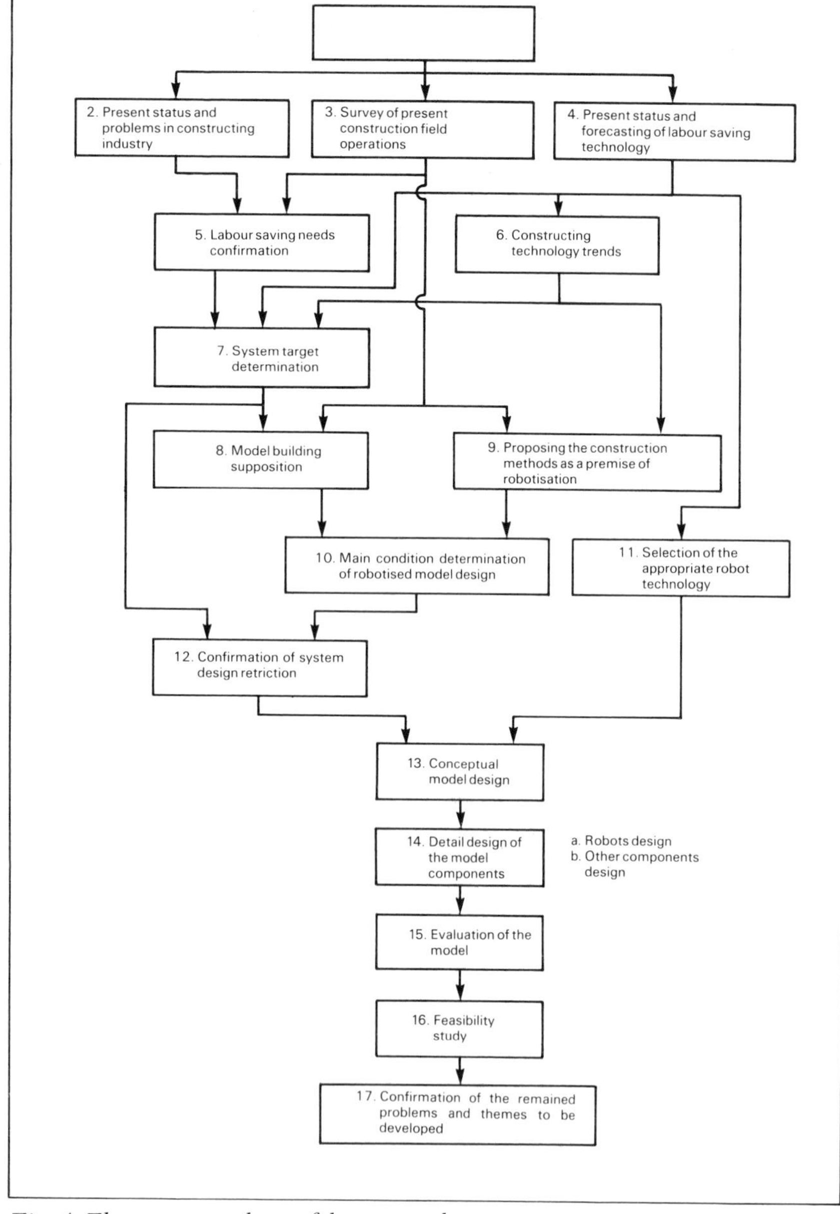

Fig. 4. Flow process chart of the research.

construction industry. Fig. 1 shows a typical scene involving manual construction operations. Today, the majority of concrete building construction work is done manually. The reasons why robotisation has not yet been introduced are as follows:

- □ Buildings use a large number of different items.
- □ Product size is large and the weight of components is great.
- □ The location of the construction site is constantly changing.
- □ A variety of operations is needed to complete a building so multi-functions are needed for robots.
- □ There are many complicated group operations.
- □ Information feedback from production people to product designers is poor; such people cannot improve the construction productivity because of many design restrictions.
- □ The stability of operation is inadequate.

Fig. 2 shows the tendency of productivity to increase throughout many types of industry. In the construction industry however there has been no increment of productivity rise in spite of the rapid increase in manufacturing industry.

Fig. 3 shows the trend of material, labour and total costs of wooden concrete form construction per square metre during the past 15 years. In this time labour costs have increased more than four times.

Because of the lower productivity today costs are increasing yet on the other hand working conditions remain poor. For that reason people in this industry have begun to consider the introduction of robots to break through the thick wall of the seriously low productivity.

Turning now to the conceptual design of construction industry robots it is important to look at the methodology for introducing industrial robots into the industry. To conquer the difficulties of introducing robotisation throughout general construction we have developed a process chart as shown in Fig. 4. The flow chart was used as a basis of the research work.

In the research emphasis was put on the following points:

- □ To maintain a consistent system approach from the beginning to the end.
- □ To make a survey of flow of technology and its forecasting.
- □ To decide the priority of operations to be robotised based on a questionnaire survey and an analysis of operations.
- □ To select an appropriate method of construction for robotisation.
- □ To decide the robotised system targets to be accomplished as a result of the research.
- □ To conceive a building model as a foundation for the research.
- □ To determine the restrictions as a premise of conceptual design of the new robotised systems.
- □ To make a conceptual model design of the robotised building construction system.

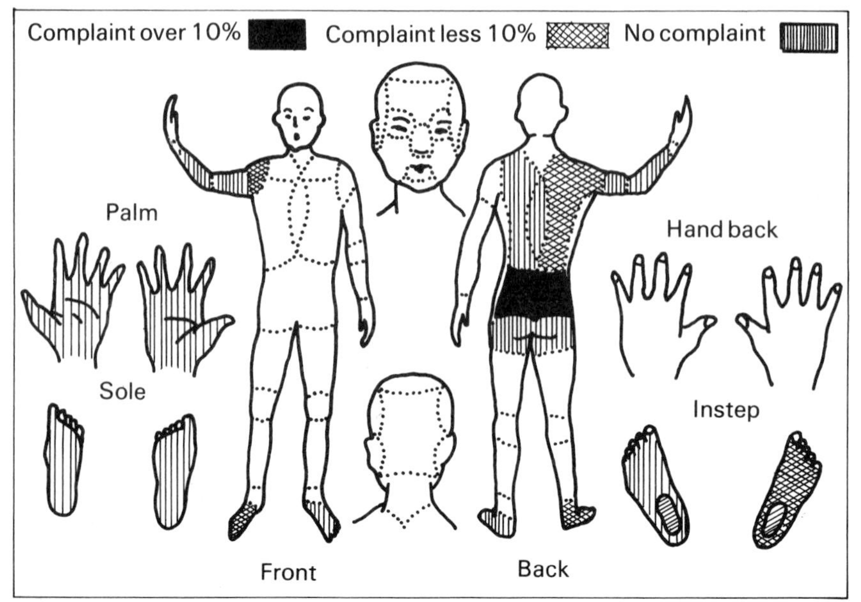

Fig. 5. Operators fatigue concentration.

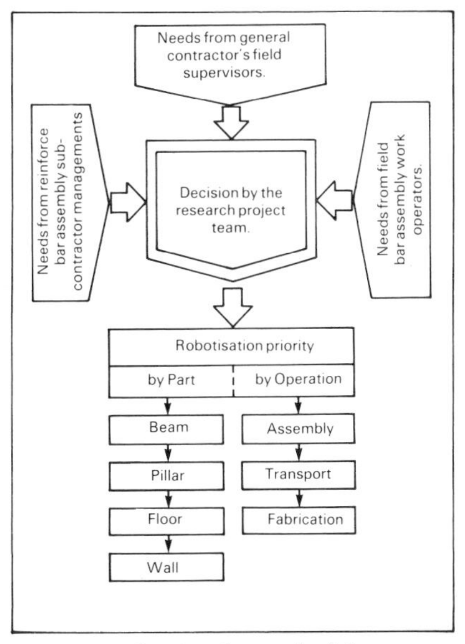

Fig. 6. Robotisation priorities.

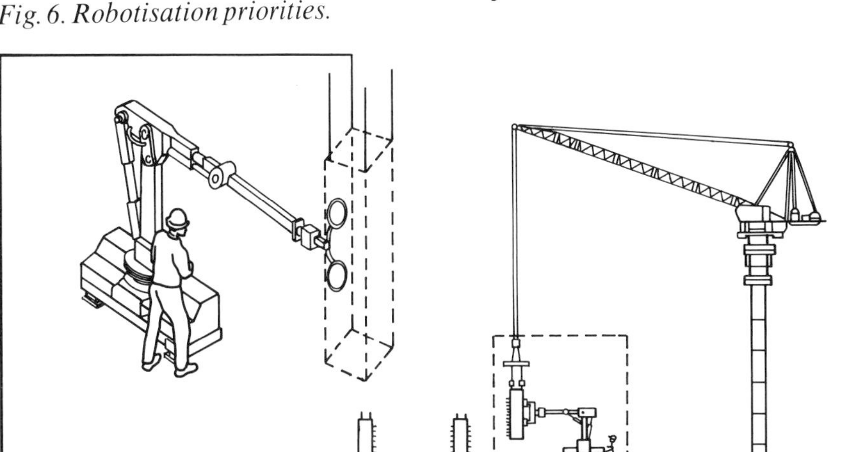

Fig. 7. Image of robotised constructing yard.

- ☐ To determine the expected specification of robots and other associated components.
- ☐ To make an evaluation of the model alternatives.
- ☐ To make a feasibility study of models.
- ☐ To determine the problems for materialising model plans and future themes to be solved.

For this purpose a questionnaire was sent to operators, subcontractors, managements and general field supervisors in the construction industry. In addition using a memo-motion film analysis method, a study was made of operator posture and fatigue.

Fig. 5. shows the result of survey of operators when handling reinforced steel bars during assembly work. Fig. 6 is an example of robotisation of priority based on the above mentioned survey.

Modifying conventional methods

What should be the design targets for a construction industry robot? To achieve a breakthrough in solving this difficult problem it is necessary to decide some fundamental policies to modify the conventional process and methods of construction. For example, it is necessary:

- ☐ To develop a new construction process which will increase the ease with which robots can be introduced in place of conventional methods.
- ☐ To reject dangerous operations from the process or use robots as substitutes for operatives undertaking dangerous work.
- ☐ To standardise and modularise each component of the building and let the new system help introduce the advanced CAD-CAM system.
- ☐ To seek the possibility of introducing prefabrication production components and sub-assemblies.
- ☐ To remove disassembling operations from construction processes which are difficult to robotise.
- ☐ To consider adopting three types of robot (automatic, remote control and hybrid) because of the flow repeatability of construction work.

So how is the conceptual design of new building construction robots likely to look? Fig. 7 shows an artist's impression of a future robotised building construction yard. In the years ahead many types of robots will be used, including those for ground digging, soil handling, steel beam assembling, form assembling, reinforcing bars assembly, concrete pouring, painting, measuring and so forth.

Building constructing operations are composed of dozens of complicated operations. It is therefore important to give a wide variety of functions to any robot used in the construction industry.

For several years many leading Japanese construction firms and robot manufacturing companies have been continuing their efforts to develop construction robots for the sake of saving constructing cost, and improving safety and working conditions. This robotisation is not yet totally synthesised in the way that factory automation is in manufacturing industry; designs shown in Figs. 8-11 suggest the future direction for robots in construction operations.

In conclusion it is worth pointing out that even today workers in the construction industry have problems of low wages, long working hours, and poor working environments because of lower productivity compared with other advanced industries. Whether we like to do so or not, people in the construction industry are expecting the invasion of industrial robots and a solution to their problems by using advanced robot technology.

For fulfilling such social needs it is important to develop and apply more systematic methodology of industrial robot application. It is certainly worthwhile, taking painstaking steps in this area of robot research and development.

Fig. 8. A concrete distributing robot.

Fig. 9. A stud bolt setting robot for atomic power plant building.

Fig. 10. An antifire insulating coat.

Fig. 11. Large concrete panel handling robot.

Japan's robot industry: Where does the future lie?

Kanji Yonemoto, the Japan Industrial Robot Association

The practical use of industrial robots in Japan over the past 10 years or so has produced a number of economic and social advantages. Future prospects are discussed.

ECONOMIC and social advantages of the use of industrial robots include an improvement in productivity, greater humanisation of working life, prevention of occupational and industrial accidents, improvement of product quality, and early return on capital investments.

These economic and social effects stemmed from the fact that the industrial robots are flexible and versatile. This has led to the automation of multi-product small-batch production, and changed the production system from a man-machine system to a man-robot-machine system. This change means that people have been released from dangerous, unfavourable work.

Efforts have recently been made to popularise the use of industrial robots in Japan through policies implemented in 1980:

- □ Establishment of a leasing system and of leasing companies designed to make industrial robots more attractive to small and medium sized companies.
- □ Special finance from the Small Business Finance Corporation and the People's Finance Corporation to small and medium companies for the introduction of industrial robots designed to ensure workers' safety.
- □ Establishment of a special depreciation system for high-performance computer-controlled industrial robots.
- □ Application of loans and leasing programmes to industrial robots by local governments to help small firms to modernise their equipment.

The use of industrial robots in Japan advanced throughout the 1970s, with a total of 57,000 machines in use at the end of 1979. However, it is predicted that substantial popularisation of industrial robots in Japan will occur after the 1980s (see Figs 1 and 2). It is predicted that demand for industrial robots in the manufacturing industries in 1985 and 1991 will be approximately Yen 290 billion and Yen 520 billion, respectively.

Table 1. Classification and Definitions of Industrial Robots
(Japanese Industrial Standards JIS B 0134–1979)

Classification by input information and teaching method	
Manual manipulator	A manipulator that is directly operated by a man.
Sequence robot	A manipulator, the working step of which operates sequentially in compliance with preset procedures, conditions and positions.
Fixed sequence robot	A sequence robot as defined above, for which the preset information cannot be easily changed.
Variable sequence robot	A sequence robot as defined above, for which the preset information can be easily changed.
Playback robot	A manipulator that can repeat any operation after being instructed by a man.
Numerically controlled robot	A manipulator that can execute the commanded operation in compliance with the numerically loaded working information on e.g. position, sequence and conditions.
Intelligent robot	A robot that can determine its own actions through its sensing and recognitive abilities.
Classification by motion	
Cylindrical coordinate robot	A manipulator which moves primarily in the cylindrical coordinate system.
Polar coordinate robot	A manipulator which moves primarily in the polar coordinate system.
Cartesian coordinate robot	A manipulator which moves primarily in the cartesian coordinate system.
Articulated robot	A manipulator which consists primarily of an articulated arm, that is, revolute joints.
Note:	
Manipulator:	A device for handling objects as desired. Without touching with the hands and it has more than two of the motional capabilities such as revolution, out-in, up-down, right left travelling, swinging or bending, so that it can spatially transport an object by holding, adhering to, and so on.
Robot:	A robot is defined as a mechanical system which has flexible motion functions analogous to the motion functions of living organisms or combines such motion functions with intelligent functions, and which acts in response to the human will. In this context, intelligent functions mean the ability to perform at least one of the following: judgement, recognition, adaptation or learning.

Furthermore, it is expected there will be substantial future demand for industrial robots in non-manufacturing such as nuclear energy, ocean development, and public works and construction as well as in service trades such as medical services and transportation.

Industrial robots at present are those devices which provide versatile and flexible moving functions similar to those of human limbs and enjoy intellectual functions through their sensing and recognition capabilities. The Terminology Standardisation Committee (TSC) of the Japan Industrial Robot Association (JIRA) worked out the classification and definitions of industrial robots in 1977 as shown in Table 1, and this was officially incorporated into the Japanese Industrial Standards in February 1979.

Along with the anticipated technological progress, however, there may come into being those robots which can perform such motional functions as for animals like snakes and crabs. Hence the classification and definitions of industrial robots will need to be modified as technology continues to progress.

There have been a number of factors affecting the diffusion of industrial robots throughout the country.

□ First, the development and practical application of industrial robots were stimulated by a serious labour shortage in late 1960s, a decade which witnessed the nation's highest economic growth. GNP showed an average 12% growth annually during this period, whilst the shortage of

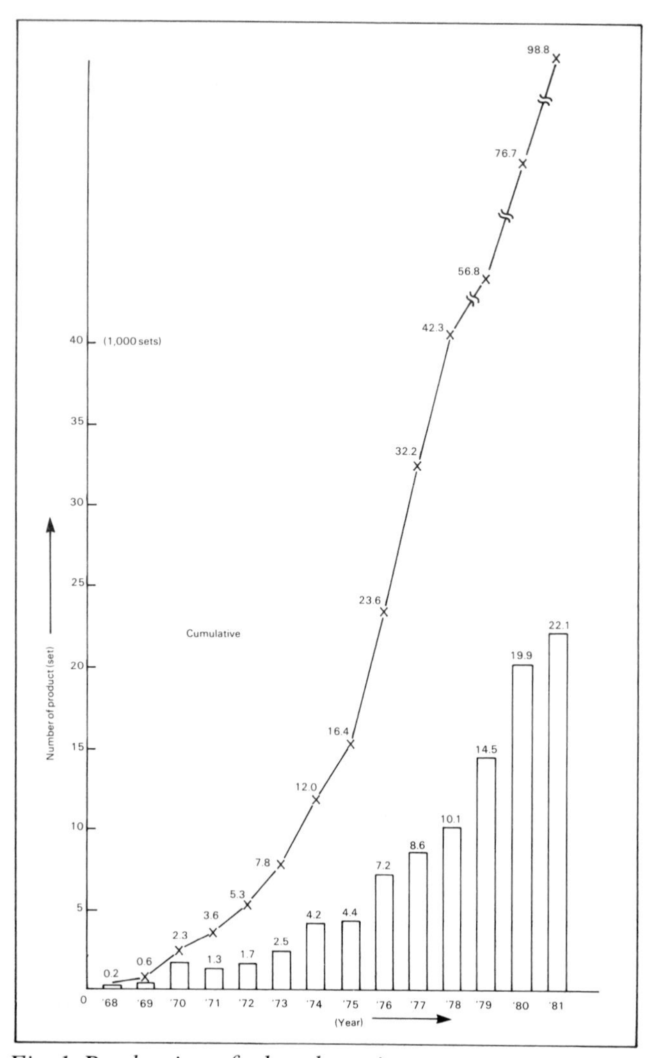

Fig. 1. Production of robots by unit.

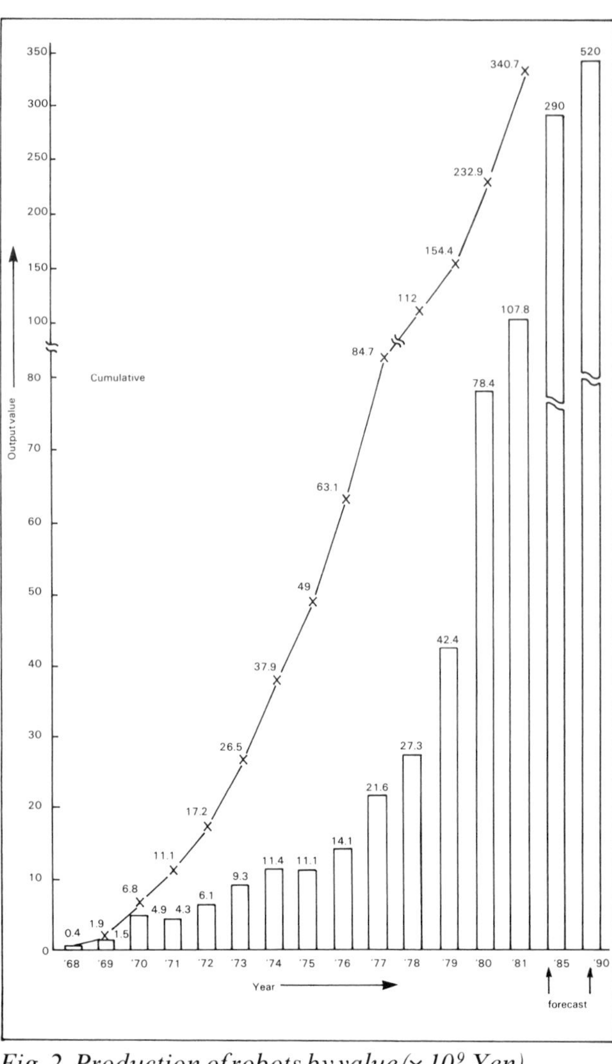

Fig. 2. Production of robots by value (× 10^9 Yen).

labour, particularly skilled labour, became more acute and widespread, reaching as many as 1,800,000 in 1965.

□ Second, the first oil crisis in October 1973 forced Japan to change direction to low economic growth. As a result, the strain on the labour market eased to a remarkable degree, but the country's highest oil and other natural resources bills pushed up commodity prices drastically and hence labour costs. To fight against such inflationary effects required a dramatic improvement in productivity, encouraging private business to invest more in automation.

□ Third, Japan's population has showed a 1.1% increase per annum, whereas the growth rate of its labour force has levelled off at 0.7% chiefly due to the mounting number of students going into higher education. This has increased the need for higher productivity to achieve the 5% annual economic growth target. The need is particularly felt in the manufacturing sector wher, unlike the tertiary industry, an increase in labour force is not expected (Table 2).

□ Fourth, shortages of skilled labour have again reached considerable proportions due to a growing number of young people moving on to a school of higher grade. In June 1980, skilled labour shortages amount to about 840,000.

Table 2. Changes in the Labour Force by Industry
(Unit: in 10,000)

Industry	*1975*	*1980*	*1985 (est.)*	*1990 (est.)*
Primary	661	577	505	401
Secondary	1,841	1,926	1,962	1,970
– Manufacturing	1,346 (25.8%)	1,367 (24.7%)	1,377 (24.0%)	1,334 (22.5%)
Tertiary	2,721 (52.%)	3,033 (54.6%)	3,269 (57%)	3,558 (60%)
Total	5,223 (100%)	5,536 (100%)	5,736 (100%)	5,920 (100%)

est. Prime Minister Office "Survey on Work Force"

□ Fifth, there is the need to prevent industrial accidents and occupational diseases due to dangerous work as well as unfavourable working conditions. To meet this need, great pressure has been placed on management to increase the use of industrial robots.

Finally, organised labour in Japan is not primarily in the craft unions but in industrial unions or unions organised on a company basis, making it much easier to introduce industrial robots and relocate workers.

The major industrial robot users are:

- The electrical machinery manufacturing industry (whose share of total distribution of industrial robots in 1981 was 32%);
- the automobile industry (30%);
- the plastic moulding and processing industry (9%);
- the general machinery manufacturing industry (8%);
- the metal working industry (5%).

In addition, the textile, chemical, steelmaking, and ship-building industries are users, indicating that industrial robots are demanded in a wide range of industries in Japan. However, demand for industrial robots in non-manufacturing industries accounted for only 1% of total demand in 1981, while exports stood at a meagre 6% of total output that year.

Industrial robots were applied in 1980 and 1981 as given below:

Process	*1980*	*1981*
plastic moulding	8000	7500
machining	3200	4200
assembly	2800	2900
welding	1600	2600
press	800	1100
casting	700	1000
painting	130	350

It is expected that automation of assembly processes in small-batch production in different sectors of the machinery industry will be possible in the future as the technological development of intelligent robots makes progress.

Exports of industrial robots are expected to reach 16% of total output in five years from now and 17% in another five years. For this to be true, however, Japanese makers will have to team up with overseas engineering firms, since industrial robots require highly sophisticated systems engineering prior to installation and also good maintenance services after they are put into operation.

Wide use of robots

Aside from the manufacturing industry, the wider use of robot technology is expected in areas such as:

- □ Nuclear – for inspection and maintenance of nuclear power plant, and radioactive waste disposal.
- □ Medical and social welfare – to help the physically handicapped (1,989,000 in 1980, of which 1,130,000 are crippled) to work as efficiently as the average person, and to assist the bedridden (530,000 in 1979) in making their living.
- □ Ocean exploitation – for a variety of handling operations needed in underwater construction, machining and geological surveys, and for multi-purpose observations.
- □ Agriculture and forestry – for a number of agricultural and forestry jobs including fruit harvesting, crop-dusting, lumbering, trimming, and lumber collecting.
- □ Construction – for assembly of reinforcing bars, painting of bridges, and interior and exterior finish work in high-rise buildings.
- □ Transportation and other service sectors.

In order to strengthen international competitiveness and profitability, in the years to come productivity improvement is an essential consideration that businesses should take into account. Particularly in Japan, which is lacking in natural resources and is thus supporting its national economy by means of processing trade, there is a pressing need to improve productivity.

Also, there is a need to prevent industrial accidents and occupational diseases and to upgrade the quality of human labour. Mankind is turning into a slave of machinery as exemplified in the conveyor system and there is a need to reverse this degradation of humanity. The provision of agreeable working environments and the humanisation of working life are an important key to promoting industrial and social welfare.

Product quality

Industrial robots provide a powerful means of attaining these goals. The application of industrial robots has a wide variety of economic and social impacts. Characteristics such as flexible movement functions enable small batch production to be automated and are causing a change in the production system from man-machine to man-robot-machine. These changes are improving the working conditions of human labour.

Industrial robots not only improve productivity and welfare but they upgrade product quality, save resources and energy based on more effective capital investments and a reduced rejection rate, improve production control, stability of employment, solution of skilled manpower shortages, creation of new technology, development of new industrial sectors, and so forth.

1.2 The personal view

Automation must be flexible to allow firms to move into the 1990s

Hiroshi Makino, Professor of Precision Engineering, Yamanashi University, Japan, and the developer of the Scara robot, answers questions about the future of assembly automation in Japan.

Professor Makino

JAPANESE INDUSTRY is highly automated, particularly in industries where volumes are high. But generally, these companies rely upon hard automation. Do you think they will switch to flexible automation? And if so, why?

It is true that up to now, most Japanese companies have been using hard automation. But they are now planning to switch to flexible automation. The reason is that owing to engineering developments and market forces, model changes are becoming more and more frequent.

The wrist watch illustrates the trend. For over 200 years, there was no model change. Then came the quartz crystal, the digital watch, the digital/analogue watch and the calculator watch. All these have come in the past ten years or so.

But what about the economics?

The big problem is price. If flexible automation were to cost the same as hard automation, then all the companies would buy flexible automation.

For example, a typical pick-and-place unit in hard automation costs about one million Yen (£2,000) in Japan, whereas robots are six to ten times more expensive. If we can get the price of the robot down to three million Yen (£6,000) the cost will be competitive.

Hard automation has a life of about two years only in Japan, whereas the company can use flexible automation for five years or more.

In which industries is flexible automation coming first?

The automotive industry is leading, followed by the electronics industry, and electrical industries. The watch industry is also beginning to look at flexible automation.

So what is the future for hard automation? Many manufacturing engineers are sceptical of the use of robots and similar devices, and point out that in many plants hard automation is faster and in any case, the changes are not so great. Do you agree? Or do you think that hard automation is on the way out?

Hard automation is very important, and it will always be so. We are working on mechanisms for hard automation as well as robots. The need is to separate those functions that will change frequently from those that will not.

Of course, human beings are still used for a host of assembly operations, many of which are very complex. So how do you think automation can compete?

Almost all assembly is very simple – I would say 80%. It is a case of either inserting something vertically, screwing it down, or pressing it in. All these can be automated quite easily. But there are jobs, such as the fitting of springs to hooks, which do require considerable dexterity.

The main elements required in automation are high productivity, flexibility and dexterity. Robots are generally dexterous, with many degrees of freedom. But they have too many degrees of freedom for simple work, so they are slow and lose productivity.

My thinking is that for this simple work, dexterity is not needed. But flexibility is necessary, and high productivity is very important. That is why we call our Scara robot a P-F type – productivity and flexibility. This may be the machine people in industry need.

So what are the top priorities?
For flexible operation, there may be NC or computer control, but it must be very fast. So we designed the Scara first and foremost for very fast motion.

How fast is it? What sort of cycles do you have in mind?
For simple work, I hope two seconds is possible, and for more complicated work three seconds. A human can operate at about a 3 s cycle, so I suppose the Scara is similar.

Can you give some examples?
As you know, we built our first robot for pick and place work, while our second, which is larger, is intended principally for screwdriving. The gripper or tool can move 50 mm in 0.2s, and 150 mm in 0.4s, so with a pitch of 100 mm, we can move from hole-to-hole in about 0.3s. using 4 mm diameter, 10 mm long screws, we need 2-2.5s for driving; we can drive six screws into place in about 18s – a cycle of 3s.

We use magazine and pneumatic feed for the screws, and feeding takes 0.3s. So if the screws have a small pitch, some waiting is involved. That is why we are currently moving in a sequence of alternate holes, rather than successive holes. Of course, this can be programmed easily.

For pick and place operations, the arm can move quicker, but a double stroke is needed.

But what sort of assembly jobs can Scara do?
Palletising, inserting parts with a light press fit, or simple pick and place. For example, we have programmed it to install bulbs and lenses in a car instrument panel. The bulbs are inserted, and then turned through 45° to lock. The lenses are a light press fit – that job cannot be done by hand. The cycle time is about 3s.

Your robot has some compliance to allow for errors in positioning. How does this work?
It is important the arm should move laterally and not rotate about a point above the component when there is misalignment. If there is rotation, the part can jam in the hole. On the Scara, the arm has two joints – with vertical axes – so that the structure is rigid.

However, for insertion, the arm is locked in position electrically – by braking the motors. If there is misalignment, the insertion force is sufficient to overcome this braking torque, and allow some movement.

Overall therefore, the Scara seems to have good capability for simple assembly work. How does it compare with other robots?
It has about the same capability as the Puma, and about the same speed. But the Puma has five axes of freedom, and Scara four, so Puma can do more complicated work. But, as I said earlier, most assembly operations are simple. And Scara costs about half as much as Puma.

How much does it cost? When will it be available?
There are now many enquiries for Scara, so I hope that it will enter production late this year. It will be made by Nitto Seiko of Kyoto, but first we need to improve the control unit – productionise it. The price is likely to be about five million Yen (£10,000), whereas the Puma costs nine million Yen.

They say that in Japan, the average investment needed to replace one man on assembly is three-five million Yen (£6-10,000), the lower figure being for the electrical industry, and the higher one for the auto industry. Our robot can operate as fast as one man, so it is competitive, with the payback period within a year.

Some people say visual or tactile sensing is necessary before robots can really take over in assembly – and that will make them even more expensive. What is your view?
As I said, 80% of assembly operations are simple, and we don't need any special sensing. However, depalletising is a problem. If parts are not positioned exactly then perhaps some form of visual sensor is needed. But for palletising visual sensing is not necessary.

But what about those very dexterous jobs that people do on assembly? What is the solution there?
That problem can be tackled in two directions. One is to redesign the product so that dexterity is not needed. The other is to use complicated robotic systems.

But are such expensive, complicated machines worthwhile?
I think this approach is very interesting, but it is not practical. It is interesting for us in research but it is not practical for manufacturing engineers!

So you see the future based on simple robots or flexible machines? That presumably means many less people being employed in manufacturing. Do you think this will be a problem?
Yes, I do see the future of flexible manufacture being based on simple machines. I think 80% of manual assembly may be taken over by robots. For the leading companies in Japan, this will happen in the next five years. Of course, that means less workers, but elsewhere, many new jobs will be created – in supervising the machines and planning and programming. Then, someone must design and make the robots. So maybe there will be no reduction in the number of workers.

Prof. Hiroshi Makino

After graduating from Tokyo University in 1956, Prof. Makino joined Matsushita Electric Industrial and worked in the machinery division of the Central Laboratory until 1966. He spent most of that time investigating and working on automatic assembly mechanisms in general, and cams in particular.

After working at Matsushita, he joined Yamanashi University, and in 1975 was appointed Professor. At the University he spends 80% of his time on research, and 20% in teaching. He specialises in cam mechanisms for automatic assembly machines, and in the development of the Scara assembly robot.

He designed a three-plate cam for indexing mechanisms, now being produced as the Tricam by Sanko Steel Wire Co., Kyoto, and also a cam for linear indexing. He is now finalising development of a cam system for installing inside a rotary table for very rapid indexing with 20 stations.

Prof. Makino has carried out a lot of fundamental work on cam linkages, and produced a universal formula for cams based on the movement of the follower where appropriate. In 1976 he published a book on this subject in Japanese, and has written a program in Fortran for profile development.

The movement of the Scara robot is based on motion curves produced by that program, the aim being to reduce vibration so that precision of positioning could be improved. The Scara robot has some inherent compliance, and has been designed for assembly operations. It is expected that it will go into production this year – at about half the price of the Unimation Puma.

Self confidence is essential if manufacturing is to improve

Mr Katuo Aoki, formerly Vice-president of Nippondenso Co. was responsible for much of the automation introduced into the company. In this interview he stresses the responsibility of management in introducing automation.

Mr. Katuo Aoki, former Vice-President of Nippondenso Co. Ltd., Japan.

Can manufacturing automation and full employment co-exist in a future industrial society?
Automation will cause employment problems in the future and labour unions may have objections to this aspect. It is a matter for management to persuade the labour unions that automation is a necessary development otherwise the company will collapse.

But how can management persuade labour unions to accept automation?
It is the problem and responsibility of management to increase productivity. If productivity is to increase by say 10% then it is essential that sales must also be increased by the same 10%. Robot applications can only be achieved by this equation otherwise workers will have to move to other jobs.

If sales do not increase and automation is introduced to increase productivity then management must look for new jobs for workers eliminated by productivity increases.

Do Japanese workers welcome automation?
Japan enjoys growth and so therefore it can accommodate robots to increase its production. Nippondenso was able to introduce industrial robots into its die casting factory because of the bad working conditions and in this situation it was welcomed by the workers.

What is the future for programmable robots such as PUMA and PRAGMA etc. in assembly automation?
Both the PUMA and the PRAGMA are very good robots and in general programmable robots have a promising future. In Nippondenso we are already using programmable robots.

In the past Nippondenso were able to use simple robots for mass produced items but now there is a demand for many different types of products and therefore an increase in demand for more flexible robots. However, the design of the product is important and it is absolutely necessary to look carefully at the design of any product when introducing robots. It is essential that design is considered before robots are introduced.

Is there a need for programmable feeders in the future of assembly automation?
Programmable feeders are very complex and wherever possible they should be avoided. It is much more preferable to link first operation work such as pressing, machining etc. to the automation line. We have done this in Nippondenso by connecting centre lathes, presses, injection moulding machines etc. directly onto the assembly line.

However, this is not practical in all cases and particularly if the volume of production from the first operation machines is very high. In these cases it is necessary to feed these into magazines and to use programmable feeders.

Is the use of sensory devices such as vision and other forms necessary in assembly automation?
In some cases we have been able to use vision in Nippondenso but generally the cost is too high for use in detecting the position and orientation of parts. It is more likely that vision and other sensory devices will find application in the inspection area.

It is best to look at the problem from the point of view of the human who has many excellent facilities which can be used for quality control. One example of this in Nippondenso is the automatic assembly line for producing meter gauges in which a visual sensor is used to determine the position of a needle and then to control the adjustment of that pointer for accuracy. In this case it has been possible to integrate the measurement and the actuation in a way that a human being could not work.

Does Nippondenso have any plans to sell its manufacturing expertise, for example, as has been done by the Citizen Watch Company by going into the automation manufacturing business?
No, there are no such plans at Nippondenso. The main business of Nippondenso is to sell automotive parts and it

only makes the automation equipment necessary for its own use.

In fact I have often been asked by our higher management whether we couldn't sell some of our experience in automation by manufacturing equipment. However, the main problem that I faced at Nippondenso was to increase the productivity of the parts manufactured by Nippondenso and not to have enough time to increase that of other companies who could be our competitors. This would happen if we sold them our own equipment.

What do you see as a key problem area in the development of automated manufacturing techniques in the future?
The main problem is the integration of quality control techniques as a method into the automation process. It is important that there is a balance between the introduction of robots and of quality control techniques.

Do you mean integration similar to that you have already achieved at Nippondenso on the meter gauge line in which measurement and adjustment is carried out?
Yes, this is the case. It is important to remember that the worker does the job and then carries out a quality control check on it. Robots on their own cannot do this and therefore if some quality control is not integrated into the line then the worker must stay on that line with the consequent loss of the productivity that would have been gained by the introduction of robots.

However, it must be recognised that it is very expensive to integrate the robot and quality control techniques and this is a problem for the future.

Can Japan continue as the leader in the use of manufacturing technology?
The important criteria is that of self confidence which can help to improve automated manufacturing methods. About fifteen years ago Japan adopted the technology that was based on foreign patents and inventions. Since then it has strived to establish its own technology.

If Japan is to stay as a world leader in manufacturing automation then it must innovate otherwise it will lose its lead. Japan had to use robots before Europe as it had no other labour available other than its own, whereas in many European countries the availability of immigrant workers enabled productivity and production to be increased.

Mr. Katuo Aoki

Mr. K. Aoki is a former Vice President of Nippondenso Company Ltd. from which position he retired in March 1981. He now acts as an adviser to the Nippondenso Company and is Chairman of ASMO Company Limited, a subsidiary of Nippondenso which is engaged in the manufacture of small electric motors for automobiles, for example windscreen motors and the like. The ASMO Company employs three thousand people and has an annual sales turnover of 55,000 million yen.

2. Robots

Over the past few years, the growth in the number and type of robots produced in Japan has been remarkable. In this section, we trace the growth, and include some significant case studies. Of the case studies, the installation of over 100 robots by Yamaha to weld up motor cycle frames is interesting partly because the robots were designed specially for the work involved, and partly because the installation is so practical – that is the hallmark of Japanese production engineering.

In November 1981 we reported the burgeoning interest in assembly robots, but it was not until 1983 that several of these robots were being installed together for assembly. The use of robots for arc welding was also confined to a number of installations of two or three robots until recently, largely owing to the lack of sensors suitable for welding thin panels. Nevertheless, sensing has improved, and a greater variety of robots have been introduced – again a typical pattern in Japan.

2.1 Robot applications

Three-year payback makes installing robots all worthwhile

One small Japanese company decided to make its own components instead of putting them out to subcontract using a new investment which includes 11 robots. The investment saves £1 million a year giving a payback within three years.

ONE JAPANESE COMPANY which has built a machine shop around the concept of robot handling is Miyanaga, manufacturer of rotary and percussion drills for rocks and concrete, holesaws and corebits.

The drills are generally long, solid cylindrical components, while the corebits and holesaws are assemblies consisting of a shank centering device and the cup-shape cutter. Thus, the first operations are generally turning, followed by milling and boring, heat treatment and grinding. In some cases, of course, there is an application of a hard coating or hard tips.

Miyanaga has its head office at Kobe, on the south coast of Japan, but its factories are some 30 miles away in small towns. It employs 220 people and its business has expanded substantially in the past few years. It now produces 1,200,000 straight shank drills, 120,000 core bits and 100,000 holesaws annually.

Until three years ago Miyanaga relied on sub-contractors for all turning operations; this is a common practice among Japanese companies. However, it then decided to build its No. 2 factory to concentrate on turning, with some other operations being done as well.

Productivity

Masaki Miyanga, the president of the company, said when they were planning the new factory, they found that if they relied on conventional machine tools with manual handling, a total of 100 employees would be needed on two shifts. So they opted for NC machines and robot handling at the lathes.

In the No. 2 factory there are 11 Fanuc Model 1 robots handling workpieces at a total of 14 NC lathes. Also in the shop are a number of NC bar autos and a small bar auto that can turn and drill as well.

Total investment in the plant, including the land and all equipment came to £2,500,000, while the total workforce numbered only 25. The plant operates on two shifts with two part-timers looking after the robots during the night shift. This fact is significant, since part-timers, who often work for a few months only in a plant, are paid less than regular workers. They do not receive the normal bonuses which boost total annual earnings by 33% to about £6,500 a year in the case of the average Miyanaga employees.

A robot depositing a workpiece in a box alongside the table for partly-machined parts.

The installation of the robots is straightforward enough. Three robots each serve two machines while the remaining eight robots each serve one machine only. Most of the lathes are small Ikegai FX20N machines with Fanuc controllers, but there are also some Warner-Swasey lathes.

Each robot is equipped with two double-ended hands, one for small diameter parts and the other for larger parts. The two ends are needed so the robot can extract the machined part while it is holding the next part, ready for insertion into the chuck.

Where one robot is used for one machine, a group of three or four robots are placed at the four corners of a circle, facing outwards to their machines. Adjacent to each robot is a table which carries workpieces. These tables are simplicity in the extreme; their tops are made from plywood in which a number of holes are drilled – generally about 80.

The workpiece, which invariably takes the form of a shouldered spindle is placed in a hole. These holes are drilled in the table at intersections of arcs and radii of the robot's main axes of movement. Thus, a number of lines are drawn towards the axis of the rotating base of the robot on the table. Then, a number of other lines are drawn as arcs cutting across these, the arcs having the robot axis as their centre.

It is a simple matter to program the

Arrangement of Fanuc robots in Miyanaga machine shop.

robot to pick workpieces from the holes in the desired sequence. In this way, not only are expensive carousel tables unnecessary, but also more of the robot's capabilities are exploited. This inexpensive tooling was necessary owing to the diversity of the product range – 10,000 different products. Batches are limited to 5,000 units.

Operation of the robots is conventional: the robot moves over the table, picks up a workpiece, and then the arm is moved to the machine. When the guard is opened, the hand enters the machine, removes the workpiece from the chuck and then withdraws. The wrist rotates, and the arm moves into the machine again to load the new workpiece. Subsequently, the robot places the machined workpiece in a basket.

In theory, the machines and robots operate automatically without any manual assistance. Automatic adjustment for tool wear is incorporated, and if there is more than 0.3mm wear, the machine and robot stop automatically.

Miyanaga said this automatic 'stop' takes care of any tool breakage. In addition, the machine also stops if the table is empty, while the robots will stop moving if the hand strikes something. Miyanaga claims that none of the robots has done any damage at all, and that downtime is minimal.

However, on the question of safety, the operation would not be considered acceptable in many large Japanese factories, and certainly not in any British factories. In all, there are five people working around the robot area; their job is to empty the baskets, load the tables, and make any necessary changes to tooling.

As mentioned earlier, in one case four robots are placed near one another, but in other cases there is one robot next to a machine. In some cases, there is just a piece of string cordoning off the robot area, and a sign saying that people should not enter.

However, the operators do enter that robot area to remove full baskets of swarf, and they seem to be quite happy to walk within 20cm or so of a moving robot arm. Miyanaga seemed quite unconcerned about the possible dangers as well.

In practice, this situation has little effect on the efficiency of the overall installation – only on safety. The men can load the tables from outside the danger area, while the baskets could just as easily be located elsewhere.

As common practice in Japan, the night shift staff is much smaller than on the dayshift – only two against 12 or so. The reason for this is that during the day any maintenance or tool changing is also carried out, whereas the men working on the night shift are merely machine minders; if a machine should stop it is left for the day shift staff to rectify.

Although the machine shop is operated on a nominal 80h week, another two hours overtime are worked daily. Since the machines also operate through the lunchbreak, they operate for 107.5 h/week.

Overall, this is a highly profitable plant. Miyanaga said that its use has reduced annual machining costs by £1 million – the sum paid previously to sub-contractors. However, since it is a new plant, it is not possible to quote a precise figure for the labour savings obtained by the use of robots. In all probability, though, the use of robots has saved 13-14 salaries or £85,000 a year for an investment of around £300,000.

Miyanaga quotes an overall investment in the plant so far of £2,500,000, including all machinery. Over the next three years a similar amount is to be spent on an extra factory. But since the first factory has shown an annual saving of £1 million, it paid for itself within three years, which is not bad going.

In the next stage, Miyanaga aims to adopt direct numerical control (DNC) for the machines and robots in order to gain greater control of production.

This application illustrates how much easier it is to install robots where there is growth, and in this case that growth has come from the cost competitiveness of the Miyanaga products. To some extent that competitiveness reflects a willingness of the management of a private company to accept small profits to help increase market share. That increasing market share then gave it the opportunity of investing in more modern plant.

Indeed, it is this faith in the future, and the recognition that if they do not automate they will lose out to their competitors in the long run that has encouraged Japanese manufacturers to automate. And it is that long term view that has done a lot in giving Japanese industry its competitiveness recently – and its increasing robot population.

Home grown arc welding robots reap dividends

Yamaha Motor has installed over 100 robots of its own design to arc weld motor cycle frames. This is a report of these developments and their economic consequences.

ALTHOUGH robots are suitable for arc welding, which is considered one of the largest potential growth areas, the robots are still expensive for this application. To overcome this problem, cheaper robots are needed, and these should ideally be used 24h a day. In this way, the capital cost can be amortised quickly, so the economics improve drastically.

Yamaha Motor has done both of these things in its Hamakita motor cycle frame factory, where there is a huge array of 145 arc welding robots – perhaps the biggest number in any one plant at present. Apart from one or two machines, these were all designed by Yamaha, and are much simpler and cheaper than the universal arc welding robot produced by robot manufacturers.

Yamaha has also produced its own assembly robots, but these are scattered around various assembly lines, whereas the arc welding robots are concentrated in a small factory that covers an area of 21,000m^2. The robots are used to weld up motor cycle frames, which are made from a number of steel tubes and pressings with wall thicknesses of around 2mm, at a rate of 150,000/month. Most of these welds are short, and many are at junctions of round tubes.

In 1974, Yamaha built its first arc welding robot, which was a Cartesian co-ordinate type, with a telescopic arm mounted on a column to give vertical motion. About 30 of these were produced, but by 1978, the company had switched to the manufacture of two different types and now the original type of robot has been superseded. It is claimed that the current types are 30% more productive and 20% cheaper than the old design.

For most applications, Yamaha developed a robot of polar type – a telescopic arm which can articulate and which can rotate on its base. For some applications, an overhead Cartesian co-ordinate robot is used. This consists of a carriage from which a column carrying the welding gun hangs down. The carriage can move longitudinally through about 1.5m, and laterally through about 0.5m on slides, while the column can extend to lower the gun to the workpiece. In addition, the gripper can rotate the welding gun.

Pivot

The polar type robot has a 900mm long arm which can pivot on a short pillar. The pillar can rotate through 100°on the base, while the arm can articulate through about 30°. There is 600mm extension of the arm, so the robot can weld within an trapezoid envelope of 500mm to 800mm in height by 600mm long. The span of the envelope, provided by the 100° rotation, is 900mm. As on the overhead robot, the wrist houses a motor which can rotate the welding gun through 270°.

Both these robots have only four axes of movement, and a much smaller envelope than is normal on an arc welding robot. This design was selected after examination of the jobs for which the robots would be used. The result is a relatively cheap and simple design, with a stiff mechanism and rapid operation. Maximum welding speed is 600mm/min.

Initially, sequence controllers were used, but now Yamaha is using electronics systems with the program steps stored in random access memory (ram). This arrangement facilitates the changing of programs, and allows 1,024 steps to be stored at one time. A battery is provided to prevent loss of memory in the event of a power failure.

Noboru Nagata, production manager at the Hamakita plant, said that the main problem in installing the robots was to provide sufficiently accurate jigs. With the thinner tubes, an accuracy of ±0.02mm was necessary, and in other places, an accuracy of ±0.05mm was sufficient. He said that in most cases, the combination of CO_2 welding and the relatively thick tubes or pressings meant that distortion during welding was not a problem.

In a typical application, six overhead robots, attended by two men, weld up steering head assemblies. In this particular design, the assembly

Yamaha's polar robot has four degrees of freedom, but the extent of each motion is less than with conventional arc welding robots.

The robots are arranged in rows usually, with the men who load then having welding booths nearby.

consists of the tubular steering head which is welded to a pair of channel section pressings to produce a T-shape box section. The assembly is approximately 300mm long by 100mm deep by 50mm wide. Welds are made between the steering head and the pressings, and along the seams between the pressings.

Supplement

To supplement the robots, there are special fixtures and removal arms. The fixture is carried on a plate mounted on a turntable which can rotate on horizontal spindles. There is a large hole in the plate through which the assembly can pass. Beneath the fixture is a tray carried on an arm.

The operator loads the assembly in the fixture manually, and as soon as he presses the start button, a pair of curtains is automatically drawn across to prevent glare. Then, the fixture rotates as necessary to present the assembly to the welding gun. When the welds have been made, the fixture releases the workpiece which falls through the hole on to the tray below. The tray is then articulated on a pivot so that the workpiece can be removed later, and the curtains are opened automatically.

In one or two places, the robot cannot make the weld easily, so the operators make a few short welds afterwards. In addition, alongside the last robot is a fine boring machine, in which the operator machines the steering head before placing it in a pallet.

Most of the polar robots are used to weld up frame sub-assemblies or complete frame assemblies. Generally, the welds are short – 25-50mm is typical – but there are many of them. In one line, eight robots are arranged in a circular path to weld frames, and in another case, 10 robots are arranged in a straight line – the layout was dictated by the space available. In both cases, three men are employed on each line, to load the workpieces into the jigs, and to make some welds.

Three types of fixture are used: in a few cases, there is a turntable fixture, with a man welding on one side, and a robot welding on the other; in many others, the fixture can turn through 180°, so that the robot can weld both sides of the frame; and in the third type, the frame is held in only one position, with the robot attacking from that side alone.

Hammer

At the turntable fixture, the operator first loads the tubes and pressings, and then tack welds them in position. Before the turntable rotates, he knocks the reinforcing panel into position with a hammer – that is the sort of job that is often needed with a welded structure, but which a robot cannot do. In most of these other applications, the operator stands at a bench opposite the robot, and in addition to loading and unloading, completes a few welds. These are always short welds, which can be made without any jigging. Nearly all these robots are equipped with automatic curtains which are closed before the robot starts welding, a simple and neat solution to one problem in arc welding by robot.

Overall, Yamaha is using 135 polar type robots and 10 overhead robots at Hamakita; it has also sold 50 robots to its subsidiaries, and is using a few at its other plants. Nagata said that the polar type is used where dexterity is needed, and the overhead type is used with simpler welds, and where space is limited. In fact, despite the flexibility of the robots, one overhead line is used only for one type of steering head assembly. But each of the other lines is used to produce three or four different assemblies. 'We have two types of fixtures: some have clamps for two different assemblies, but in other cases, we need to change the jigs completely,' said Nagata. 'We do this every other day. It usually takes 30-40 minutes to change the jigs, but we are trying to reduce this to 20 minutes,' he said.

Usually, the cycle time is about 2min, and each robot makes 30-40 welds per cycle. In fact, the actual number of welds on the assembly may appear to be only 15-20, but to allow

The overhead type of robot is used for simpler welds and is equipped with a turnover fixture.

the robot to make the welds without the need for too much dexterity, one weld is often broken down into two or three short runs. 'For example, we divide a joint between two tubes, where a full 360° weld is needed into three welds,' said Nagata. These tubes are usually of 20-50mm diameter.

To make the robots pay, and to meet the increasing demand for its motor cycles, Yamaha operates the Hamakita plant on three shifts, employing 200 people as direct labour. In addition to the 145 Yamaha robots, there is also one Yaskawa machine, while in another plant, Yamaha is using two OTC robots for arc welding. 'We are using these to see how well they perform,' explained Nagata.

Since these Yamaha robots are not for sale, their cost has not been revealed, but it is probably less than £12,000 or half the cost of a proprietary arc welding robot. Nagata says that on average one robot replaces half a man per shift. 'If the welding is simple, then one robot replaces a man, but at the other extreme, a robot may only replace one-third of a man,' he said. The workforce at Hamakita has been reduced in the past two years, as the amount of automation in arc welding has increased from 30 to 80%. The robots are now doing the work of 220 people, so the payback period is probably less than two years.

Diversific tion continu s

More robots aimed at different jobs on offer as 'robot fever' subsides; more flexibility in assembly and welding.

ALTHOUGH robot sales are still strong in Japan, the 'robot fever' generated last year has subsided. To some extent, of course, this results from the recession which is being felt in Japan as elsewhere. Thus, companies, especially the smaller ones, are finding it more difficult to invest in new plants or new robots.

Nevertheless, manufacturers are still introducing new robots – Aida Engineering, Mitsubishi Electric and Kobe Steel are among those involved – and the uses are increasing. For example, Fanuc has now started to assemble servo motors by robots on a big scale.

Despite the temporary setback, with some makers of arc welding robots evidently cutting prices to clear stock, Kawasaki Heavy Industries, which manufactures the Unimate robots in Japan, is still expecting annual growth of 25-30% in the next few years, although growth will not be completely linear.

Kawasaki is now building the 2000, 3000, 4000, 6000, 8000 and 9000 series robots, while in the summer it started to build the PUMA 500 and 700. By next January it will be producing the PUMA 200 model as well. According to Naohide Kumagai, senior manager, industrial robot sales, Kawasaki is now making 60/month of the larger models, and 20 PUMAs each month. By early next year, output of PUMAs is expected to increase to 50/month, so that total volume will be more than 100 monthly.

So far, 140 orders for PUMAs have been received, but most of these have been for trial use, mainly by the automotive, electrical and precision machinery industries. Most applications are for handling, but Japanese companies are also just curious to see how well the robot performs it seems.

Surprisingly, Kumagai said that the visual sensing system described by Maurice Dunne of Unimation at last year's Robot Symposium in Tokyo, is still a long way from production. This was developed jointly by Unimation and Kawasaki, who implied that it was close to production last autumn. One

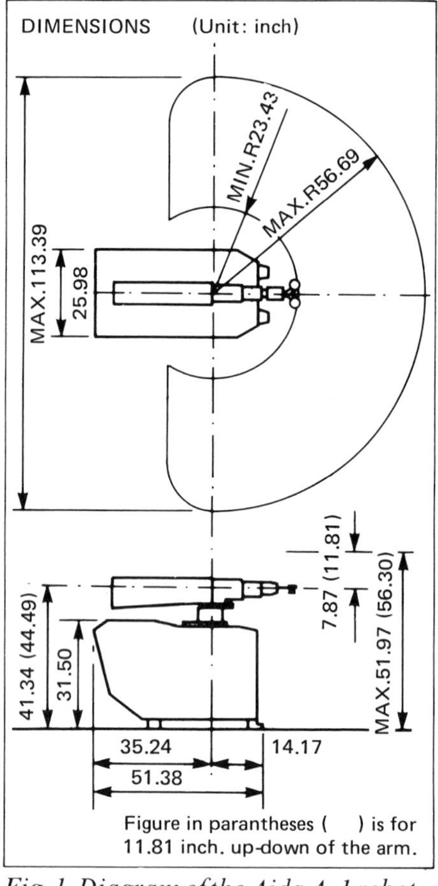

Fig. 1. Diagram of the Aida A-1 robot and its working envelope.

was shown operating at the Robots Show in Tokyo, but a year later, this is still far from developed. Kumagai said: 'The system is all right for laboratory use, but it is very difficult to manufacture. The problems are cost, which is too high, and the operating speed, which is too low.' He thinks a one-to-two year programme is needed to overcome these problems.

Kawasaki sells Unimates principally in Japan, although it has sold about 50 robots in the USSR, an action that aroused the displeasure of the Reagan administration. But it does make all the 6000 arms – in fact, these have all been 6060s – wherever they are sold. However, Unimation makes the controllers for export models. So far, about ten 6060 arms have been exported, the biggest single order coming from BL for four, which will be installed at one station on the LM10 welding line in January.

Incidentally, Kumagai confirmed that under the agreement with Unimation, robots built by Kawasaki and sold in Europe or the USA will be marketed by Unimation, not least because Kawasaki has no sales or service organisation. Currently, the 9753 paint spraying robot, which can

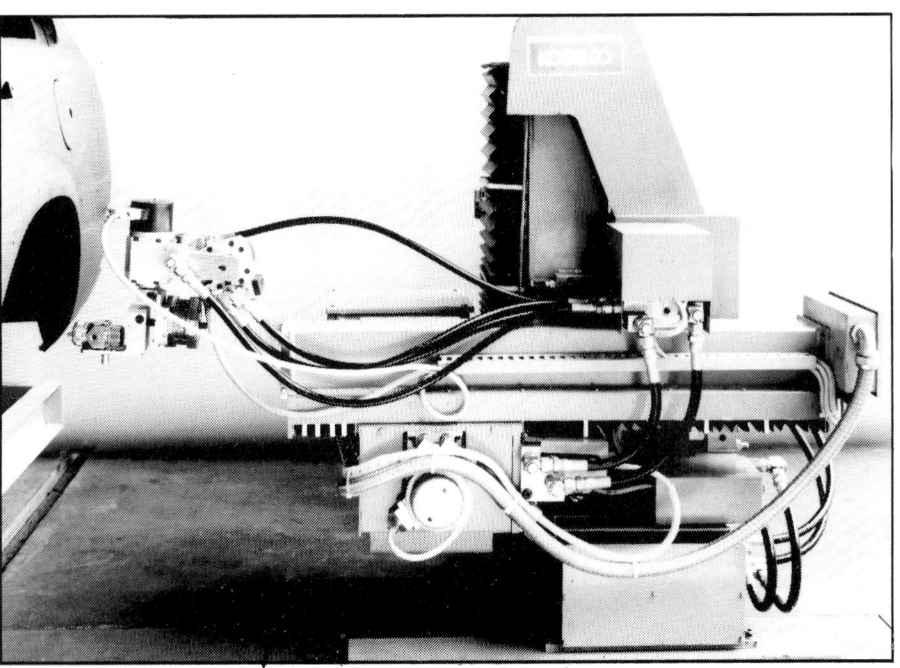
Fig. 2. Kobe Steel's compact new paint spraying robot.

be supplied on a linear slide so that it can move along with the conveyor, is attracting a lot of interest among European and American auto companies, Kumagai said. In Japan, Nissan, Toyota and Toyo Kogyo have embarked on programmes to install robots in paint shops. Nissan has started in its Murayama plant, with what appears to be an arrangement close to the 'NC Painter' developed by General Motors, while Toyo Kogyo has installed full robot painting at the new Hofu plant, which came on stream in August. Various makes of robots are used in these applications.

Elephant trunk

Kumagai claims that Kawasaki has had orders for 30 of the 9753 robots from Japanese car companies, and that one is on trial in Detroit, and another in Germany. He also said that by the end of the year an 'elephant trunk' wrist, which allows the robot to spray easily inside small housings, will be available.

In the arc welding business, Kawasaki is trailing Yaskawa, largely because the 2000 series robots are suitable for large assemblies only. Most users are in the construction equipment industry. However, Kumagai hopes that the PUMA will change all that. Within Kawasaki, ten or so robots are in use, and five of these are PUMA 760s – four being used to weld up motor cycle frames, and one to weld rolling stock. In addition, two 2852 arc welding robots are in use in the construction machinery division. There is still only one robot being used in the machine shop of the hydraulic division, although this machine has been in use for about seven years now.

Fig. 3. Mitsubishi Electric RH-211 assembly robot.

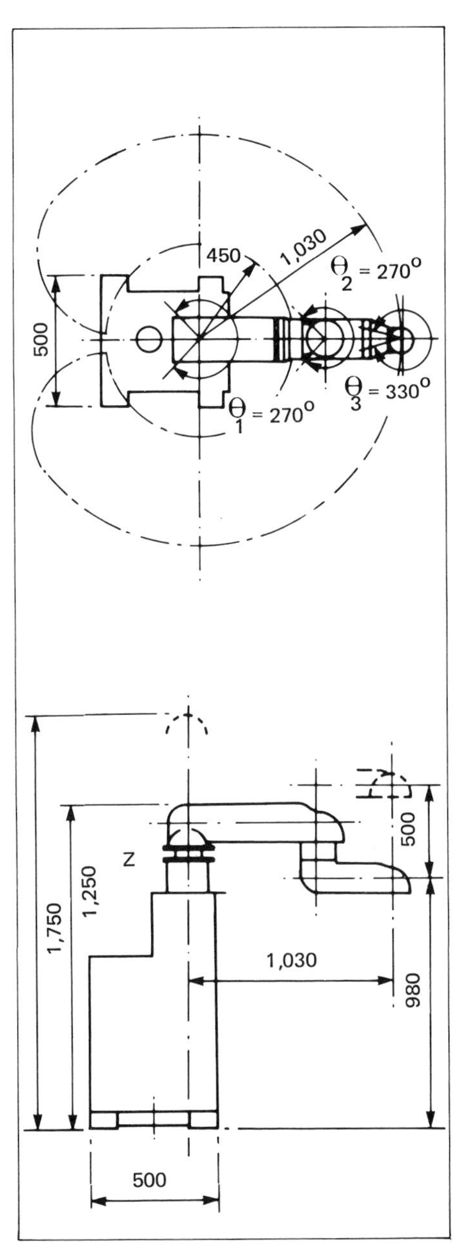

Fig. 4. Working envelope and dimensions of Mitsubishi Electric RH-223 robot.

The fact that so few robots are being used, in such a big group that makes robots, is rather startling; after all, here is Kawasaki trying to sell robots to industry, yet in its own companies it has generally found that hard automation is cheaper! This is doubly surprising, when Kawasaki's operations are considered; it is generally involved in small batch production, such as in its hydraulic, construction equipment, rolling stock and aerospace divisions. Production of Kawasaki motor cycles is in relatively low volumes, too. Clearly, big Japanese companies are just as conservative as their European counterparts where return on investment is concerned.

Kawasaki claims to be the biggest robot maker in Japan, with sales of £16 million last year, and expectations of £20 million sales this year (March 1982-3). Kumagai is hopeful that the PUMA will boost sales, as will the reorganisation of Kawasaki's organisation. Now it has a technical division to develop new robots; a mechanical-electrical division to develop engineering systems, including those involving robots, while robot manufacture remains part of the hydraulics division.

A much smaller company that is now hoping robots will boost sales is Aida Engineering, one of the larger Japanese makers of presses. It switched from the manufacture of robots to simple manipulators for presses some years ago, but has now produced manipulators with NC controls as well as a cylindrical co-ordinate robot, the A-1. Actuated by electro-hydraulic servos, the A-1 has four or five axes of freedom and can carry workpieces weighing up to 3kg. The robot is competitively priced at around £15,000 in Japan.

Mayuki Iwamatsu, vice president of Aida Engineering, said that the cylindrical co-ordinate configuration was chosen because it allows much faster movements than the jointed-arm or polar type. The robot is aimed at palletising and handling at presses and machine tools, where great speed is needed in linear motion – in this case the telescoping of the arm – as the gripper is moved into, and away from the machine. In this mode, the operating speed is 1,500mm/s, and Iwamatsu claimed that this is far faster than would be possible with a polar or jointed arm robot. He also claimed that the A-1 can operate at a press to a cycle of 2s, which is fast.

The base rotates at up to 180°/s, while the arm is raised at up to 800mm/s. Wrist bending speed is 240°/s, and accuracy is put at ±0.25mm. The arm can be raised vertically 200mm with an option of 300mm, it can rotate through 200° and can extend 600mm. The wrist can bend through 180° while there is an optional wrist rotation through 180°.

The electro-hydraulic control and actuating mechanism are all housed in the base of the robot arm which is carried on a pillar. Iwamatsu claimed that the electro-hydraulic actuation system was more suitable for factory use than an all-electric machine, especially where the environment is bad. The robot incorporates a new Aida hydraulic servo valve, which is compact, and through which flow can be controlled with great precision, owing to the use of a microprocessor system. The fluid flow is controlled in pulses, each pulse being 0.1cm³, so that 1,000 pulses are needed for the delivery of 100cm³ of fluid, for example. In addition, acceleration-dwell-deceleration curves are used to ensure smooth operation without vibration.

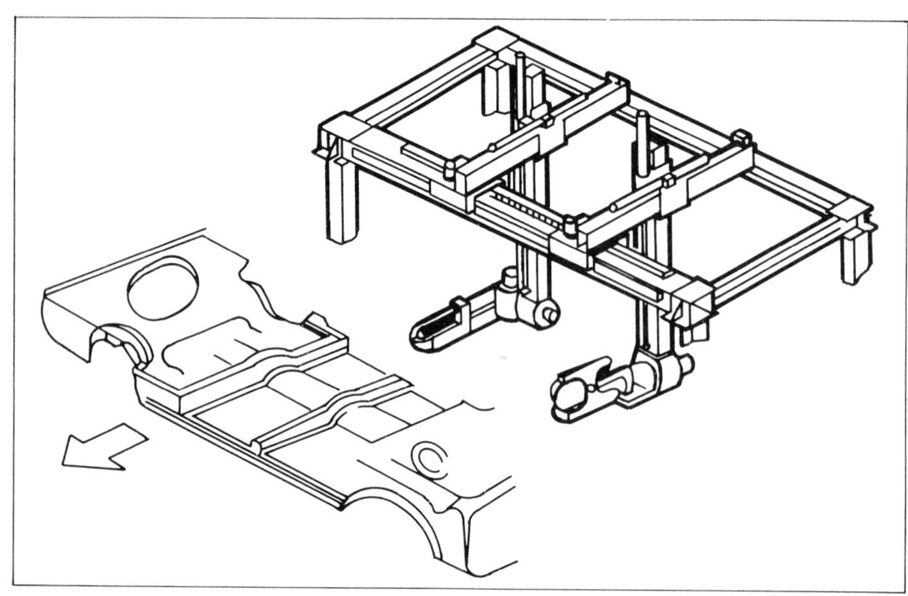

Fig. 5. Diagram of Mitsubishi Heavy Industries' modular robot

Since a new software system has been developed, the A-1 can be taught more quickly than most robots. The operator manually leads the robot gripper from point to point, and presses a button on the teach-box when each point is reached. Between points, he can follow any path he likes. This should be quicker than the normal method in which the path has to be keyed into the teach-box, movement by movement.

Teach-box

Aida has also introduced a range of NC programmable arms for installation in press lines, instead of the arms controlled by limit stops. The arm, which is mounted on a console, can move vertically and horizontally, so it is placed between a pair of presses as an alternative to a shuttle mechanism. The arm has vertical movement of 25-65mm, and a feed stroke of 750-900 or 1,000-1,200mm according to the model. Maximum operating speed is 2m/s, and the device can operate to a cycle of about 3s. The basic machine costs about £6,500 in Japan, but the cost of ancillary equipment can double this.

Daido Steel, which makes a number of special-purpose robotic devices, has now produced a range with capacities of 100-1,000kg. Although these robots have long, jointed arms, Daido says that the articulation is similar to that of a cylindrical machine. Actuated hydraulically, these machines fall into two categories – manipulators and robots, the former being controlled by an operator.

The largest robot can work in a rectangular envelope 3m long by 3m high. At the lowest point, the wrist joint is 950mm above the floor, and the arm can articulate so that it comes to within 1.5m of its body. The arm and wrist can each rotate through 240°.

The upper and lower arm each consist of two parallel links, there being a three-point hinge at the elbow joint. The robots have been designed principally for palletising.

Mitsubishi Electric (Melco) which plans to sell 100 robots/month within a year, has shown how it hopes to achieve that volume – it has introduced a range of assembly robots and a double-arm welding robot to supplement its neat arc welding machine. This is the second Japanese electrical company – the other was Matsushita – to mount a comprehensive attack on the robots business in a span of only three months.

There are seven machines in the assembly robot range, similar in principle to the Scara robot, with the arm pivoting on vertical joints, and actuated by dc servo motors. First, there are the RH-111, 211 and 212, which are similar except for their size and capacities. They can lift 5, 10 and 20kg respectively, and total arm lengths are 310, 650 and 1,030mm. The shoulder and elbow joints can both move through a total of 270°, and the gripper through 370°. Positional accuracy is put at ±0.05, 0.05 and 0.1mm respectively. The arm can also be moved up and down on its column. Three areas of movement can be controlled simultaneously by the point-to-point/continuous path controller. Prices are 5.5-7.5 million Yen (£12,000-£17,000), and the production plan is 30 units/month.

Basically similar, are the 'modular' RH-122/221/223 models except that four axes can be controlled simultaneously. The movement of these robots is rapid, the speed of the arm movement being quoted as 1,300-2,500mm/s according to model. There is also the RL-223 model, which is similar to the RH-223, except that it is mounted on a slide giving 3m longitudinal movement.

Small unit

Before long, Melco will introduce a PUMA style robot with a capacity of 2kg and six axes of freedom. This is a small unit, with the shoulder pivot 455mm above the base, and upper and lower arm effective lengths of 300mm each. Also in the pipeline is a 'flexible parts feeder', consisting of vibratory bowl feeders, an optical system to

Fig. 6. In the Mitsubishi Okazaki plant, the underbody travels down first half of line tranversely, and some robots are carried overhead on gantries.

check and identify the part and check orientation and an assembly robot, all mounted on one table. The robot has two horizontal arms, which normally both grasp the tool.

These will be supplemented by a much larger jointed arm robot with a large operating envelope. The robot will reach down to within 90mm of the floor, and to a total height of 3,540mm, and 2,800mm forwards from the axis of the base. The forearm is 1,765mm long, and when horizontal, it is 2,500mm above the floor. Capacity is 50kg.

Also on the massive side is the RW-251 double-arm arc welding robot which has two arms projecting from one side of a 2m high tower. They are carried in a circular block in a housing which can move up and down in the tower. The block can also rotate, and the arms are telescopic. In addition, the arm has an elbow joint, and can rotate on its axis on the main arm through a total of 180°. In addition, the whole tower can run along rails up to 20m long. Clearly, the answer for welding up railway rolling stock, aircraft, truck frames and even hulls of ships. Melco expects to make five of these a month, and to sell them for about £110,000.

Profiting from its experience as the Japanese licensee of Trallfa, Kobe Steel has developed its own paint-spraying robot. This is a cartesian co-ordinate machine, which is unusual for a spraying robot. This construction was adopted to keep the robot as small as possible. It has 400mm movement in the x, y and z axes, while the wrist can articulate through 210° vertically and horizontally. It can also twist through 90°.

Kobe Steel says that the robot is intended principally for spraying in small spaces, and identifies wheel-arches of cars as a main potential use. It claims that the new robot can do anything the Trallfa robot can do. The production plan is for four units/month, which will be sold at about £28,000 in Japan.

Applications

Fanuc has introduced a new range of robots, the S Models. These are jointed arm robots on the ASEA, and are intended for special-purpose use. Therefore, some are mounted on horizontal slides to increase flexibility.

Fujitsu, just going into production with the Farot robot, has produced a second prototype of the Micro Arm. This unit is equipped with a number of infra-red leds which sense the positions of objects. However, they are not arranged so that distance can be measured.

Intended to be used for assembly, the robot has six axes of freedom, and is controlled by a 16-bit microprocessor. The controller is such that after the robot has found a component, it can progressively move faster as it learns the path it needs to follow. The prototype has a capacity of 3kg, and a positional accuracy of ±0.1mm.

Fanuc has started machining and assembling servo motors by robot at a new factory. There are 60 workers and 100 robots in the plant, which is alongside the FMS plant where machine tool and robot parts are machined. Capacity is put at 10,000 motors of 40 kinds monthly.

The plant is a two-storey building, with cells comprising NC machines and robots on the ground floor. In all, there are 60 cells with 52 robots, and about 900 different parts are produced. Lot sizes are from 20 to 1,000.

There is a large automatic warehouse which extends from the ground floor up through to the first floor, and when parts have been machined, they are stored in the warehouse before being withdrawn for assembly.

On the first floor, there are four rows of assembly robots, with a total of 25 cells incorporating 49 robots – some M Models, but mostly A Models. The robots are arranged linearly, and workpieces are transferred between cells by conveyor. On both floors, unmanned trolleys are used to transfer workpieces between the stores and cells.

Nissan has started to use some robot assembly at the Murayama plant, where Skyline and Laurel models are built. Robots are now used to install the spare wheel, windscreen and rear window. To enable the robot to position the components correctly, the body is carried on a shuttle conveyor through these stations. Significantly, these robots are of the cartesian co-ordinate type and were made by Nissan. It seems that for security reasons, Nissan prefers to build its own robots now that it has decided on which types it should use in various applications.

Mitsubishi has started to use some robotic devices on car assembly as well, and is concentrating on jobs involving heavy lifting. First, the battery is lifted from a store to a position just above the body by crane resembling an x-y-z robot. The operator presses a button to identify the battery, which the crane then selects from the store, and moves to within a few inches of the battery tray in the car. The installation is currently completed manually.

The front seats too, are moved from a position alongside the track into the car by robot. The body is lifted above the normal moving conveyor, and locked in a stationary position while a robot installs the seat. There are two robots, one on each side of the line, and these travel along angled rails, at about 60° to the line. The robots have four axes of movement.

First, the robot lifts the seat, by squeezing the cushion in a big gripper, then moves along the rail towards the car, before tilting down slightly, and turning to insert the seat in the car. All

Fig. 7. To get to the centre of the underbody, some robots have long jaws, while forks hold the panels in position.

these developments reduce manning level a shade, and move towards automated assembly. Wheels are fitted with a semi-automatic machine, the machine carrying the wheel, but the man checking that alignment is correct. Nut tightening is automatic.

Squeezing

Mitsubishi Motors Corporation (MMC) has been using robots made to its requirements by Mitsubishi Heavy Industries for several years, and it has recently refurbished the Okazaki plant to produce the Starion, Tredia and Cordia as well as the Sigma. Robots are usually associated with big volume models, such as the Toyota Corolla and Nissan Sunny, but MMC is using robots to build cars in relatively small volume – about 200/day of each model.

Owing to the way the robots are used, Okazaki is to some extent the plant of the future, with models going down one line in random order. MMC uses 103 Mitsubishi Robitus robots and 200 people to build Sigma, Starion, Tredia and Cordia models to a 1min 30s cycle. Two shifts, each working 1h 30min overtime daily, produce about 650cars/day. MMC claims an automation rate of 80%, of which 60% is by robot.

MMC's Tredia and Cordia are front-wheel drive saloon and coupé versions sharing a common underbody, whereas the Sigma and Starion have common engine bays and similar rear ends. Basically, therefore, the lines had to be designed to deal with three underbodies and four body shapes, but with the added complication of front- and rear-wheel drive.

MMC has built massive four-post structures all the way down the lines to carry the robots. The Robitus robots are small, but they are generally of the cylindrical or cartesian co-ordinate type with five axes of freedom. Most hang down from gantries. Typically, a robot can slide along a rail across the sub-assembly, move horizontally on a small slide, and also move vertically. The wrist can twist and rotate. With this layout, the transformers are usually carried on the gantry.

MMC's side assembly line is L-shape with six stations and 16 robots. The side assemblies are laid on their sides and pass along a shuttle conveyor. There are two, three or four robots are each station, and these do not seem to be well matched; at some stations one robot had finished welding when the other ones were only half way through their cycle. The complete cycle was about 1min 15s, and doors for all four models, left and right hand, go down the same line.

MMC has an interesting way of cleaning the electrodes at these and other stations: near the robot's rest position are elecrically driven wire brushes. When the robot approaches, the wire brush starts to rotate, and the robot moves the welding electrodes into the brushes for cleaning.

Underbody

MMC's underbody line is an equally massive structure, with 17 stations where 35 robots and 204 'multiguns' are used. What MMC terms a multigun is a welding gun which may make just one weld, or six, moving along slides to do so. Thus, multi-welders as such are not used; instead, these guns are built into welding fixtures along the line.

There are two 'first' stations, one for the engine bay of the Tredia/Cordia, and the other for the Starion/Sigma. Here, loading is manual, and each station resembles a mini-multiwelder. The next station is used for all engine bays, but two men make the necessary welds. At the following stations, the underbodies are passed along one set of jigs, except for the last fixtures on the line, which are duplicated. The engine bay is carried along a conveyor above the line between these first stations, but once the floor comes in from the side, the underbody travels along a normal shuttle. Generally, there are two or three robots, and two to four multiguns at each station. At most stations, the robots have long jaws on the welding guns –300-600mm. In this section of the line, the robots are welding for about 60s, the shuttle is moving for about 15s, and when I visited the plant, the line was idle for about 60s in each cycle.

For framing, MMC uses two main fixtures which resemble Honda's butterfly fixture. When the complete underbody comes to the station, the fixtures for the sides are horizontal, and the side assemblies are lowered into position. Then, the side fixtures are turned to the vertical, and move on slides into the floorpan, clamping the sides in position. To cope with the different models, there are four sets of clamps built into the fixtures, but not many welding guns. In fact, only six welds are made each side along the edges of the floor. At the next station, similar fixtures, but fewer movements are used, and another 70 welds are made. Although this is rather an elaborate framing method, its use simplifies the jigging.

There is a pair of unusual robots on this line. Each can rotate on its base, move along on a slide, and move an enormous pair of jaws, each about one metre long. One jaw is hinged halfway along so that it can get through the front door aperture. When the body arrives at the station, the robot slides in towards the body, one welding gun jaw going beneath the floor, and the other through the front door aperture. Then, the front or upper jaw articulates about its hinge so that the electrode is moved down to the cross member at the rear seat. Meanwhile, the other electrode is in position, ready to strike the panel opposite. Elaborate, but according to the engineer involved, easier than bringing the robot down from above the body.

Unusually, a special-purpose multi-welder is used to weld the roof to the body. It consists of a turntable each side of the body. The turntable is rotated to align the correct set of guns to the body, and then moves in to make the welds. In the respot lines, the Mitsubishi robots are generally suspended from gantries above the line. At the end four robots make short arc welds at one station. The arc welds are short, such as at joints at the corner of the door apertures.

With actual output of 13,000-14,600 in the past few months, output is running at 68-76% of the real total capacity, based on a 1min 15s cycle. But MMC prefers to operate at a cycle of 1min 30s, and on this basis, the utilisation rate is a more creditable 78-88%.

Elaborate the MMC line may be, but it is certainly flexible. When the new models were introduced, neither the robots nor basic structures needed changing. Nor would MMC have been able to produce these models economically if separate multiwelder lines had been needed for each model, especially since the model mix between Cordia and Tredia has not been as expected. Certainly, the Okazaki plant is a good example of how robots give significant advantages in body manufacture; and these advantages are just as significant in other applications.

2.2 Assembly robots

Assembly by robots: Is this the next growth area by Japan?

Dominating the Robots 81 in Tokyo were several types of assembly robots, many based on the Scara robot. Are the Japanese out to launch into robot assembly in a big way?

THEY were talked about at the Syposium – in theoretical terms; they could be seen at the Robots 81 exhibition – in the "flesh"; they were assembly robots – from NEC, Nitto, Pentel, Hirata, Sankyo and Fanuc. All these, with the exception of Fanuc's, are similar to the Scara design, developed by Professor Hiroshi Makino of Yamanashi University.

The Scara robot is a simple arm with an elbow joint carried on a pillar. The shoulder and elbow joints both pivot on vertical axes, so that the arm can cover quite a large area. This type of robot has some compliance built into the joints, which are braked electrically. Thus, the electrical braking force holds the hand steady, but allows some compliance.

All are small robots, generally with a capacity of less than 5 kg, and a total arm length of 250 mm to 1050 mm. They are designed to insert parts in housings, with a light press fit if necessary, and to feed and tighten screws. Clearly, this type of machine will play an important part in assembly automation. For example, Hirata intends to integrate its machines into its automated assembly lines. Hirata is already producing an x-y table to give flexible operation in assembly, and this can operate with the same repeatability as the robots – ± 0.05 mm.

Sales target

Although the exhibition was the first public showing of most of these commercial versions of Scara, many have already been ordered and are in use. There are 50 Sankyo Skilam SR-3 arms now in operation and a further 70 on order. In fact, Sankyo are confident of reaching their sales target of 300 for 1981, doubling this in 1982 and doubling it yet again in 1983! This must mean a significant number of

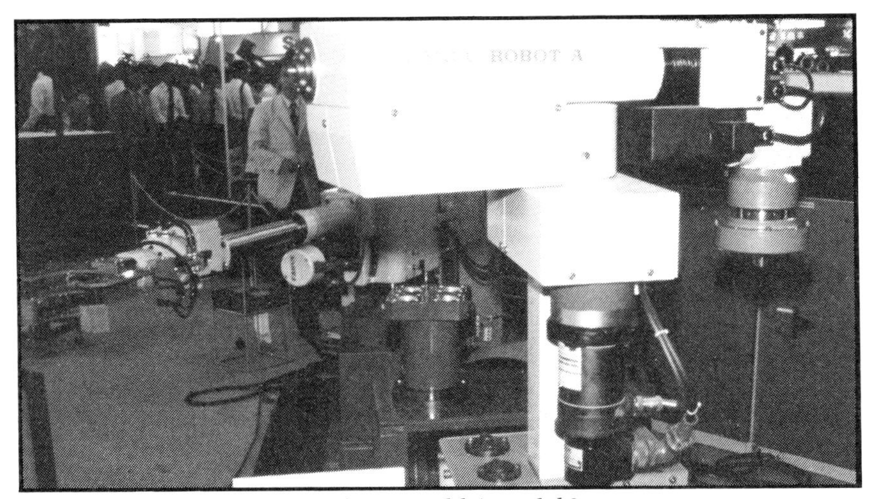

Fig. 1. Fanuc's new robot A (for assembly) model 0.

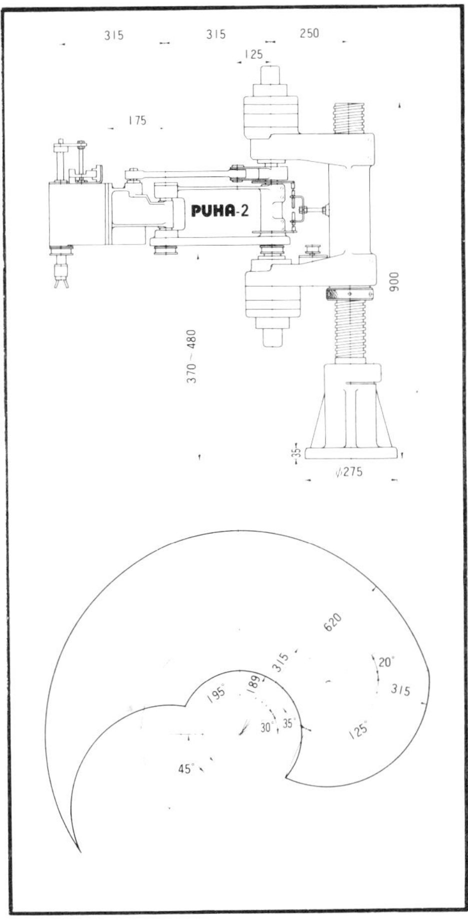

Fig. 2. Layout and assembly envelope for a typical Scara type robot – in this case the Pentel Puha-2.

Fig. 3. Hirata's new assembly robot is designed for integration into automated lines.

export orders and to ensure this, agency arrangements are to be set up in Europe.

Nitto's Picmat-Scara is available in 3 models – the 220, 422 and 620 with capacities of 6, 10 and 20 kg, respectively. They are now building 10 of these every month. The cost of the Picmat-Scara, in Japan is Yen 4-5 million (£10-12,000) and these prices are typical of all the Scara versions.

Pentel's Puha range neatly dovetails in with Nitto's. The Puha 1 has a capacity of 3 kg. and the Puha 2, 6 kg. Within its working envelope the Puha takes a maximum of 0.8 s. between any two points. With a maximum working speed of 800 mm/s. and an accuracy of ±0.05 mm. Pentel claim it can perform more effectively and faster than human hands. It has been developed through Pentel's in-house experience gained on assembly of small precision products such as pencils, rollers, etc. So far 20 have been sold.

NEC's model-B robot is somewhat similar to the Scara, but the model-A, designed to operate with a repeatability of ±0.008 mm, is quite different. It consists of a pair of small arms hanging down from a frame above a table, similar to but smaller than the Olivetti Sigma. Each arm can move in the x and y axes, and can be moved vertically in the z axis, and it can also be rotated. There can be up to 10 axes of movement, all controlled by an Intel 8085 cpu.

IC assembly

Interestingly enough, the linear movements are all acutated by linear motors, while stepping motors are used for rotary motion. Each arm can move 500 mm longitudinally, 270 mm laterally, and 150 mm vertically. NEC is already using this machine to mount integrated circuits on substrates, this being a complex operation taking 2 min. 20 s. Clearly, a special purpose machine, but one of a type that is likely to become increasingly common in the future.

Designed to work as a flexible tooling robot with the model-A is NEC's cylindrical coordinate *handling robot.* There are two versions available, one with an 0.5 kg. capacity and 180 mm horizontal reach and the other with 2 kg. and 400 mm, respectively. Each type is micro-computer controlled with six independent axes.

NEC has been using all three robots, a total of 60, in its own factories for two years now so has built up experience of their use. Its plans for commercial sales are almost frightening – 3,500 over the next three years! Initially, sales will be restricted to Japan with exports scheduled to start in 1983.

This was a significant exhibition. Many robots that are likely to be used for assembly purposes were being displayed, even if these do lack the sophistication that many researchers feel is essential for assembly. But the Japanese are past masters at getting just the right level of technology to match the situation. The indications are that they have their sights on robot assembly and on past expeerience they will make it happen.

Fig. 4. Sankyo assembly robot.

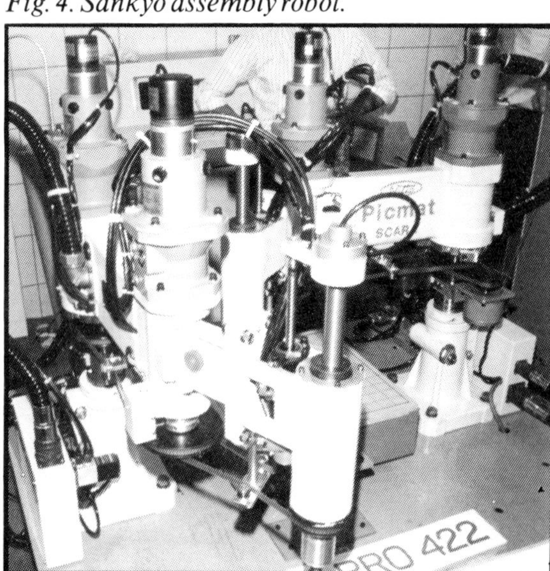

Fig. 5. The Nitto Picmat Scara robot picking-and -placing.

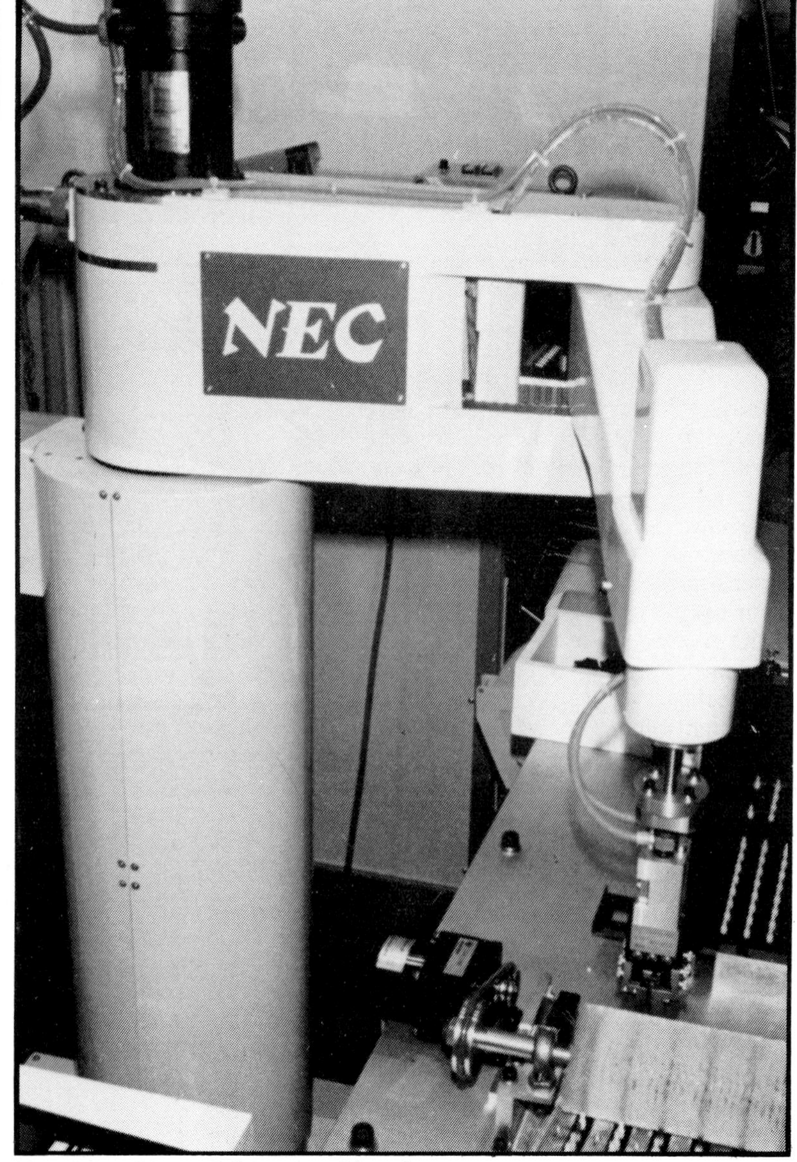

Fig. 6. NEC's model B assembly robot.

MITI lends a hand

While more robot models are introduced, the Japanese government starts a long-term research project.

AS PART of its aim of pushing Japan into the role of the foremost high-technology nation, Japan's Ministry of International Trade and Industry (MITI) is to start a vast research project, which will surely make Japan's efforts so far seem small fry. It is to spend £40 million on research into 'intelligent robots'.

Meanwhile, a number of new robots have been introduced, although these are improvements more than anything else. But coming from another MITI project – one to produce a totally flexible manufacturing system – is an assembly robot of considerable interest. Meanwhile, Matsushita has made a determined bid to enter the assembly robot business with a range of new machines.

MITI, no doubt alarmed by the fact that so far Japanese robot makers have tended to look overseas for help developing sensors for robots, has decided to invest this big sum in a seven-year project starting in 1983. The project will be handled by the Electrotechnical Laboratory of the Agency of Industrial Science and Technology (AIST) and some 10 robot manufacturers, the main aim being to develop new sensors and systems and robots for specially difficult environments.

Specific aims include the development of:

- □ sensitive tactile and visual sensors;
- □ an ultra-small processor to control the vision sensor;
- □ an arm capable of moving heavy objects with great precision;
- □ a highly efficient electric motor for robots;
- □ lightweight robots.

Evidently, most of the effort is to be placed in a project to develop a universal robot with these new sensors, even though there are signs that robots are becoming more specialised. On the other hand, though, MITI is also including in this project special robots that can operate in space, in nuclear reactors, and on the seabed, as well as robots to help handicapped people.

In the first year, about £5 million will be spent on fundamental research, the plan being to give the Japanese robot industry the sort of lead that the Japanese semiconductor industry gained from the vlsi – very large scale integration – project. And to think that the British government, after discussions with MITI about a year ago, was saying that it was confident that Japan would not be sponsoring any major projects in the future! Apart from this one, there are projects to develop new materials, including composite and ceramics, the fifth-generation computer, and the semiconductors of the next century.

Hitachi's latest Process robot has a 16-bit processor and crt mounted in the top face of the control console.

All big projects, are all inter-related in some way or another. For example, a project to develop a robot with vision sensing will gain from the fifth-generation project's hardware and software, with new ultra-fast processors and the ability to respond to voice commands. Then, new companies will help in the development of a lightweight robot – and so it goes on.

In contrast with that homegrown effort, Dainichi Kiko, lacking the necessary technology, has enlisted the aid of GCA Corp. of Bedford, Mass, USA to develop robots with sensors and CAD/CAM systems. In fact, the two companies have already agreed that GCA, which specialises in the production of manufacturing equipment for semiconductors, should sell Dainichi Kiko robots in the USA and Canada. The agreement will last five years, and the plan is that GCA will start building robots in the USA by 1986, under a joint venture. In the first year, GCA expects to be able to sell 240 robots, with an increase in sales of 50% pa thereafter – implying that there will be sufficient business to justify a 50 units/month plant within three years. Clearly, a substantial volume.

Homegrown

But on top of that, the two companies will work together on CAD/CAM systems and visual and tactile sensing. Significantly, GCA will enlist the aid of US academic institutions in these projects, so it looks as if Dainichi Kiko will gain access to a lot of new technology which it cannot find in Japan.

Not content with that news, Dainichi Kiko introduced a cheaper articulated robot for arc welding. It has a 16-bit processor, and positional accuracy of ±0.1mm, the company claims. However, to stimulate demand, the price has been reduced by some 40% to about £14,000.

Meanwhile, Fujitsu, anxious to use robots in assembly of electronics parts – it has already announced that it plans to use robots on assembly in a new factory for small computers and peripherals – has developed a robot with a claimed positional accuracy of only four microns. This is considered to be sufficiently accurate for assembly of small electronic components and watch parts, and has been tested in joining optical fibres.

A tiny robot, but this time intended for educational purposes, has also been introduced by Mitsubishi Electric. The 'Move Master' is a five-axis jointed arm robot and is only 245mm high. The robot can be controlled by a personal computer, using Basic or a new language devised by Mitsubishi called M-Roly.

It has a metal-plate structure and the joints are actuated by stepping motors claimed to give repeatability of ±0.3mm. The robot weighs only 10kg, and can lift objects weighing up to 500g in weight. The body can rotate through 240°, the shoulder 150°, the elbow 120°, and the wrist 360°. The wrist can also bend through 180°, while the hand can open to grasp an object up to 80mm in width. Mitsubishi Electric plans to sell the machine, with its controller for about £1,400.

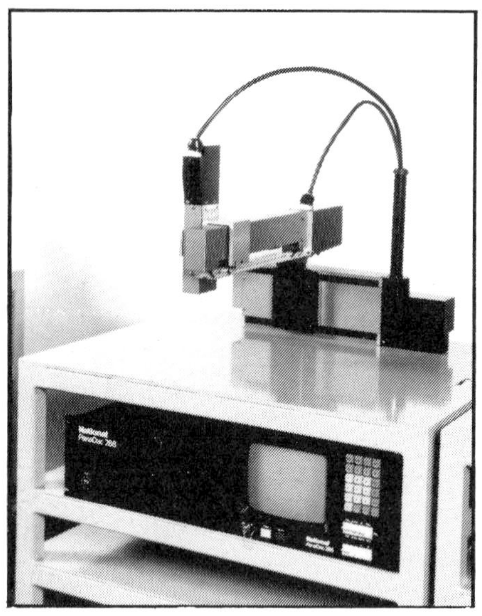

Matsushita Pana-Robo A-1 assembly robot.

The Arcman robot, a rather ponderous device produced by Kobe Steel, has been joined by an altogether much more modern and compact arc welding robot – the Arcman S. This is a compact jointed-arm five-axis machine actuated electrically, the motors being concealed in neat housings.

It is a 5kg capacity machine, but is intended primarily for arc welding, its main features being a self-tracking control and automatically optimised welding parameters. Kobe Steel has collaborted with Shinko Electric in the development of suitable electronics for the control of the robot and of the welding parameters.

The positions and length of welding runs are taught, and the electric current, voltage, speed, torch angle and number of passes are optimised automatically by the controller. Also new is a sensing system that detects the position of the start of the weld, using the welding rod as a probe. Then, during welding, the controller monitors the welding current, and can thus detect any divergence between the path being followed and the required path – this seems to be similar to the system shown by Yaskawa at the last Robots Show in Tokyo.

Accuracy

The Arcman S is claimed to have a positional accuracy of ±0.3mm, and a memory with 700 steps. The arm can revolve through 270° on its base, the lower and upper arm can articulate through 93° and 78° respectively, while the wrist can bend through 135°, and twist through 400°.

The Arcman S will cost about L23,000 with sensor. Production is expected to build up to 30 units/month by October, and by that time, the table, is a circular drum-type tool carrier. These have vertical axes, and the one near the horizontal arm carries grippers of many types – including one with remote compliance – while the other carries a variety of tools.

Hitachi has updated its Process Robot, by adopting a 16-bit processor and magnetic bubble memory. The new processor results in faster handling of data, and the potential for more precise computation. To simplify programming, there is a 9in crt built into the controller, while the magnetic bubble memory doubles the capacity from 1,000 to 2,000 program points and job steps. The price of the robot in Japan is about £24,000 and Hitachi expects to be able to boost output of its Process Robot to 70/month by the summer owing to this development.

Meanwhile, Hitachi is concentrating its research effort on developing its robots for assembly. It is already using some experimentally in its factories, relying on the precision of the robot, together with ancillary equipment to overcome the lack of vision. The Hitachi marketing people are evidently not keen to try to sell the vision system developed by Hitachi yet owing to its cost so this project is now on the back burner. For welding, though, Hitachi is developing its sensors used in the Mr Aros robot further as a cheap alternative to vision sensing.

Nachi-Fujikoshi, maker of the Uniman robots, claims to have developed a new cheap but precise sensor. It will be used in the 7000 Series arc welding robots first, and then in the 5000 Series spray painting robots. The price in Japan of the robot with new sensor will be £23,000. The company is building a new robot assembly shop in its main factory, and the new robots will be built there starting in the late summer at the rate of 15-20/month, it says.

Big push

On the commercial front, there is news of a Japanese company deciding to import robots into Japan. Murata Machinery has concluded a deal to sell Prab robots in Japan, and if volume rises sufficiently, it will also assemble the robots locally. Murata is already building the Digitron Robotrailers under licence, so it is concentrating on the larger machines which will be used in handling, complementing the trailers.

Matsushita has decided to enter the robot business in a big way – and demonstrated 70 machines at an exhibition it held in Tokyo in June. Its main effort will be in assembly and fastening.

Just over a year ago, the company introduced its Pana-Robo jointed arm robot – an ASEA lookalike – for arc welding and spraying, and it is already using 100 of these in its own plants. Now it has added four more types of

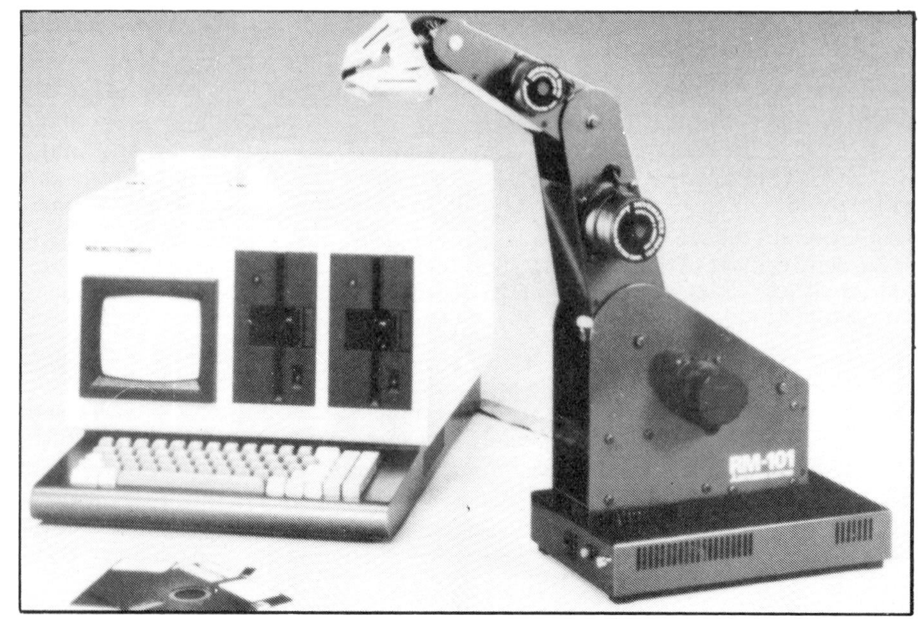

Mitsubishi Electric's tiny Move Master robot intended for educational purposes.

robots: the Pana Robo T and A types are Cartesian-type arms/tables, while the H and V series are jointed arm robots.

The Cartesian coordinate machines are available in three sizes, giving motion covering a table 220 × 150mm, 400 × 300mm and 600 × 400mm. Load capacities are 2-12kg for the arms, and 35kg for the x-y table. They are designed to move at speeds of 400mm/s, while positioning accuracy is put at ±0.05mm.

Similar in concept to these arms are the nut-running models, which incorporate automatic nut/screw delivery from hoppers through chutes. These operate at speeds up to 200mm/s, and repeatability is put at ±0.15mm.

Matsushita plans to build the A and T models and the nut-running units into assembly systems, some examples being shown at Tokyo. For example, one row of nut-runners were shown tightening the screws in a video cassette. In another line, an assembly line for small components was simulated. At the first station, a robot placed a plastics pallet on the line, and at the second, a small component was inserted. Subsequently, robots carried out sequences simulating sealing, wire bonding, and insertion of screws.

Shown separately was a 'conveyor module' a chain driven system for small parts. Stoppers are provided at each station, and at the corners are 90° turntables. Matsushita also makes a range of multiple-head screwdrivers whose positions can be set to suit different products.

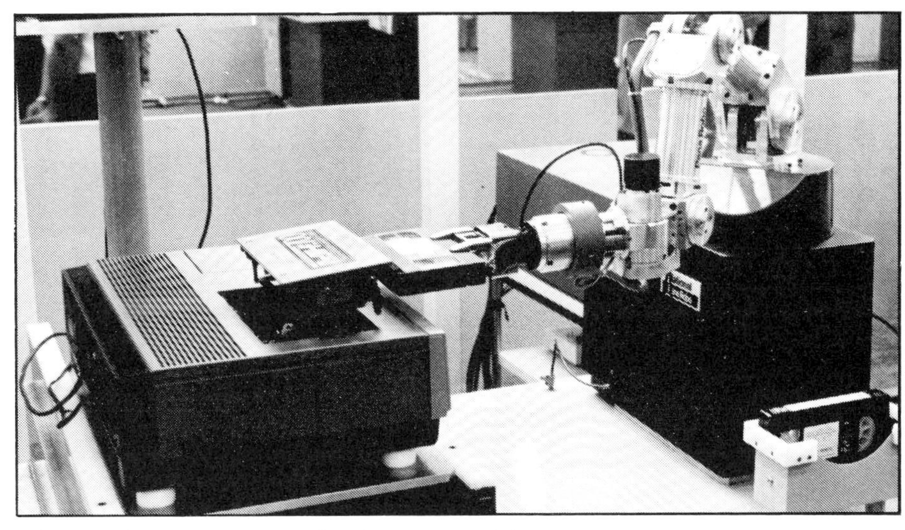

Matsushita vertical type robot inserting a video cassette into a recorder.

Matsushita screw loading/fastening robots.

But the main interest, of course, centres on the assembly robots many of which will be sold as stand-alone models. The Pana Robo H or horizontal robot is like the Scara machine, with a double-jointed arm with vertical pivots. The largest of these machines has total arm lengths of 500 and 700mm, and can normally carry up to 8kg. The smaller models are available with total arm lengths of 400, 600 and 800mm, but can carry only 2kg. They also have 100mm vertical movement at the wrist for operation of a tool. The Pana Robo H model NM-6650, for example, has a capacity of 5kg, and can rotate through 270° on its pillar. The 'upper' arm is either 200 or 400mm long, and the forearm is 300mm long. Maximum speed of movement is 1,000mm/s, and both continuous path and point-to-point control is available.

Matsushita horizontal type robot demonstrating pick-and-place movements.

There are also three models in the V range, with up to six axes of freedom. These are built on a modular system, so that a wrist gripper can be used alone, or can be built on to a forearm, or a complete arm and base. Matsushita has developed a new range of 16-bit controlled systems for all these robots.

Currently, Matsushita claims that its factory automation division, which also sells motors, a host of components, and industrial cameras, has a turnover of £35 million pa. It is expecting growth of 40% pa to give it a turnover of £230 million in five years' time. As yet it has not fixed the prices of the jointed arm robots, but these are expected to be under £10,000 each. The A and T type are in the region of £4,500-£6,500 each.

Assembly project

In 1977, Japan's MITI started a project to develop a completely flexible manufacturing system, which included a complex assembly robot. The prototype system has now been built by the various companies involved, and the plan is to assemble the various units together soon. The system includes powder metallurgy, a general purpose NC forge, laser beam heat treatment cutting and welding, NC machining and robot assembly and inspection. The budget, spread over seven years, was set at 30 million, and the project is due to be finished by March 1984.

Toyoda Machine Tool was chosen to build the assembly robot, which may seem odd since it has no robot experience. However, it seems that Toshiba, Japan's largest machine tool builder, was given the NC machining project, Toyoda as the second biggest maker of machine tools, was in a strong position to press for the robot project.

Be that as it may, the robot has now been built, and a complex device it is. Indeed, it is so complex that Dr Ryuji Wada, head of R&D at Toyoda, considers it a trifle impractical. The system consists of three robots: one to mount the parts on a pallet, one for assembly, and one for bolt/nut tightening.

At the first station, an arm of the polar type is mounted alongside a fixture that can carry two pallets. This is a four-axis robot, and adjacent is a tool drum carrying six grippers. The robot arm loads the workpieces on the pallet in various positions.

At the second station, there are two arms mounted on a massive four-post structure approximately 4m × 2m × 3m high. Both arms are of the cartesian type, and each has four axes of freedom. In addition, the pallet carrier can rotate to give a ninth degree of freedom.

On the side of the structure are a pair of horizontal rails along which the arm and its carrier can slide. In addition, the carrier incorporates a pair of vertical slides, so that the arm, which is telescopic, can move vertically. The overhead arm hangs down from a beam so that it, too, can move in the x and y axes, while the arm can also move vertically. The wrists of both arms can be rotated, and actuation is by recirculating ball screws to give a suitable combination of accuracy and strength.

Adjacent to each arm, but outside the table, is a circular drum-type tool carrier. These have vertical axes, and the one near the horizontal arm carries grippers of many types – including one with remote compliance – while the other carries a variety of tools.

There is a separate small fixture with a gantry-mounted Cartesian arm which is to be used to tighten screws and nuts.

Circular

The system has been designed to assemble gearboxes and spindle/bearing/housing assemblies weighing up to 30kg. In the sample gearbox, there are two parallel shafts, each carrying a gear, while the housing consists of a central portion with end plates.

The principle of operation is that a trolley, following cables in the floor, takes parts to the mounting robot, and then carries them to the main assembly robot and fastening robot. Subsequently, the assembly would pass to an automatic inspection machine. There is a tower carrying about 20 pallets and assemblies serving as a buffer store, and the trolley can take the assemblies to and from the store to the various robots to allow assembly in random order.

Clearly, this is an elaborate system, designed to allow almost unmanned operation in all stages of manufacture including assembly. But seven years is a long time, and as the project has progressed, so ideas on flexible manufacture have changed. Now, the system seems too universal and grandoise to be practical and saleable. However, as Wada pointed out, this project has encouraged many companies to think about unmanned operation, and so has been beneficial.

In addition, of course, Toyoda itself has learned a lot from the project, and so plans to proceed from the manufacture of DNC machining systems, which it has been producing for some time, to assembly robots. 'We will start making assembly robots next year,' said Wada, and the first will be used by Toyota Motor Co.' The Toyoda assembly robot will be a modular design which will be built into flexible assembly systems. The first machines will be designed to handle quite small parts.

So one more company, with the most solid backing possible in the form of the Toyota Motor Co., is about to enter the assembly robot business. Of course, it will concentrate initially on systems for the motor industry, but eventually it will build equipment for other industries as well – pointing the way to the special-purpose assembly robot.

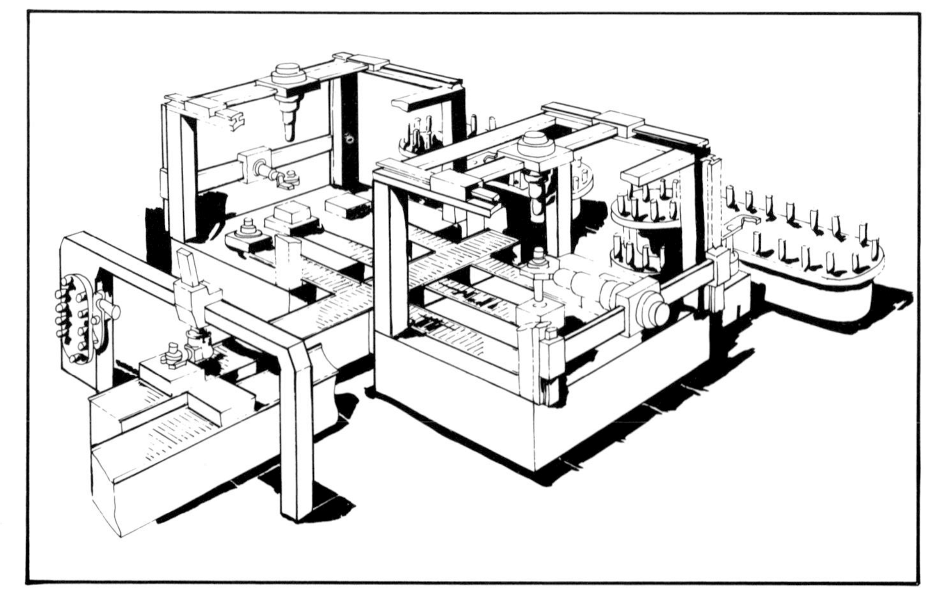

Concept of the assembly system developed by Toyoda Machine for the MITI Flexible Manufacturing System project.

More assembly robots

Seiko is all set to introduce robots to assembly, and the accent is on assembly at the 'Mechatronics' Show.

MANY of the manufacturers of robots in Japan have developed robots partly for their own use, and Suwa Seikosha, part of the Seiko watch group, is a case in point. The main area in watch production where automation is not used is in the final assembly of the vement, face, hands and glass to the casing. It is interesting therefore that Suwa Seikosha sees robots as the solution, despite the fact that great precision is needed.

Thus it has developed the Seiko SSR-H Scara-type robot, and plans to make 40/month by the end of the year, when exports will start. Prices will be in the £10,000–14,000 range in Japan. These are the smallest Scaras, with total arm lengths of 250mm and 410mm, and Suwa Seikosha claims they are the most accurate, with a repeatability of $\pm$0.01mm, compared with the more usual figure of 0.05mm. The arm can move the gripper or tool at 1.2m/s, while the arm can articulate on the base at 385°/s.

Fig. 1. Seiko robot set up to transfer chips from tapes to substrate

Movements

So far, these robots are being used to palletize watch movements at the end of the line only. However, trials of a number of other applications are at an advanced stage, so 50 or so should be in use by the summer.

First applications include the insertion of the battery in the movement of a quartz watch, palletizing of the tiny minute and hour hands, and palletizing of the watch faces. For battery insertion, 10 watches are mounted on their sides in a magazine which is passed along a conveyor to the robot station. The robot picks the battery up, drops it in a jig for centering, picks it up again, and then turns it through 90° so that it can insert it in the watch horizontally. However, since the battery must pass round a contact, the robot arm moves in at an angle, and then straightens up afterwards.

The palletizing operations are preliminaries to more complex jobs, currently under trial in Suwa Seikosha's development department. For example, the palletized watch faces will go by conveyor to another robot station, where a robot will pick the face from the pallet, place it on a fixture and mount the bars that indicate the hours. The assembly fixture is immediately in front of the robot, while there are vibratory bowl feeders to its left, and the pallets for completed faces on its right. Therefore, the complete working area of the

Fig. 2. Ancillary equipment has been produced to allow another Seiko robot to place tiny bars on the face of a watch

robot is used.

To pick up the tiny bars that indicate the hours on the watch face, Suwa Seikosha has developed a pneumatically actuated gripper which has a circular pad with 12 recesses for the bars. Each bar has two tiny pegs on one side to fix it to the face, so they must be carefully orientated before being picked up.

To check whether the bars have been installed, an optical fibre directs beams of light at the holes. If the light passes through the hole, the face is placed in the reject pallet.

Also under development is an application in which 30 different chips are picked from a tape and mounted on the substrate of a watch. Alongside the robot is a magazine of 30 reels of tape, each carrying tiny chips less than 1mm square. A sprocket engages with holes in the tape to index the chip forwards to the pick-up point. With this arrangement, the chip is located with sufficient precision for the robot to pick it up.

Meanwhile, the substrate for the watch ic, at this stage a rectangle about 20mm square, is fed from a pallet along the conveyor to the robot station. The robot takes 2–2.5s to place each chip, prior to bonding.

At the 'Mechatronics Show' in Tokyo in February, there were a few new robots on display, mainly aimed at assembly. For example, both Sankyo Seiki and Nitto were showing x-y-z machines – with Sankyo claiming a repeatability of ±0.008mm for one model.

Sankyo's range includes an x-y table as well as two x-y robots. In all cases, there are models giving 300 × 200 and 500 × 400mm movement. Drive is by dc servos and ball screws, and there are optical encoders.

Sankyo has adopted the Hewlett Package GPIB program interface to compensate for deflections in motion, and therefore claims repeatability of 8μm for one model, and 20μm for the other. The more precise model is supported on three legs, and to obtain such precision, it moves slowly, at up to 100mm/s. The other machines are carried on two-legged gantries, and move at up to 450mm/s. These robots have capacities of around 4kg, and the table has a capacity of 20kg.

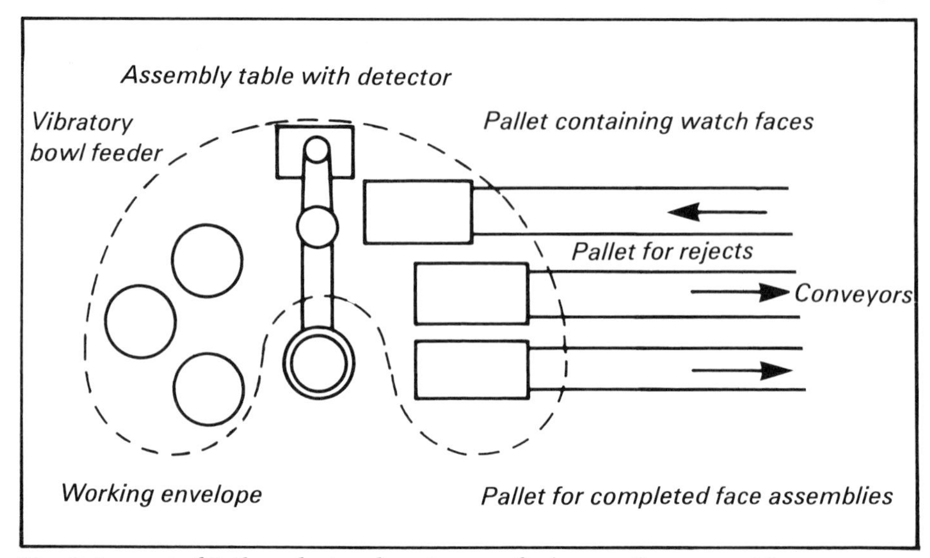

Fig. 3. Layout of Seiko robot and equipment for bar-to-face assembly

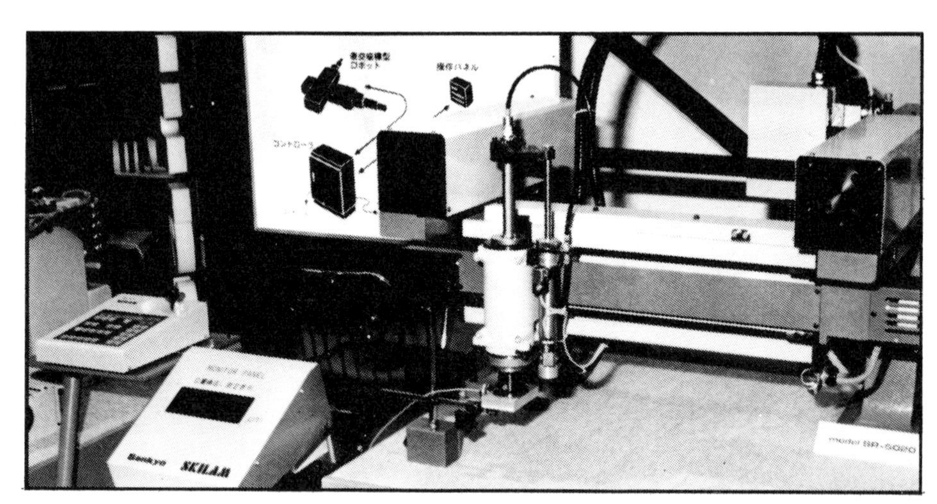

Fig 4. Sankyo Skilam x-y-z robot SR-5020 has a claimed repeatability of only 8μm.

Four different models are available

By contrast, the Nittoman-M x-y robots can be carried either on a column, or one of the slides can rest on the workbench. Four different models are available, to cover working areas of 300 × 300 to 600 × 600mm, and repeatability is put at ±0.05mm. Capacities are 10–15kg for the robots, and 25–35kg for the table, while the arm moves in the x and y axes at 400mm/s.

However, as the work at Suwa Seikosha shows, the use of robots for assembly is still in the early stages in Japan. Moreover, the reasons are not the lack of performance of the robots so much as the amount of ancillary equipment that has to be developed to complement the robot – and that all takes engineering time and skill. Of course, 'intelligent robots' would need far less equipment, but even if available, their cost would almost certainly rule them out.

Assembly and greater precision in view

While Toyoda and Seiko are getting to grips with the use of assembly robots, Fujitsu is making a belated entry into the robot business.

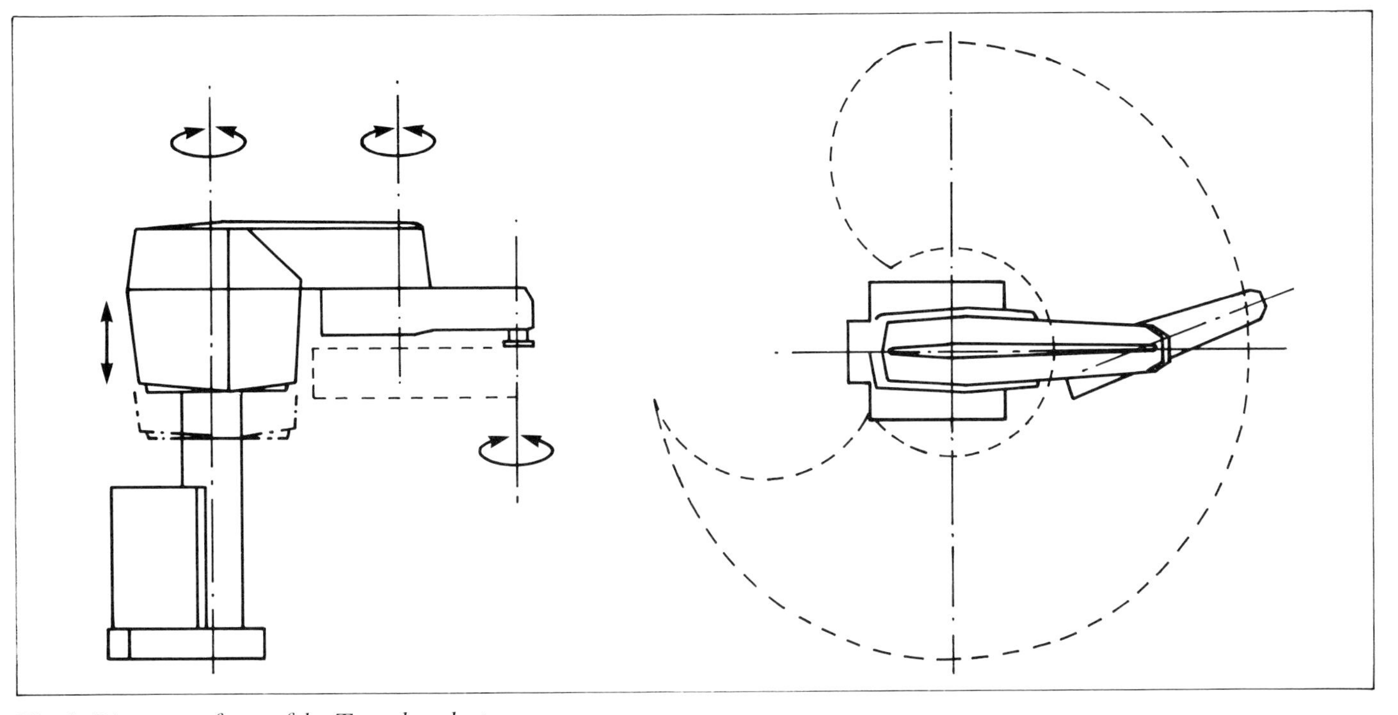

Fig. 1. Diagram of one of the Toyoda robots.

THE USE of robots for assembly, though attractive, and an essential feature of any true FMS, is a complex business requiring a progressive approach. Recently, Toyoda Machinery Works and Suwa Seikosha, a member of the Seiko group, have made serious moves towards robotic assembly with robots of their own design. The Toyoda robots are not particularly conventional in design, but are being used in assembly systems; the Seiko machines are small versions of the Scara, and are gradually being introduced for a number of jobs where small components must be handled.

The fact that Fujitsu, which has revealed a couple of robot prototypes in the past, is now producing its own robot is further evidence that large Japanese companies want to make their own key equipment, rather than simply buy it from a vendor, and robots are key manufacturing equipment anywhere these days.

Toyoda has developed three types of robots, and has also produced two systems to be installed in its own factories. The systems will be used to assemble control valves and power steering pumps for Toyota Motor Corp in high-volumes.

The three types of robots are: an articulated arm type; a modified Scara type, and an overhead cartesian co-ordinate type. A variety of grippers are available to handle the different parts. In fact, there are several variations, with the RA4 articulated arm unit being built in sizes with 2.5 and 6kg capacities. It can be mounted upright, or upside down on a gantry.

Since the RA4 models are intended for assembly, they are fairly small. The smaller RA4-1 has both upper arm and forearm lengths of 335mm, while both arm sections of the RA4-2 are 500mm long. These machines have four axes of freedom, both arm joints being able to articulate through 90°. There is also the RA6-2, which has two extra axes of freedom at the wrist.

All three RA robots, and the RC Scara type, have quoted repeatabilities of ±0.05mm, which is sufficient for most assembly jobs. However, the RA models are designed to operate faster than the other designs: the RA4-1 at 2.5m/s on point-to-point control (PTP), and the RA4-2 at 3.0m/s. Under continuous path (CP) control, both are limited to a more conventional 1.0m/s.

Two RC models are also available, the RC2-2 being a simple two-axis machine, while the RC4-2 has four axes of freedom. Of course, with this type of robot, the pivots of the arm have vertical axes, so the gripper can be moved across a large area. The basic model has articulation at the shoulder and elbow only, while the RC4-2 arm can move up and down on the column 125mm, and also have 405° rotation at the gripper.

In both cases, the flange of the tool holder at the wrist is 675mm above the ground, while the upper arm and forearm are respectively 500 and 315mm long. The gripper, which has a capacity of 6kg, can be moved at 2.3m/s under PTP, and 1.0m/s under CP control.

Toyoda's RR4-2 is more like an assembly machine, and is clearly a much simplified version of the highly-complex robot Toyoda developed for the Japanese 'Flexible Manufacturing with Laser' project. The robot arm hangs down from a beam on a four-post structure which has an area of 1,220 × 1,490mm. The structure is just over 2m tall.

The arm can move over a table 800 × 630mm, and move 250mm vertically. In addition, there is a fourth axis of movement – rotation of the tool holder through 360°. This machine is

Fig. 2. Seiko robot set-up to transfer chips from tapes to an integrated circuit.

Fig. 3. Ancillary equipment has been produced to allow another Seiko robot to place bars on the face of a watch.

designed for more precise work than the others, and it has a claimed repeatability of ±0.03mm. However, it does not operate as fast as the others, at 1.0m/s under both PTP and CP control.

Of course, all these robots share common control and actuation systems, the arms being actuated by servo motors. Main electronic units are a microprocessor controller and a cmos (complementary metal oxide silicon) random access memory (ram) with a capacity of 40 or 62K byte according to model. The cmos memory, which has a lower power consumption than other semi-conductor memories, has a battery back-up, to ensure that programs are not lost. It has enough capacity for 1,000 points in 99 programs to be kept. However, an external memory, in the form of a magnetic bubble cassette with a capacity of 128k byte is also available. For each robot there is a conventional teach box and control console.

Toyoda has designed two systems around these robots, the larger one consisting of 15 robots arranged in a row to assemble the 35 parts of hydraulic pumps for power steering systems. The system operates at a cycle time of 45s, to give an output of 20,000 units on an efficiency of 75%.

Toyoda Machinery claims that the system, now being installed in its Okazaki plant, cost about £1 million, with the individual robots priced at £14,000 to £22,000. It replaces a total of 20 men working on two shifts, so the nominal payback period is four to five years. Although this seems a long period, the system can remain in operation for many years, irrespective of design changes, while it can handle five different assemblies.

In the second system, three robots are grouped together to carry out 21 processes in 15s; they assemble control valves. In this case, the three robots are arranged in a system to make a flexible assembly machine. Pallets are supplied to a conveyor running beneath two inverted robots, suspended from the roof of the structure. Also on the line are a number of other special purpose devices to feed parts, and to check that operations are complete. The third robot is used to palletise the completed assemblies. Clearly, a step forward, although to some extent, one that demonstrates the limitations of robots in assembly.

Watch assembly

The work Suwa Seikosha has been doing with robots for assembly also shows what a long way we have to go before robots can be used for the large proportion of assembly, as a routine matter.

Suwa Seikosha is part of the Seiko watch group, and has a number of factories near lake Suwa, some 150 miles west of Tokyo, near the Japanese Alps. It produces watches, and also the Epson range of printers, personal computers, and peripherals such as floppy-disc drives.

It set out to develop robots to assemble watch components, and so great precision was needed. Assembly of the watch movement is largely automated by orthodox means, but there is little automation in the final assembly of the movement, face, hands and glass to the casing. It was decided, that here robots were the answer.

Thus Suwa Seikosha developed the Seiko SSR-H Scara-type robot, and plans to make 40/month by the end of the year, when exports will start. Kazuo Abe, assistant manager of production engineering, claims that the Scara robot is suitable for 70% of assembly work, and that it has better repeatability than the Unimate Puma. These were the reasons why the Scara was adopted.

There are four models altogether – two sizes and two levels of accuracy. The SSR-H253 and SSR-H414 have total arm lengths of 250mm and 410mm respectively. Abe claims they are the most accurate, with a repeatability of ±0.01mm on the vertical movement of the tool, ±0.02° on angular rotation, and ±0.015-0.03° on positioning in a horizontal plane. The speed of operation varies according to model, linear motion of the tool ranging from 1.0m/s for the SSR-H 253-H, which has no vertical motion at the tool to the 1.4 and 2.0m/s for the larger SSR-H414-H and SSR-H414, respectively. Capacity is nominally 1kg for the 253 models, and 2kg for the c414s. The arm can articulate on the base at 385°/s.

An interesting feature of the robot is that the Epson HX-20 hand-held computer is used as a teach box, the keys normally used to move the cursor are used to move the robot arm. It operates on PTP control, and there is normally as cmos ram with a capacity

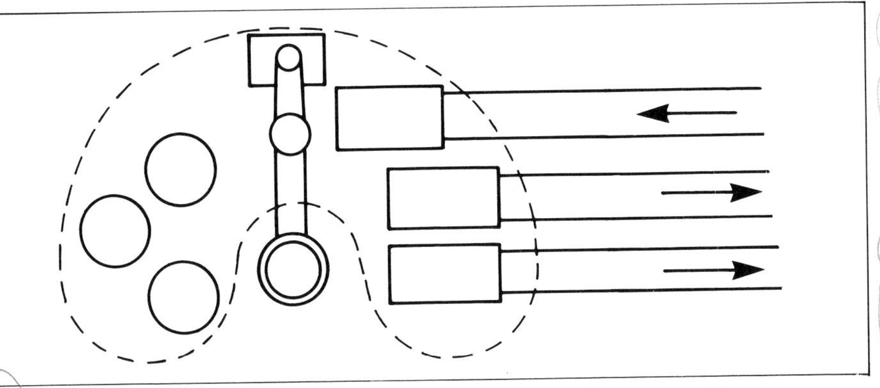

Fig. 4. Layout of Seiko robot and equipment for bar-to-face assembly.

of 16k bytes for program data storage. This memory can hold up to 200 points, but there is an option of a 24k byte memory while an external bubble memory cassette of 32k byte can be added. Prices will be in the £10,000-£14,000 range in Japan.

So far, these robots are only being used in production to palletise watch movements at the end of the line. However, trials of a number of other applications are at an advanced stage, so about 50 should be in use within a few months.

Quartz watch

These applications include the insertion of the battery in the movement of a quartz watch, palletising of the tiny minute and hour hands, and palletising of the watch faces. For battery insertion, 10 watch movements are mounted on their sides in a magazine which is passed along a conveyor to the robot station. The robot picks the battery up, drops it in a jig so that it can be centred precisely, picks it up again, and then turns it through 90° so that it can insert it in the watch horizontally. However, since the battery must pass round a contact, the robot arm moves in at an angle, and then straightens up afterwards. The line for this job was undergoing trials, within the production shop when the company was visited, so it was almost ready to be installed.

In the palletisation job, the robot picks watch faces from the tray, which they come on from another shop, and transfers them to a conveyor, which delivers them to a pick-up fixture. There, the robot picks up the face and places it on a pallet. Each pallet can hold five faces, and when full, the robot pushes it into a housing, and pulls the empty pallet from another stack.

Meanwhile, robots are currently under trial in Suwa Seikosha's development department for some other applications. For example, the palletised watch faces will go by conveyor to another robot station, where a robot will pick the face from the pallet, place it on a fixture and mount the bars that indicate the hours. The assembly fixture is immediately in front of the robot, while there are vibratory bowl feeders to its left, and the pallets for completed faces on its right. Thus, the complete working area of the robot is used.

To pick up the tiny bars that indicate the hours on the watch face, Suwa Seikosha has developed a pneumatically actuated gripper which has a circular pad with 12 recesses for the bars. Each bar has two tiny pegs on one side to fix it to the face, so they must be carefully orientated before being picked up.

To check whether the bars have been installed, an optical fibre detects beams of light at the holes. If the light passes through the hole, the face is placed in the reject pallet.

Also under development is an application in which 30 different chips are picked from a tape and mounted on the tiny printed circuit board (pcb) of a watch. Alongside the robot is a magazine of 30 reels of tape, each carrying tiny chips less than $1mm^2$. A sprocket engages with holes in the tape to index the chip forwards to the pick-up point. With this arrangement, the chip is located with sufficient precision for the robot to pick it up.

Meanwhile, the pcb, at this stage a rectangle about $20mm^2$, is fed from a pallet along the conveyor to the robot station. The robot takes 2-2.5s to place each chip on the pcb prior to bonding.

In Japan, most of the big computer manufacturers are wings of large organisations, with other electronic and mechanical engineering divisions. Therefore, several are already involved in robot manufacture, combining their mechanical and electronic experience. Typical examples are Hitachi, Toshiba, Mitsubishi Electric, and NEC. The exception is Fujitsu, whose main business apart from computers is telecommunications.

Now, however, after various experimental projects, Fujitsu has come into the robot business with the Farot, which is rather a disappointment. It might have been expected that Fujitsu would come up with its own design,

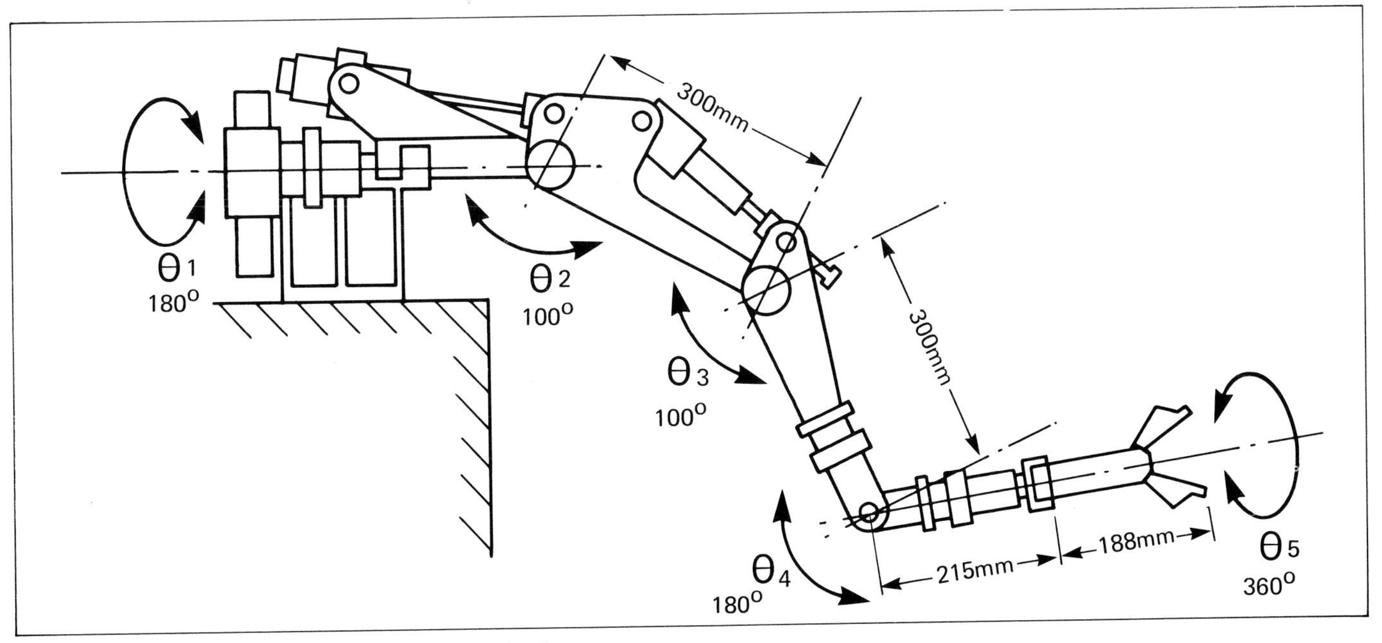

Fig. 5. Arrangement of Fujitsu experimental robot.

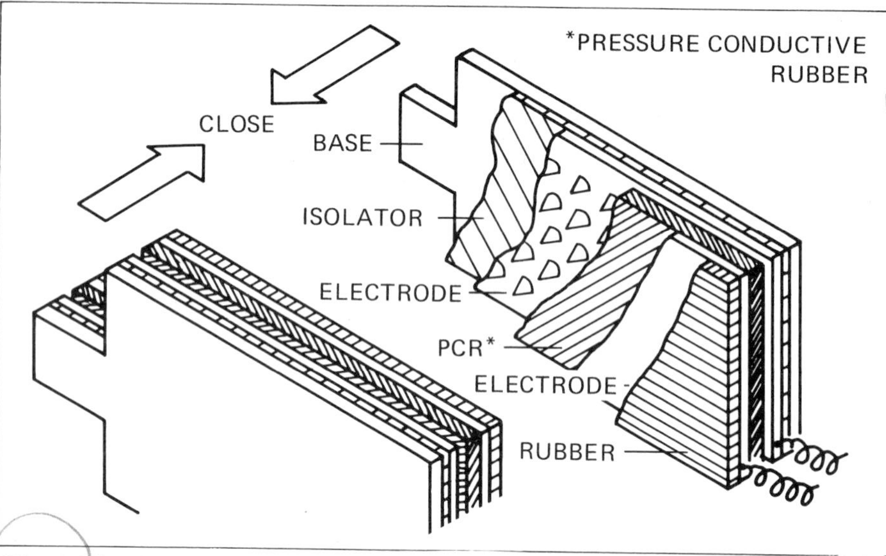

Fig. 6. The Fujitsu tactile sensor consists of laminations of rubber and electrodes.

but instead in has produced one more Scara robot. On the other hand, though, Fujitsu is involved in some interesting research on sensing, and has already produced a prototype with a repeatability of 4µm. The Farot was, therefore, obviously produced so that Fujitsu could get some experience of robot production and usage quickly.

The latest device to come from Fujitsu's laboratories is an articulated-arm five-axis robot with optical position feedback system and tactile sensing. To cut the time taken to compute the data regarding positions of workpieces, and to cut costs, many reseachers are looking for simpler devices than solid-state television cameras. Fujitsu is no exception; it has opted for psds – pin silicon photo diodes – instead of a camera. These devices are sensitive to light at a wavelength of 960nm, and so in conjunction with light emitting diodes (leds), they can be used as optical position sensors. (The pads are supplied by Hamamatsu TV Corp, 1126-1, Ichino, Hamamatsu).

The psds are used to locate the positions of the target light – a led is attached to the robot's hand – and the positions of up to 8 leds can be simultaneously measured in 3.3mm. Thus, the control system can measure two-dimensional positions in real time.

Pressure-sensitivity

To provide a pressure-sensitive hand, Fujitsu has built up layers of electrodes, pressure sensitive rubber, electrodes and rubber on an isolated base. Fujitsu claims that the gripper can sense a force of only 2g, while the gripping force is controlled by computing the dc motor currents needed to lift the object.

In practice, the psds indicate distance between the object to be transferred and the target, and this information is used to move the hand to the target. Thus, because the controller is measuring the relative distance between the two objects, and not their absolute positions, minor inaccuracies in positioning the robot at the beginning of the cycle, and deflections in the arm structure, do not affect the accuracy of the handling operation.

The two major joints of the arm are actuated by ball screws and levers, while the others are actuated by harmonic drives – the motive power being a 10 or 20W dc servo motor. Angular movement at the joints is detected by optical encoders and these are transformed, with the addition of a time base, into velocity and acceleration by the digital controller. The optical encoder has 4,096 pulses/rev.

Whereas it is normal for one central processor unit (cpu) to control all servos, Fujitsu has developed a multi-processor system. A minicomputer and five microprocessors are linked to a multi-processor bus (mp bus), an arrangement that simplifies the control software, according to Fujitsu's researchers. Repeatability is put at ±0.02mm.

Fujitsu has experimented with this robot in conjunction with a machine tool carrier, powered by a small linear induction motor. The carrier operates at up to 7m/s, and can be positioned within ±0.7mm. The robot moved the tool from the carrier to a tool box,

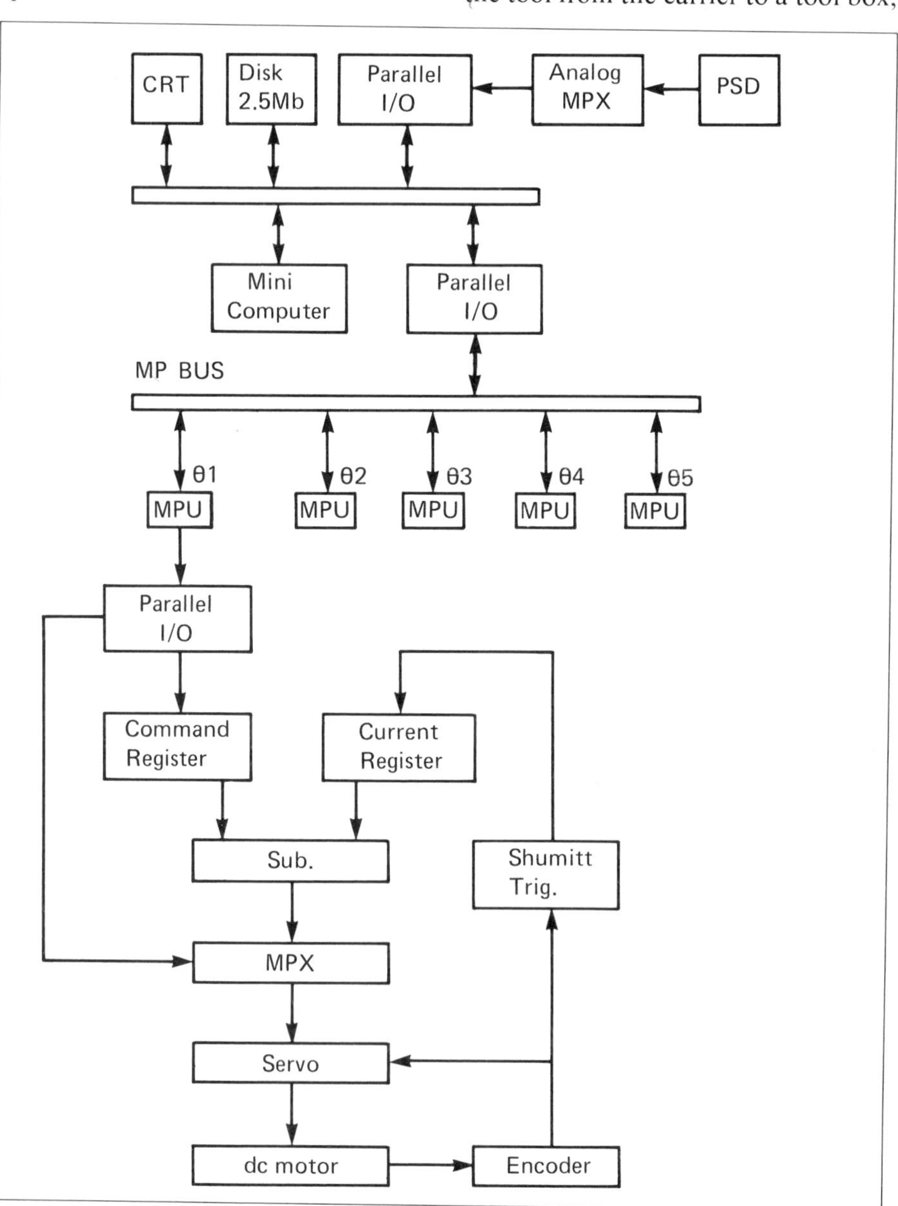

Fig. 7. Fujitsu multi-processor system.

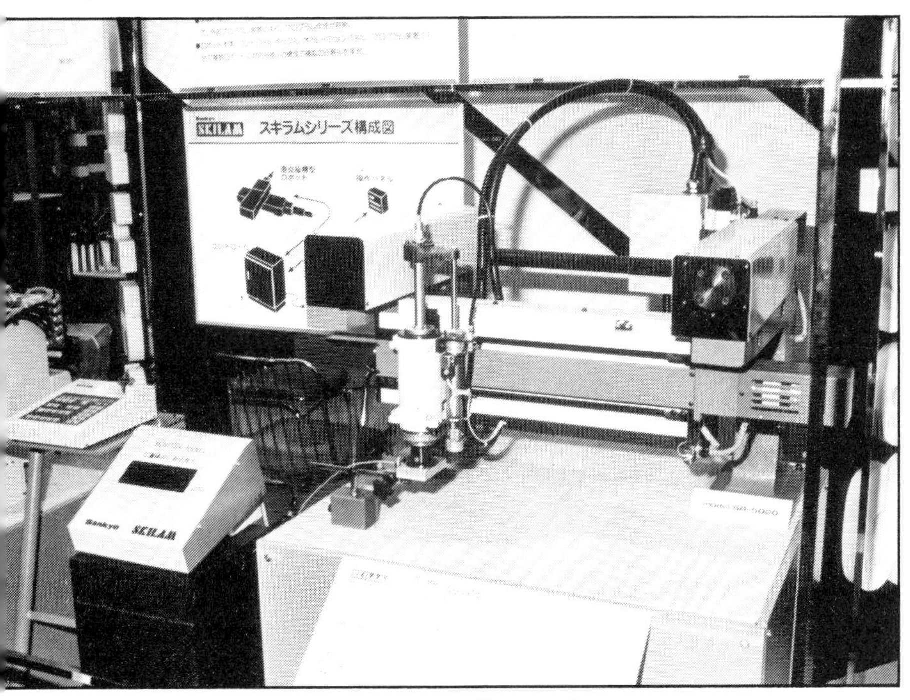

Fig. 8. Sankyo Seiki's new cartesian co-ordinate robot.

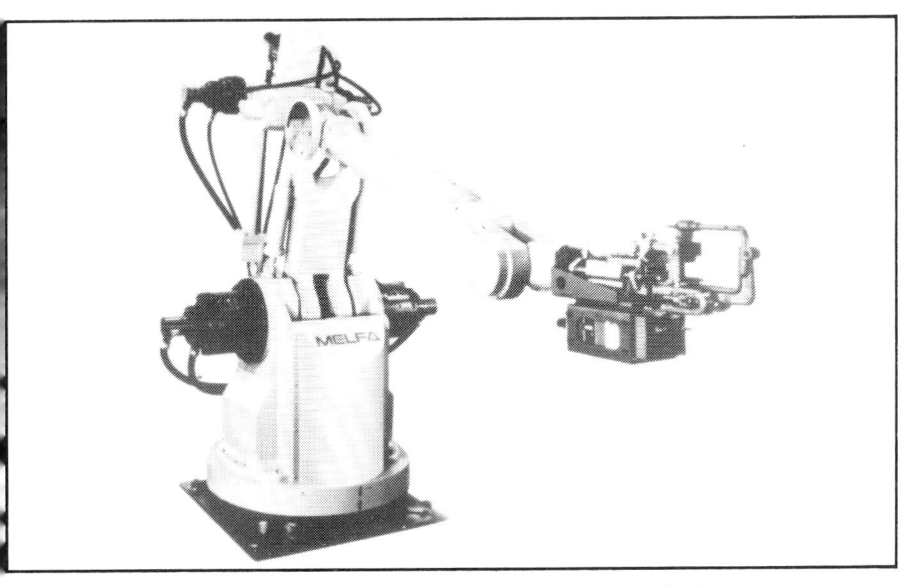

Fig. 9. Mitsubishi Electric's RV-242 robot has a capacity of 60kg.

under guidance from a psd above the track, and perpendicular to the bench on which the equipment was mounted. At the first attempt, the robot took 50s to load and unload the tool. At the second attempt, it learned from the first one, optimised the path, and so reduced the time to 18s.

Locus and position control

Thus, Fujitsu has developed a method of locus and position control. It claims that simpler algorithms can be used for this control system than when movements at articulated joints are converted to cartesian co-ordinators. In addition, the system is also suitable for tactile and optical sensors. Next, Fujitsu is to develop a control algorithm for an arm with six degrees of freedom.

For its Farot Scara-type robot, Fujitsu has trod a far more familar path. The Farot has three or four axes of freedom and a capacity of 3kg, but a normal working capacity of less than 1kg, and repeatability is put at ±0.05mm. The first and second portions of the arm are 225 and 175mm long respectively, so this machine is between the two Seiko robots in size. However, with a speed of 570mm/s at the tool, and an articulation speed of 90 and 180°/s, it is considerably slower than the Seiko robot.

The Farot has the ubiquitous Z-80 microprocessor, while the user's programs are stored in 10 or 16k bytes of cmos ram. In either case, 100 programs can be stored, but the smaller memory is limited to 850 steps, against 1,000 steps for the larger one.

Fujitsu has installed about 15 of these robots and is using them mainly to install switches, connectors and bypass condensers in pcbs.

Recently, both Sankyo Seiki and Nitto Seiki, two of the original Scara group have extended their robot ranges – with the addition of x-y-z machines. Surprisingly, Sankyo claims a repeatability of ±8μm (0.008mm) for one model.

Sankyo's range includes an x-y table as well as x-y robots. In all cases, there are models giving 300 × 200 × 500 × 400mm movement. Drive is by dc servos and ball screws and there are optical encoders.

Sankyo has adopted the Hewlett Packard GPIB program interface to compensate for deflections in motion, and therefore claims repeatability of 8μm for one model, and 20μm for the other. The more precise model is supported on three legs, and to obtain such precision, it moves slowly, at up to 100mm/s. The other machines are carried on two-legged gantries, and move at up to 450mm/s. The robots have capacities of around 4kg, and the table has a capacity of 20kg.

There are PTP controllers with program storage in cmos ram. Memory capacity is 6k bytes, which can be extended to 24k bytes. In Japan, prices are in the region of £7,500-£10,000.

By contrast, the Nittoman-M x-y robots can be carried either on a column, or one of the slides can rest on the workbench. Four different models are available, to cover working areas of 300 × 300 to 600 × 600mm, and repeatability is put at ±0.05mm. Capacities are 10-15kg for the robots, and 25-35kg for the table, while the arm moves in the x and y axes at 400mm/s.

At the other end of the market, Mitsubishi Electric, which before it came into the robot business a couple of years ago had a lot of experience as a maker of CNC, servo motors and computers, continues the proliferation of its robot range. Its latest model is the RV-242 six-axis jointed arm machine which has a capacity of 60kg. In addition, it has made some improvements to its arc welding robots, the RW 111, 211, and 213.

Actuation of the 60kg machine, as on the other Melfa robots, is by dc servo motors, while all six axes can be controlled simultaneously. Repeatability is quoted as ±0.5mm.

Competitive market

Mounted on a base that can be rotated through 270°, the first and second portions of the arm are 800 and

1,300mm long, respectively, and the gripper can be moved at up to 1,500mm/s, which is rapid for such a large machine.

Mitsubishi Electric plans to make 10 RV-242s monthly, and has priced the robot at Yen 17 million (£47,000). The production schedule for the three new arc welding robots is 50/month, while prices are from Yen 7.5 to 9.7 million (£20,000-£27,000).

With all these developments, robots are spreading through Japanese industry at a steady pace, and with the inevitability that drives innovation along in a competitive market. Now that Toyoda has unveiled its assembly robots, Nissan's engineers will be following Fujitsu into the use of robots for minor pcb assembly – if they are not already ahead. And so it goes on.

2.3 Welding robots

Now the emphasis is turning to robots for arc welding

A combination of new machines and new companies to make them is signalling a switch in emphasis from spot welding to arc welding robots in readiness for the next wave of increased sales, though this does not mean spot welding robots are by any means dead.

AFTER A SLOW START, sales of robots for arc welding in Japan are beginning to accelerate. Indeed, forecasts of sales are so optimistic that several new robots aimed purely at this operation have been introduced.

This year, sales of arc welding robots in Japan are expected to reach a value of 5-5.5 billion Yen (about £11 million), implying sales of around 600 units. But then, the industry forecasts that sales will accelerate rapidly to 21 million Yen (about £45 million) by 1985. That implies that 6,000 robots will be sold for arc welding in five years, over 2,000 of them in 1985.

It is not surprising therefore, that many companies are turning their attention to this potentially lucrative market, especially since there must be an equal number of sales opportunities outside Japan.

But the main reason there are expectations the market will grow so quickly is the catalytic effect that is a feature of Japanese industry. The Japanese love new gimmicks and new machines, and in business they are all afraid of being left behind.

In other words, many companies spend a lot of time 'keeping up with the Joneses'. This situation is made more intense by the fact that the shop-floor workers and middle managers have a bigger say in policy matters than in European companies.

Arc welding is generally a job done by what are thought of as small-to-medium size companies in Japan – 200-500 people. These are the makers of agricultural equipment, sub-contractors to the car industry, and a host of other sub-contractors making small fabrications. Once one of these small companies gets a robot, then its competitors will feel obliged to do the same – even though they may not be sure whether it is needed.

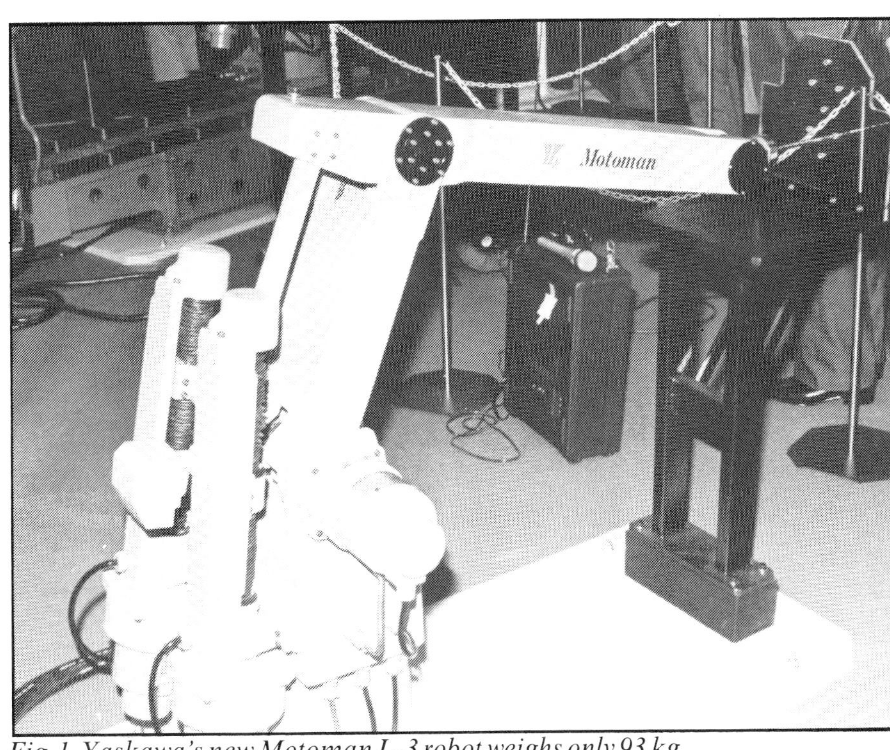

Fig. 1. Yaskawa's new Motoman L-3 robot weighs only 93 kg.

In that sense, the robot revolution in Japan could have dramatic effects; it may lead to an increase in bankruptcies in small companies – the rate is already high – as some unwisely invest in robots. But it may help the better managed companies to move up the ladder, by taking business away from the less-efficient, and those that are not prepared to invest in the future.

In other words, because the market is big, it does not follow that all the users will find that robots cut their costs.

Snowball effect

For all that, the robot makers know that once they can get the ball rolling, the snowball effect will guarantee them sales. These in turn will cut their manufacturing costs, so that they can attack the export market aggressively.

There is the added carrot that a robot designed for arc welding is generally suitable for simple assembly and pick-and-place operations such as palletising. Thus, the companies making efficient arc welding robots are likely to find other business as well.

Many of Japan's new robots were displayed at the Welding Fair in Tokyo in the Spring. Amongst notable absentees were Kawasaki, and Toshiba which appears to be making no inroads into this market at present.

New robots came from Shin Meiwa and Yaskawa among established companies, while Mitsubishi Electric

(Melco; not to be confused with Mitsubishi Heavy Industries, which makes spot welding robots) and Matsushita, better known under its National and National Panasonic brand names, were two big companies who have decided to break into the new market.

Not that Kawasaki, which is now making 60 robots a month, has forgotten the arc welding business. It recently introduced the A2850 machine, a five-axis robot working in combination with a three-axis table, to give eight axes, all operated from one control system. In the first application, on heavy fabrications for construction equipment, the robot is installed between two tables, so that one assembly can be welded while the other is being set up on the fixture.

Fujitsu Fanuc, which is making almost as many robots as Kawasaki now that it has started to operate in its new factory, has so far ignored the welding field. It is still concentrating mainly on robots for handling at machine tools, but is using some of its existing robots, in modified form, in the old factory, to assemble electric motors. It is planning to launch an assembly robot in the autumn, and is also working in conjunction with Fujitsu, its associate company in the Furukawa group, on a robot with visual sensing.

Market leaders

Yaskawa, Shin Meiwa and Hitachi are the leaders in sales volumes in welding robots in Japan. Yaskawa claims to have sold 550 of its Motoman L-10 robot, which is of the jointed arm type resembling the ASEA design. About 80% have been sold to the auto industry, 10% to the electrical industry, and 10% for handling. What is more 100 have been exported in the last couple of years, 40 to Scandinavia, 30 to Germany, and biggest disappointment of all, only a handful to the UK. Whether this indicates that British businessmen are too hard-headed to spend money on welding robots, or that the price is too high, or that GKN is learning how to develop another robot, is not clear.

In any event, the L-10 is available without welding equipment for around £21,000, and Yaskawa also supplies a small table that can be actuated by the controller.

For some applications, however, the L10 is too large, so Yaskawa has just introduced the dainty L-3, a smaller replica, with a gripper capacity of 3 kg instead of 10 kg. The layout is identical to the L-10, with four exposed dc servo motors, two driving the arm joints through ball screws.

Fig. 2. The Shin Meiwa Robel has a long reach, and can weld near ground level.

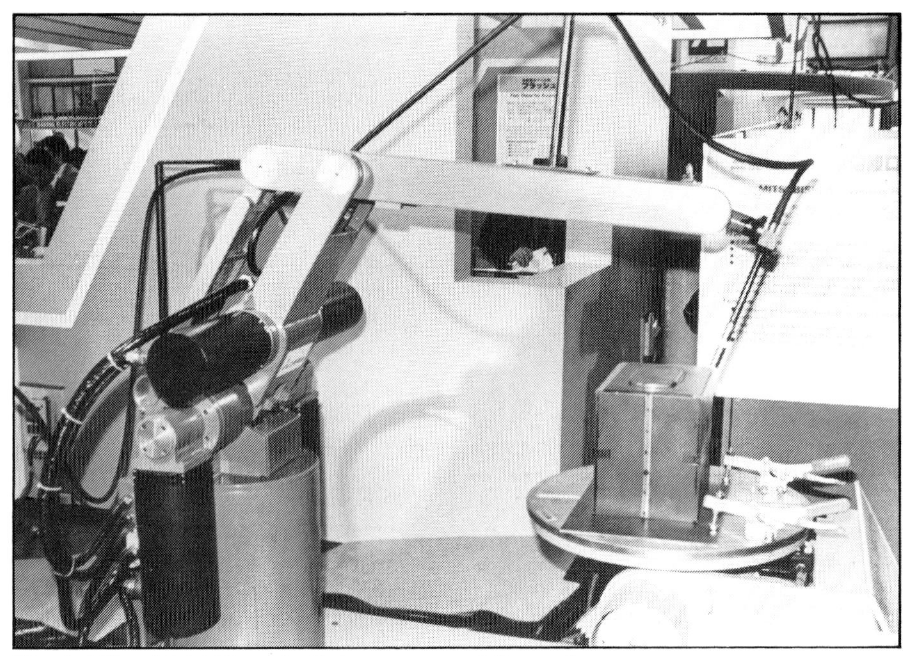

Fig. 3. Mitsubishi Electric Co's new arc welding robot

The base can rotate through 240° and the maximum reach is 919 mm compared with 1186 mm for the L-10. The umbrella-shape working envelope has maximum horizontal and vertical dimensions of 616 and 725 mm, against 820 and 929 mm for the L-10. Significantly the weights of the machines are 93 and 400 kg – an enormous difference. There is also a model with a longitudinal slide to carry the complete robot – this is Motoman L-3C – and this gives 800 mm movement.

Yaskawa claims a repeatability of ± 0.1 mm for the L-3, against 0.2 mm for the latest L-10, which is an improvement over the ± 0.3 mm of a couple of years ago. Both the robots have a new controller, the Yasnac 6000RG which replaces the 6000RB.

The main change is from hard wire to integrated circuit memory, which has a battery back up to prevent loss of memory in the event of a power failure. The memory is 8 x 16 k to give a total of 128 k byte, which means 1,000 points and 600 instructions can be stored.

Yaskawa expects the L-3 robot to be used principally to weld small components in the electrical and automotive industry, but production volume has been set at only three-to-five units a month. On the other hand, it has recently increased output of the larger robot from 25 to 40 a month, and expects this to reach 45 units this year. The L-3 is priced at £18,000 in Japan, so that it is cheaper than any of the other new models.

Try-out lines

Hitachi's process robot is also being

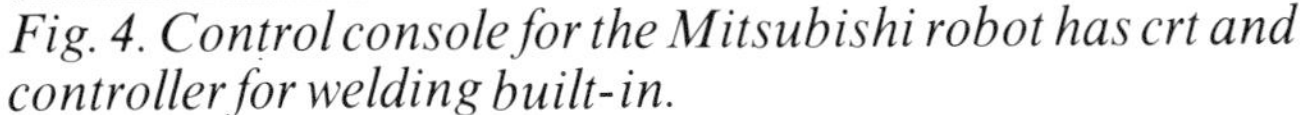

Fig. 4. Control console for the Mitsubishi robot has crt and controller for welding built-in.

Fig. 5. The Thor-K robot is similar to the ASEA machine.

used mainly for arc welding, which is not so surprising, since its design is also close to that of the ASEA. About 30 are being built monthly, with 80% going into the auto industry and its sub-contractors.

Significantly, though, about half of the robots sold are being used in try-out lines, to see where the robot can best be used. In other words, as yet, the Japanese are not convinced of the advantages of robot arc welding, but they are setting about solving the inevitable problems in the new applications. But because industry has got this far, all the new companies feel it is worth entering the field.

Shin Meiwa, whose robots are handled by Grundy in the UK, has concentrated so far on Cartesian co-ordinate column type robots that are unusual in that worktable has many axes of movement. Indeed, these are more like programmable welding machines than robots, and this may have helped them gain some popularity.

Not surprisingly therefore Shin Meiwa has been selling 20 units a month, even though these machines are very slow to program. Now however, Shin Meiwa is starting to produce a jointed arm robot, the Robel RJ-65. Production is expected to reach 30 a month this year, but at the same time, output of the other robots is expected to fall to around 10 a month.

With its combination of dc servo motors and harmonic drives, the five-axis Robel is typical of the new breed of jointed-arm robot. However, it looks more like a Puma on a stand than an ASEA machine, the dc servo motors being enclosed in a body on which the main arm pivots. The forearm is offset from the main arm, and the result is that the welding gun can cover a wide area.

Maximum and minimum horizontal reach are 1,357 and 440 mm respectively, while the gun itself can stretch down more than 700 mm below the arm pivot, and 1,297 mm above it. There is an ic internal and cassette tape external memory, the capacity being 500 steps in 8 blocks. What is interesting is the speed range with this machine – it can inch at 0.2-5 mm/s, and has speeds ranging from 12 to 20, 50 and 1800 m/s, the fastest speed being much higher than is general.

Teaching

Also interesting is the way in which the robot is taught. The software is such that throughout teaching, the welding gun is maintained at the otpimum angle. In addition, the robot can be taught either by operating each axis individually, or with them all operating together to give certain movements.

Thus, in the first case, the base might be rotated through 90° to move from one jig to another. Then, the operator can move one switch to change the mode of operation so that the gun is moved in the x, y or z axis.

Thus, the gun can be lowered vertically to the start of the weld, and moved horizontally in a straight line to follow the desired path. This is clearly easier than having to move the gun by rotating joints in sequence. It is also possible to operate two switches simultaneously to move the gun upwards at an angle, for example.

As on the other Shin Meiwa robots, by teaching the robot three points, an arc can be programmed, with four points identifying a circle. There is also the SLS sensing system available, in which the welding gun can find the starting point of a weld. Also of interest is the fact that the lower arm can turn through 180° so that it could move from a weld on the floor to one on the roof, keeping the gun in the right position.

An important innovation in the Robel is the use of a 16-bit microprocessor, whereas most companies are using 8 bit devices. However, it is reported that Hitachi has also moved into the use of 16 bit devices to enable data to be processed more quickly. Shin Meiwa is to build up production in the next couple of months, and plans to sell the Robel in Japan at £21,000. In the first month following its introduction, 20 orders were taken. It is expected that the first Robel will appear in the UK towards the end of the year.

Resemblance

Both the Melco and Matsushita robots resemble the ASEA design, and since both companies are involved in a big way in the welding business and in the manufacture of electric motors, this is a logical development. In fact, Melco insists that it alone is self-sufficient in all the requirements for the new robots – microprocessors, ics, CNC controllers, dc servo motors and precision machinery. Actually, Matsushita is lacking only in CNC controllers.

Despite its 8-bit chip, the Melco welding robot can be taught with all its axes moving simultaneously, like the Shin Meiwa Robel. In addition, during teaching, the welding gun is maintained at the optimum angle.

There are the usual five dc servo motors, one mounted in the base to rotate the robot, and the drives generally being harmonic. Two motors hang down from the bottom pivot, these two actuating the two arm joints – shoulder and elbow. The two motors mounted on the lower arm drive through chains to control the wrist axes. It will be interesting to see whether these chains are capable of maintaining the precision needed for the claimed repeatability of ±0.2 mm.

Melco adopted the ASEA-like construction because this allows faster movement than the column and arm type – in this case 1,000 mm/s. The robot arm can be rotated through 300° on its base, and the envelope is typical for this type of arm. Maximum reach is 1,280 mm horizontally, and the gun can be used to weld only 280 mm from the vertical axis of the base. The maximum vertical and horizontal dimensions of the envelope are 1,129 and 1,000 mm respectively.

Equivalent envelopes for two more robots of this type, the Matsushita Pana-Robo AW 1000, and OTC Thor-K are: maximum horizontal reach and angle of rotation, 1,320 mm and 300°; 1,445 mm and 320°; minimum distance, axis of base to welding point, 270 and 390 mm; maximum vertical and horizontal dimensions of envelope: 1,270 and 1,050 mm; 900 and 1,055 mm. By comparison, the Robel has vertical and horizontal envelope dimensions of 1,947 and 1,297 mm. Of course, in many applications, a small envelope is more than adequate.

Features

In any case, there are some worthwhile features of the Melco robot that will appeal to users. First, there is one hand-console for teaching, and a separate one for operation. Then, the controller is incorporated in a console with a 9in. crt and a CNC controller for the welding sequences built in. The welding conditions are shown on the crt, and if necessary these digits can be magnified, although then less steps are shown.

Overall, this seems to be a neat package, especially since Mitsubishi can also supply the welding equipment to suit. However, despite the ability of being able to move the gun in the x, y, and z axes, the interpolation is not so good. For example, to teach a circle it is necessary to identify a whole series of points at intervals of around 5 mm.

Mitsubishi has set the price of the machine at £21,000 or £23,000 with welder, and is planning to make 30 units a month.

Matsushita's Pana-Robo AW 1000 and Osaka Transformer Co (OTC)'s Thor-K are very similar to one another except for the way in which the dc servo motors are mounted. Both have the conventional point-to-point controllers, ic memories and a capacity of 800 points. Both these companies have also introduced column-and-arm robots as well, the OTC Thor-T having been in production for some time. The Thor-T is a neat machine, with the controller mounted on the end of the base. It provides up to 1,000 mm horizontal movement in the x axis. There is 500 mm vertical (z axis) movement, and 500 mm horizontal movement of the arm. There are two axes of movement at the wrist.

Matsushita's RW 2000 robot is intended primarily for spot welding. It has three axes, movement being 800 mm in the x axis, 500 mm in the y axis, and 300 mm in the z axis. The controller is separate.

Another company manufacturing robots on a modular basis is Osumi Sangyo, which has three models – light weight, heavy weight, and gantry type. These are all Cartesian co-ordinate machines, actuated by dc servos, with point-to-point controllers, and ic memories with 256 steps. These are not the only companies in the welding robot business in Japan, of course, but the main new models. They indicate just how the industry is growing.

Fig. 6. Matsushita Pana-Robo AW1000 being demonstrated as a pick-and-place device.

Spot welding

If arc welding robots are about to proliferate, then spot welding robots are into the steady expansion phase. Here in Japan, the auto makers are gradually expanding the use of robots, and are using machines that are closer to the actual requirement of the job than hitherto.

In body welding, there are various operations. First, there are the number of small sub-assemblies that are welded together prior to the production of the main sub-assemblies, such as the doors, bonnet, bootlid, the floor, the engine bay and the wheel-arches. Then there are the major sub-assemblies such as the underframe, roof and side assemblies. These come together at the framing stage, and then the final welds are made.

Final welding is the preserve of robots in most modern plants, but on the other hand, doors, bonnets and bootlids still seem ideal for hard automation, with their simple shapes, and combination of adhesive bonding, clinching and welding.

But the trend is to an increasing use of robots in all other areas, although the problem in some cases is that few of the capabilities of the robots are actually used. For a start, the same panels remain in production, running down the same line for as much as four-six years – often with no program changes once the system has been debugged.

Therefore, the robot is being used by the auto industry as a method of overcoming teething troubles in the early stages – with press welders, changes in positions of welds involve serious delays in production – an insurance policy against the car having to be replaced with a new model earlier than expected, and a rapid way of getting a new model earlier than expected, and a rapid way of getting a new model into production.

With these requirements, the call is for simpler robots. At the same time, robot use is bound to increase simply because it is the best way of ensuring that the manufacturer can cope with changes in the market place – either by running two models down one line, or enabling a model to be phased out quickly, while a new one is phased in.

In fact, the Japanese have not yet solved these problems, but as is their practice, they are slowly making progress. The main beneficiary of this programme is Kawasaki, whose new 6060 and 3000 series robots were designed specially for spot welding.

Building block

Nissan Motors in fact lays claim to the initial design of the Unimate 6060 robot, which it calls BBS – building block system. The first installation of these robots was at Nissan's Tochigi plant, where the Cherry car is produced at the rate of 1,100 a day. According to Eiichi Yoshida, manager of the engineering section, his team's original idea for the BBS was a Unimate 2000 robot suspended from a gantry, completely upside down.

Nissan was looking for quicker welding, and Kawasaki's eventual design consists of a head, normally free to slide along on overhead rail, a pillar hanging down and a horizontal arm. The pillar can pivot on the head through some 70° on an axis parallel with the slide. The arm can also pivot on the pillar, while the wrist can pivot and rotate.

Fig. 7. Osumi modular robot.

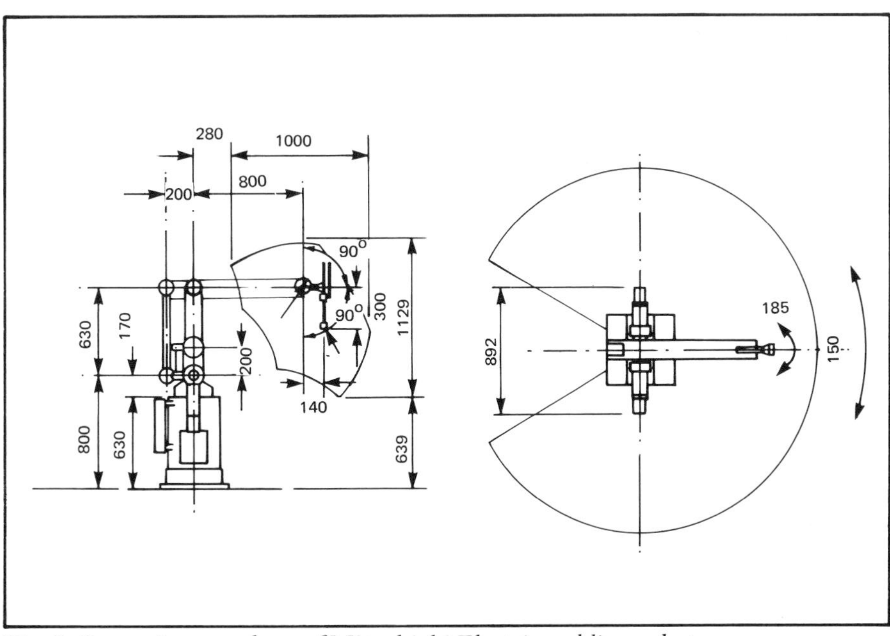

Fig. 8. Operating envelope of Mitsubishi Electric welding robot.

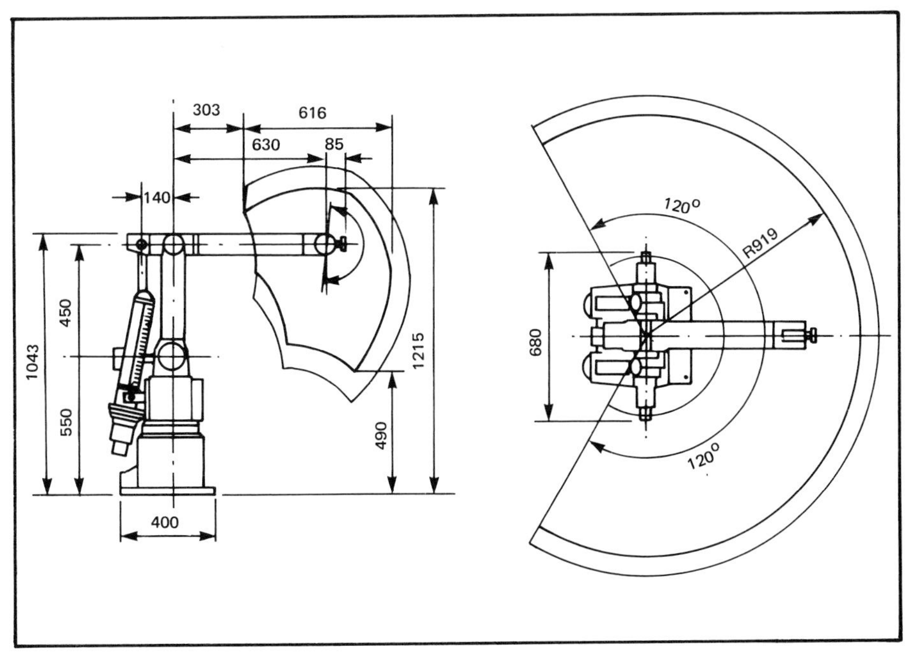

Fig. 9. Operating envelope of a small Yaskawa Motoman L-3.

Fig. 10. Kawasaki 6060 robots welding up Datsun Cherry side assemblies.

Fig. 11. Some of the Kawasaki 3000 series robots used by Toyo Kogyo have very long jaws so that they can weld up the underframe assemblies.

Actuation is hydraulic, as on other Unimates, but one hydraulic system can serve three robots, while one controller is sufficient for up to 6 robots. Yoshida pointed out that the problem with the Unimate 2000 was that all the capabilities could not be expolited, while it was too large.

Nissan has already developed a simple robot for straight welds with Toshiba, but this takes 3 s/weld, whereas the BBS can do one weld in two seconds. Nissan's target is one weld/s.

In the first installation, six of these 6060 arms are used at one station – the first on the line – to weld up side assemblies for Cherry cars. Four different models are produced, the two-door coupe, two- and four-door saloons, and the four-door estate car. Following the 6060 station there are four more stations where 10 Unimate 2000s are situated.

The 6060 arms are carried on a four-post structure, there being a pair of rails, parallel with the track on each side of the structure. Each rail, which is inset a little from the side of the structure, carries three arms. The controller and power source are set on top of the gantry.

During welding, the side is laid with the outer panel face down on the fixture. To accommodate the different models, the fixture has a number of different clamps, which can be moved in and out of position as needed. The arms are very close together on the rails, and at some stages during the welding they almost touch one another.

In addition to tack welding, the 6060 arms weld around the corners of the door apertures, where the movements between welds are small, and where the corner radii are tight. One advantage of the 6060 is that it can operate on point-to-point or continuous path control, CP being used around the corners at the bottom of the pillars. Evidently, the 6060 arms can be taught more quickly than the 2000 models, Yoshida saying that one program can be taught in 2 h.

Nissan baulked at the use of the 6060 arms for the complete line owing to the need to get to full production quickly, and their concern about reliability. But Yoshida confirmed that they would use them for the complete line now, and in fact are keen to use 6060 arms for underbody and framing welding.

The economics certainly favour the 6060. In this case, the total investment was around £150,000, whereas the five Unimate 2000s and equipment would have cost over £200,000. In some applications, the 6060s are even more cost effective.

For example, Yoshida quoted one case where eight 6060s could replace ten 2000s. He also pointed out that two stations would have been needed to replace the one on the side assembly line had 2000 robots been used. In fact, the 6060s each do about 8 welds in 20 s, the cycle time being 45 s.

New equipment

Toyo Kogyo also installed new welding equipment for the 323 model, there being three- and five-door hatchbacks, and also four-door saloons. There are two underbody and framing lines, operating at a cycle of 72 s, and one final welding line at 36 s. Each line up to framing has been designed to handle all variants, but at present three- and five-door models go down one line, and three- and four-door models down the other, to meet demand.

Toyo Kogyo uses small multiwelders in conjunction with robots for the sub-lines where the engine bay, floor and rear floor are built up. There are two robots on each sub-line.

In the main underbody line, there are no multiwelders, the main sections being joined together by robot welding. There are 14 robots/line, these being ten Unimate 3040s and four 4000s. The choice of robot depended on the reach needed, a critical factor with underbodies, and the dexterity required. Signals from sensors at the beginning of the line are used to switch the robot from one program to another, giving maximum flexibility.

The 3040 robots are simpler and cheaper than the Unimate 4000, with four axes of freedom. Each consists of a telescopic arm on a telescopic column. The robot can also traverse on a slide parallel with the track, and the wrist is also articulated.

Some of the 3040 robots have long jaws, as do all the 4000s, so that they can reach the middle of the floor. The 4000s are used where there is a curvature in the welding path.

At the framing stage, a small multiwelder and some special purpose welders are used. But from then on, all welding is done by robot, apart from a small amount of manual welding. In the final welding line, there are 15 Kawasaki 3000s, and 20 Unimate 2000s.

The 3000 arms have three, four or five axes of freedom according to the requirement, and in some cases, they are mounted side by side at common stations. Each robot makes 18-20 spots, which involves rapid movement. This is possible because all the welds are outside the body, so long movements between welding stages are not needed.

As the Japanese use more robots to spot weld car bodies, so they are installing new types aimed at matching the job more precisely. Of course, the actual programming capability of the robot is not fully used yet, but once robots are being used for all spot welding, true random order production will be practical, with many models going down one line. Whether that will in fact be an advantage remains to be seen.

New designs add to Japan's growing robot population

The emphasis at the Tokyo Robot 81 Exhibition was on welding and assembly robots. Japanese manufacturers are continually bringing out new designs.

A NUMBER of new robots were on show at the exhibition held in conjunction with the 11th ISIR in Tokyo, and all of these looked practical machines ready to go to work. There were three machines with new sensing systems for welding, several special purpose robots for welding and cutting, and a group of new assembly robots. And since the exhibition was held in Japan, the hall was jam-packed solid, many of the visitors using the opportunity to get away from the office or school. As a result, it was a very poor atmosphere in which to collect information on new products; but that is Japan.

On the arc welding side, Kawasaki and Mitsubishi Electric were both showing the sensing systems described at the Symposium while Yaskawa was also showing a new sensing system of considerable merit. It can be used for plates of 4.5 mm thickness upwards, and the welding gun is arranged to weave as it welds. When the gun tracks precisely along the true seam to be welded, the electric current measured at the end of each weave will be constant. But if the gun diverges from the seam, the currents differ. The controller continuously monitors the current, and when a difference is detected it calculates the discrepancy in the path, and adjusts the movement of the robot accordingly.

Curve following

At the beginning of the weld, the gun need only be within 20 mm of the correct starting position, and it can follow curves without programming. Yaskawa cites the example of an approximately U-shape section being welded to an end plate. In the first open wall of the U, there are three small changes in direction, while the corners of the U are tight right angles. In the second and third side of the U are curves. In this case, only 8 points need be programmed: the starting point, points before, at and after the apeces of the sharp corners, and the finishing point.

This system, which promises to simplify programming a lot, will be

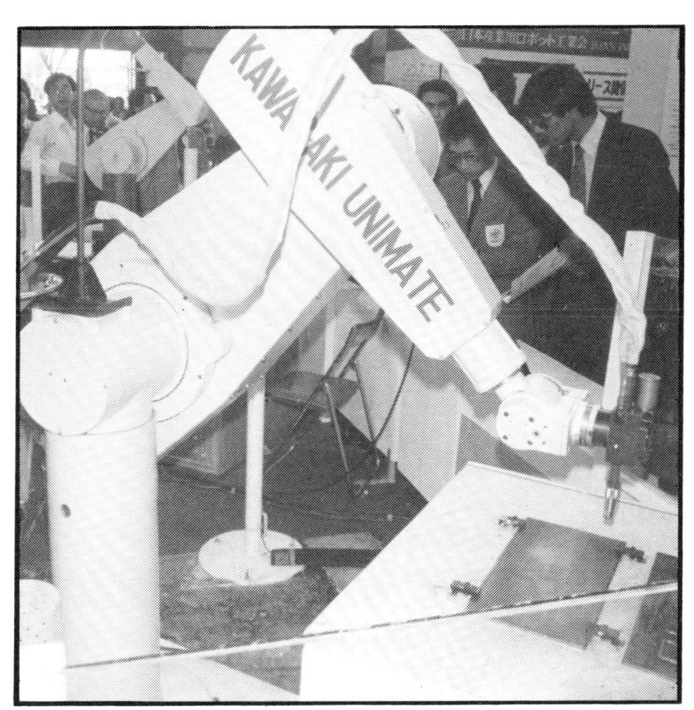

The Unimate Puma can be fitted with vision sensing for welding.

Kawasaki Unimate intended for painting can move along with conveyor.

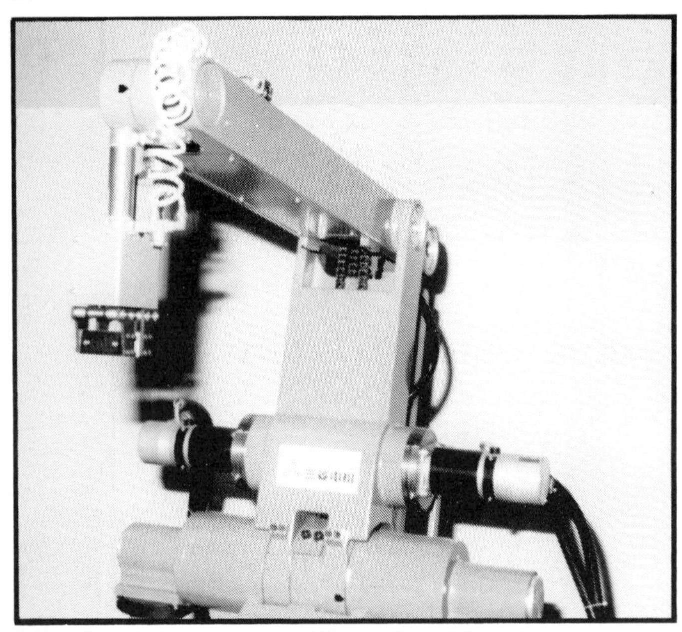

Mitsubishi Electric arc welding robot with vision sensor attached.

Shin Meiwa RC190 plasma cutting robot cutting a car body panel.

Matsushita's Pana-Robo in unmanned arc welding cell; the robots handle the parts and weld them.

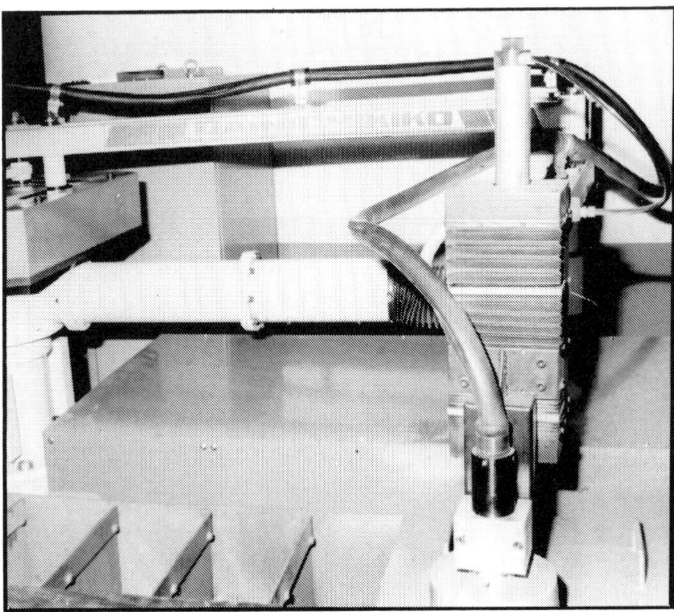

Dainichi Kiko has developed this plasma arc cutter in conjunction with O-A Machinery.

available in the Spring, at a cost of around £5,000; the robot costs about £25,000.

Matsushita, which has recently come into the robot business with its Pana-Robo, demonstrated two of them operating as an unmanned arc welding cell. One robot acts as a handling robot, while the other does the welding. A turntable completes the package.

Yaskawa introduced a new handling robot for use with standard spot welding machines, the Motoman S-30. This is, in effect, a robot turned on its side, the jointed arm being mounted so that it moves horizontally. The mechanism is similar to the normal Motoman, except that there are only three degrees of freedom. The shoulder and elbow joints produce the normal angular movement, while there is a device providing 150 mm vertical movement at the wrist.

The S-30 has the normal Yasnac controller, and is controlled on a point to point basis, with a claimed repeatability of ± 0.3 mm, while the capacity, including gripper is 30 kg. This machine was being demonstrated moving a pair of pressings through a routine at a pedestal welder – a common application in many limited production run shops.

Linear interpolation

Nachi was also showing its Uniman 8000 spot welding robot, a new machine reminiscent of the KUKA 601 series, with long arm. It is a six-axis machine, developed specially for spot welding, with the capability of carrying the welding gun and transformer – the capacity is 35-50 kg. It is actuated by dc servo motors, and is equipped with linear interpolation, so that it can move linearly between taught points.

Mitsubishi Heavy Industries, not to be confused with Mitsubishi Electric, also introduced two new robots – the Robitus RD, and the ARB version of its painting robot. The RD is a small prosthetic robot, reminiscent of the Unimate 6060 arm. Like the 6060, the RD can be mounted on an overhead gantry, on the floor or on a wall. It can also be mounted on rails so that it can slide across, above or alongside the workpiece. In that case, there are six axes of freedom.

This is a hydraulically actuated robot, a variable displacement pump being claimed to reduce energy consumption. One hydraulic unit can drive three arms. Aimed at the auto industry, the arm has a capacity of 25 kg, and a claimed repeatability of ± 1 mm.

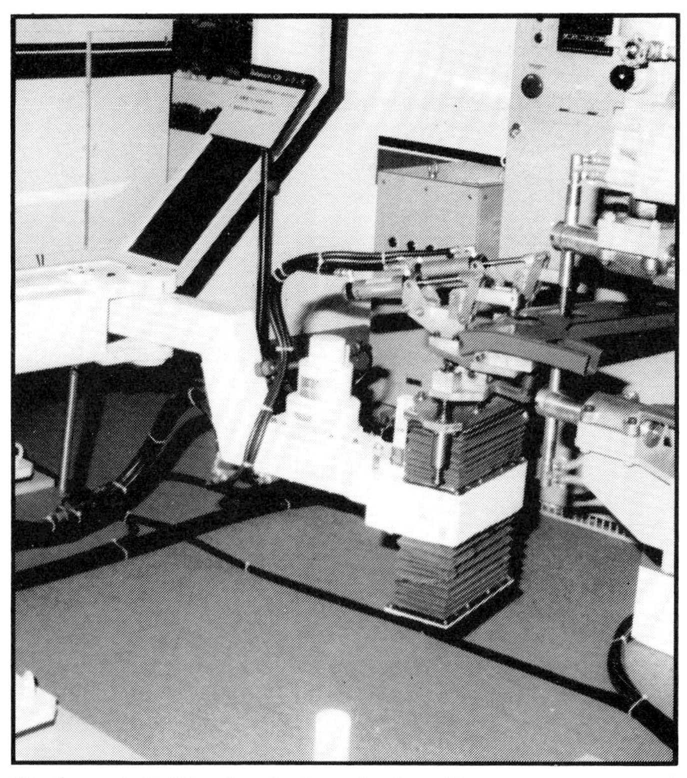

Yaskawa's S-30 robot designed to handle parts at a pedestal spot welder.

Thor robot designed to apply sealant.

Mitsubishi's ARB painting robot is microcomputer controlled, and is claimed to be able to cover almost twice the area of the previous machine with quicker motion. Five or six axis machines are available, the machines being actuated hydraulically. A conveyor follow-up function is provided, and the speed of operation is up to 2000 mm/s.

Kawasaki was also showing its new painting robot, which can be mounted on a similar slide so that it can follow the track – clearly a requirement in the Japanese auto industry – as was Tokico.

Trimming

Two plasma cutting robots were being shown, the Shin Meiwa device intended for trimming car body panels and the device developed by O-A Machinery and Dainichi Kiko. Intended to replace NC cutting, the O-A machine is designed to cut plates mounted horizontally, so the arm is designed to be able to traverse over a plate approximately 1.5 m square. The machine has three axes of movement as standard, but two more can be incorporated in the wrist.

Dainichi Kiko has extended its model range, so that now there are four Part-Time models, from the small 300 to the large 800. These are all jointed arm robots, actuated by dc servo motors, the repeatability ranging from ±0.1 mm for the 300 models, to ±0.5 mm for the 800, which has a capacity of 30 kg.

Osaka Transformer Co, which has a range of welding robots, is now to produce a robot intended specifically for the application of sealant to components such as engine pans and covers. Complete with sealant supply system and gun, the machine costs about £28,000 in Japan.

But the one clear trend at the Exhibition was the proliferation of assembly robots. There are the two models from Fanuc, and also small machines from Nitto, Pentel, Hirata, and Sankyo. These are all very similar in concept, that concept having been developed by Professor Hiroshi Makino of Yamanashi University. He calls his version, now being produced by Nitto, the Scara, and this can be caled the 'Japanese assembly robot' since it is quite different in structure from robots developed elsewhere.

There is a simple arm, mounted at the top of a pillar on which it can pivot on a vertical axis. The elbow joint also pivots on a vertical axis. Thus, the arm can cover quite a large table, despite being very small. In addition to the two joints in the arm, there is usually a third axis of freedom at the wrist – this is often vertical motion or rotation, or it can be both.

Compliance

This type of robot has some compliance built into the joints, since these are braked electrically. Thus, misalignment is sufficient to overcome the braking force, giving compliance. These are all small robots, generally with a capacity of less than 5 kg, and with total arm lengths from 250 to 1050 mm. Repeatability is ±0.05 mm, and prices are in the range of £10,000-£15,000 in Japan. They are designed for simple assembly jobs, such as insertion of parts, and tightening of nuts and screws.

NEC, the electrical and computer company, has also started to market its own robots, and these were on show for the first time. The most unusual of these is the Model-A, which has two arms hanging down from slides to give a total of 10 axes of freedom. Linear motors drive the arms vertically and horizontally, while stepping motors are used for rotary motion. Each arm can move 500 mm longitudinally, 270 mm laterally, and 150 mm vertically. Since this robot was designed for the assembly of integrated circuits on substrates, the repeatability is claimed to be a microscopic ±0.008 mm!

NEC's Model-B robot is not unlike the Scara, except that the column can raise the arm 200 or 300 mm, according to model. Capacities are 1 and 5 kg, the machine with the larger capacity having a longer arm – a total length of 1100 mm, against 900 mm.

Overall, the number of new robots indicates how seriously the Japanese are taking robots. They realise that the industry will now expand very rapidly, and they want to be there at the beginning to gain a market share. Certainly, the result is a number of highly-competitive products – in price and performance.

The pace quickens

Matsushita leads a number of companies coming into the robot business, while Yamaha installs over 100 arc welding robots, and international ventures increase.

SPECTACULAR robot installations are still few and far between in Japan, but one has surfaced at Yamaha Motors. Meanwhile, the big battalions of Japanese business are beginning to put a lot of effort into production facilities so that they are ready for the boom in robot sales – which has already started.

Production is certainly booming. According to a recent survey conducted by the Nihon Keizai Shimbun, the business newspaper, the biggest 'robot' producer by volume is Sailor Pen, which is selling 80 robots/month for mounting on injection moulding machines. Among the true robot companies, the survey showed that Kawasaki and Hitachi were both selling 70/month, Osaka Transformer and Fanuc both 50. Astonishingly, Sankyo Seiki, the maker of the Scara robot, is now claiming sales of 50/month too, yet last year it averaged less than 20. The most ambitious companies as far as expansion in the next year is concerned are Kobe Steel, looking to increase output from 10 to 140/month, and Mitsubishi Electric, aiming to go from 20 to 100/month. Of course, Mitsubishi has only been in the business for just over a year. Since the Nihon Keizai Shimbun is read principally by investors, it is to be expected that companies would tend to exaggerate their projections, but even so, these figures are dramatic.

Among the new models, the range produced by Matsushita is the most significant, largely because it is so comprehensive, but Kobe Steel's new arc welding robot is a marked advance on its original design. Fujitsu has shown that incredibly small tolerances on repeatability are practical – it has produced a robot with a claimed repeatability of $4\mu m$ (0.00016in). Also, Sumitomo Electric has come up with a robot that has two visual sensors and a carbon fibre arm.

On the applications front, Yamaha's use of 140 arc welding robots in one plant is perhaps the most exciting. They are mainly of the polar type, and are being used to arc weld

Matsushita Pana-Robo welding/handling robot operating as an unmanned arc welding station, with the aid of camera and positioning table in the background.

Estimated sales rates of major Japanese robot makers

Company	*1982*	*1983*	*1987*
		Robots per month	
Sailor Pen	80	120	—
Kawasaki HI	70	100	430
Hitachi	70	80	—
Fanuc	50	100	500
Osaka Transformer	50	—	—
Sankyo Seiki	50	100	300
Mitsubishi Electric	20	100	1000
Mitsubishi HI	20	30	—
Tokico	20	30	150
NEC	20	30	200
Nitto Seiki	18	22	—
Kobe Steel	10	140	500

Source: Nihon Keizai Shimbun, Japan.

motor cycle frames. This application is covered in a separate article.

Also of interest is Toyoda Machine's assembly robot built as part of a government-sponsored flexible manufacturing project. Significantly, a lot of research went into establishing just what tolerances were needed for robot assembly on positioning.

To show just how serious it was about the robot business, Matsushita demonstrated 70 machines at an exhibition it held in Tokyo in June. Nearly all these were new, and are aimed principally at assembly and fastening. Just over a year ago, the company introduced its Pana-Robo jointed arm robot – an ASEA look-

Matsushita x-y robot for assembly.

alike – for arc welding and spraying, and it is already using 100 of these in its own plants. Now it has added four more types of robots:

□ The Pana-Robo T and A types are cartesian-type arms/tables; and
□ The H and V series jointed arm robots.

Like the arc welding robot, these are all actuated by dc servo motors.

The cartesian coordinate machines are available in three sizes, giving motion covering a table 200 × 150mm, 400 × 300mm and 600 × 400mm. Load capacities are 2-12kg for the arms, and 35kg for the x-y table. They are designed to move at speeds of 400mm/s, while positioning accuracy is put at ±0.05mm.

The Pana Robo H (horizontal) robot is like the Scara machine, with a double-jointed arm with vertical pivots. The largest models in this range have total arm lengths of 500 to 700mm, and can normally carry up to 8kg. The smaller models are available with total arm lengths of 400, 600 and 800mm, but can carry only 2kg. They also have 100mm vertical movement at the wrist for operation of a tool.

There are three models in the V range, with up to six axes of freedom. These are built on a modular system, so that a wrist/gripper can be used alone, or can be built on to a forearm, or a complete arm and base. The models shown at Tokyo had tubular aluminium arms and simple grippers consisting of a pair of fingers which are moved together or apart by an electric motor mounted on the wrist.

Currently, Matsushita claims that its factory automation division, which also sells motors, a host of components, and industrial cameras, has a turnover of £35 million pa. It is expecting growth of 40% pa to give it a turnover of £230 million in five years time. As yet it has not fixed the prices of the jointed arm robots, but these are expected to be under £10,000 each. The A and T type are in the region of £4,500-£6,500 each.

It is not just the wide range that makes the Matsushita robots significant, but also the extent of the ancillary equipment – everything to make the robotic assembly line is available from one source. In the future, of course, many companies are likely to want to be able to buy such systems rather than separate robots and other devices from different suppliers.

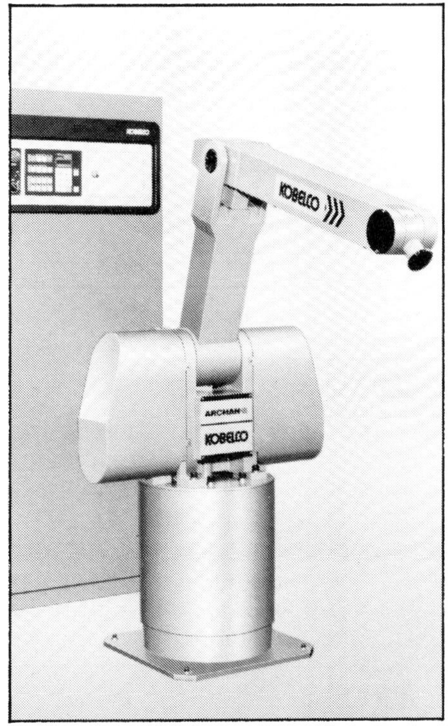

Kobe Arcman S robot.

Matsushita has also developed some new ancillary equipment for its Pana-Robo AW 2000 arc welding robot. This includes visual sensing for parts identification and orientation, and for sensing the position of a welding path in a confined space such as inside a large-bore pipe. For unmanned arc welding, the robot is combined with an automatically operated clamp on a table for welding; a separate table, which can be rotated and moved in the x and y axes, and above which is a camera; and a chute to supply parts to the orientation table; there is also a conveyor on which completed parts are placed.

Ancillary

In operation, the part slides down the chute to the table, the camera detects the position and orientation, and responding to these positional signals, the controller rotates and moves the table as necessary to place the workpiece in the middle of the table, and correctly orientated. This operation happens almost instantly.

The robot, which has a gripper, picks up the workpiece and places it on the welding table, where it is gripped by a pair of clamps. Then the robot picks up a welding gun and welds the workpiece, following the path dictated by the controller, which has received information as to the part to be welded. After welding, the robot replaces the welding gun on its stand, and transfers the workpiece to the conveyor.

This is certainly a neat installation, in that the robot really does replace the operator; in most arc welding applications, an operator is still needed to load and unload the workpieces. The ancillary equipment may be expensive, but since the camera is remote from the welding gun, it should be reliable. Currently, this is an experimental application.

Meanwhile, the Arcman robot, a rather ponderous device produced by Kobe Steel, has been joined by an altogether much more modern and compact arc welding robot – the Arcman S. This is a compact jointed-arm five-axis machine actuated electrically, the motors being concealed in neat housings.

Self-tracking

It has a capacity of 5kg, and an accurcy of ±0.3mm but is intended primarily for arc welding, its main features being a self-tracking control and automatically optimised welding parameters. Kobe Steel has collaborated with Shinko Electric in the development of suitable electronics for the control of the robot and of the welding parameters.

The positions and length of welding runs are taught, and the electric current, voltage, speed, torch angle and number of passes are optimised automatically by the controller. Also new is a sensing system that detects the position of the start of the weld, using the welding rod as a probe. Then, during welding, the controller monitors the welding current, and can thus detect any divergence between the path being followed and the required path – this seems to be similar to the system shown by Yaskawa at the last Robots Show in Tokyo.

The Arcman S will cost around £23,000 in Japan, with the sensor. Production is expected to build up to 30 units/month in the next few months, which seems optimistic. By that time, Kobe Steel expects to be able to offer ancillary equipment including a workpiece positioner to form a complete welding system.

Sumitomo Electric's new robot on a trolley with vision – not for production before 1985.

Hitachi has updated its Process Robot, by adopting a 16-bit processor and magnetic bubble memory. The new processor results in faster handling of data, and the potential for more precise computation. To simplify programming, there is a 9in CRT built into the controller, while the magnetic bubble memory doubles the capacity from 1,000 to 2,000 program points and job steps. The price of the robot in Japan is about £24,000.

Nippondenso, the auto electrical company that tends to make robots for its own use only, has developed some cylindrical coordinate machines for use in assembly – and each has two grippers. It has installed six of these robots in a new assembly line, for car air conditioning units, to install evaporators and printed circuit boards in the housings, but is now rebuilding the line to increase the level of flexibility with another four robots.

The robot has a rotating base, and an arm that can be raised and lowered on the column, and which can be extended telescopically. The two grippers are at 90° to one another, and at 45° to the axis of the arm. They can be rotated together as a unit on the axis of the arm, and have simple grippers.

Conditioners

Although the assembly job is fairly simple, many different types and sizes of air conditioners go down the same line, so the robots must be able to pick and place different units. In some units, the printed circuit boards are mounted horizontally, and in others vertically. The robot also checks the presence and security of spring clips that hold the two halves of the casing together. Owing to the different functions involved, more than one design of gripper is used, and the robot changes these itself. The line operates to a cycle of 9s.

Making a belated entry into the robot arena is Sumitomo Electric, and it seems to have bitten off rather a lot. The jointed arm robot is carried on a trolley, and is fairly small with a height of 900mm. It incorporates vision sensing. Actuated by dc servo motors, the arm is constructed from reinforced carbon fibre to cut weight. The two cameras are mounted in a housing at the top of a pillar on the trolley. They face down towards the work table, and are placed some distance apart so that they can detect distance by parallax, as well as position. There is an additional visual sensor in the front of the trolley, which not only acts as a proximity control, but also as a charcter reader so that the trolley can be sent to a series of stations in sequence.

The arm has six axes of freedom and a capacity of 1kg. The control is extraordinarily complex, with three 16 bit microprocessors and three 8 bit microprocessors, in addition to a mini computer! Not surprisingly, this machine is at the experimental stage only, although Sumitomo says that the first application will be to insert printed circuit boards in electronic control units. This seems a rather simple job for such a machine, but Sumitomo adds that it is also developing the unit for use in nuclear reactors and hospitals. The robot is said to be three-to-five years from production.

On the commercial front, Dainichi Kiko, the fledgling Japanese robot maker, has been busy, concluding deals with both GCA and Cincinnati Milacron of the USA, while it has also sold a robot to a Chinese university. GCA specialises in the production of manufacturing equipment for semiconductors, and it has gained sales rights for all Dainichi Kiko robots in North America. The agreement will last for five years, and the plan is that GCA will start building robots in the USA by 1986 under a joint venture. In the first year, GCA expects to be able to sell 240 robots, with an increase in sales of 50% pa thereafter – implying that there will be sufficient business to justify a 50 units/month plant within three years. The companies are also planning to cooperate on new sensing systems.

Dainichi Kiko's agreement to supply Cincinnati Milacron with robots is less clear cut. The Japanese company says that it will develop models specially for Cincinnati to sell under its own brand name, but Cincinnati talks of jointly developed robots for sale worldwide, which seems likely in view of the experience the American company already has. These are likely to be small and inexpensive machines.

2.4 Robots – General

Keeping up with momentum in new robot designs

A clutch of new robots has emerged from Japanese firms. They are aimed at assembly, loading and unloading machine tools and arc welding.

JAPANESE robot makers have been busy recently introducing new models and trying to increase sales, and some of this effort has been directed at overseas markets. The aim is to boost output and ensure that those interested in using robots can see them working; in Japan it is not so easy to take customers to users' factories.

Among new models the main impact has been handling parts for machine tools. Several of the machine tool companies are now making their own machine tool handling systems.

On the sales front, Yaskawa and Kobe Steel have entrusted the majority of overseas sales to Japanese trading houses. In addition, Yaskawa is to sell its robots through Hobart Brothers, a welding specialist in the US.

In Japan, Hitachi, Yaskawa and Tokico are increasing their robot demonstration facilities, while Hitachi has set up a new company to develop robot software, Hitachi Keiyo.

Although sales of robots in Japan are increasing, many of these are simple handling devices and some are not robots even at all – just handling arms with adjustable limit stops.

It is important to stress that most of the articles published in the mass media about robots in Japan include almost every automated device in its tally of Japanese robots. At the same time, it is clear the Japanese are beginning to develop some high-technology robots, as well as some that are specialised and therefore ideal for certain applications.

The automotive industry continues to be the big user, although several of the electronics companies have produced their own robotic handling devices for the assembly of small parts.

It is Hitachi that takes pride of place for new robots in this review, even though only one of its three designs is ready for production. All three were shown at Hitachi's Technology Show held to celebrate the company's 70th anniversary.

First there was the 'learning robot' designed to insert components in housings to close tolerances; then an assembly robot with visual sensing; and finally, an enlarged version of the process robot introduced at the Robot Show at the end of 1979.

Prototype

Hitachi's 'learning robot' is still a laboratory device only and as yet no production programme has been fixed. But it is an interesting design, and the technology certainly will be used in due course.

The robot is an x-y device with a horizontal arm on a column. Adjacent is a sensing table which can also be moved. The base on which the robot is mounted allows a small amount of movement at right angles to the arm while the arm can move vertically and extend. In the prototype all these movements are controlled by ballscrews.

However, it is the table that is more important for it can be moved in three directions. There is also a force sensor between the table and a lower table which is actuated by rams to create the necessary movements. The table can move in the normal x-y axes, but since there are four rams beneath, it can also be raised or lowered vertically, or tilted.

The force sensor consists of a circular band, from which four strips hang down, equally spaced, and then turn horizontal to extend inwards to the base of a spindle on the axis of the band. There are strain gauges on the vertical and horizontal parts of the

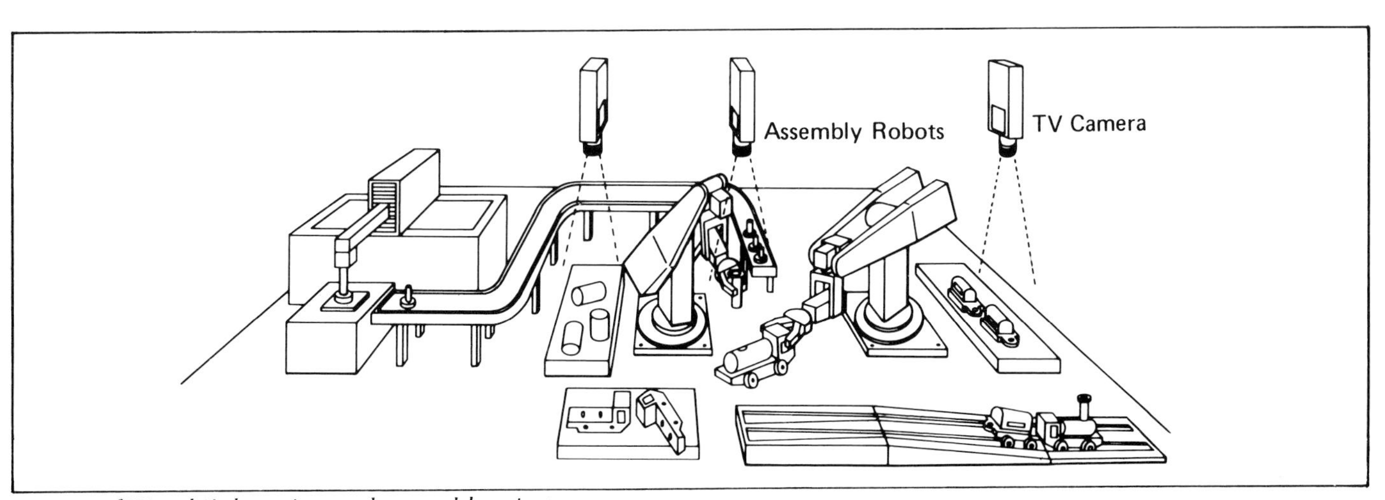

Layout of Hitachi's learning and assembly robots.

strips, so the forces in x, y, z planes, as well as the moments can be detected.

To control the motion of the robot and table, five microcomputers are used. From the forces and moment, the misalignment is calculated, and this is fed back to the electric motors that control the position of the table.

In fact, the table is designed to cater for small movements only. Corrections are made at a frequency of 20 Hz. From the first insertion, data are continually stored and computed, so that a data bank is built up to speed up further centring operations.

It can take 20-30s to install the first component precisely, but after only seven or eight operations the time can be reduced to 1.2-1.3s. The force sensor can detect forces to 20N, while the minimum resolution of forces is 2 x 10^{-2} N.

In production, the learning robot could be far simpler than this prototype. For example, a robot arm as such would seem unnecessary in many applications. If the table could be incorporated in a conveyor line, a very simple device could be used to pick-and-place. Of course, this would have to respond to control signals from the table to enable it to wait until the correct time to install the component. But such a learning table would seem to be a very important feature in any robot used for assembly.

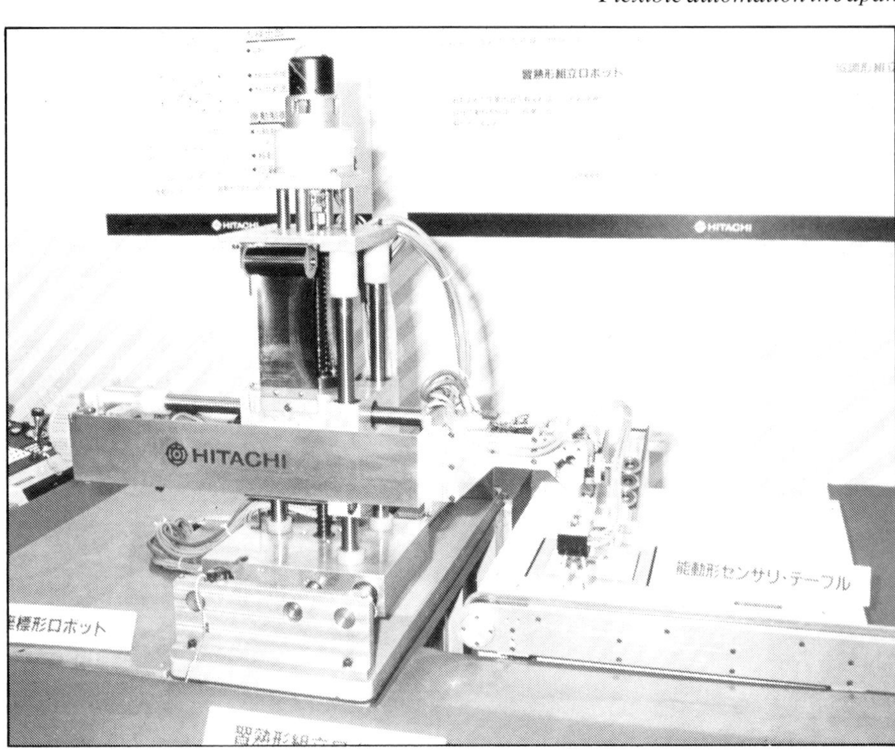

Hitachi's learning robot makes use of a sensory table.

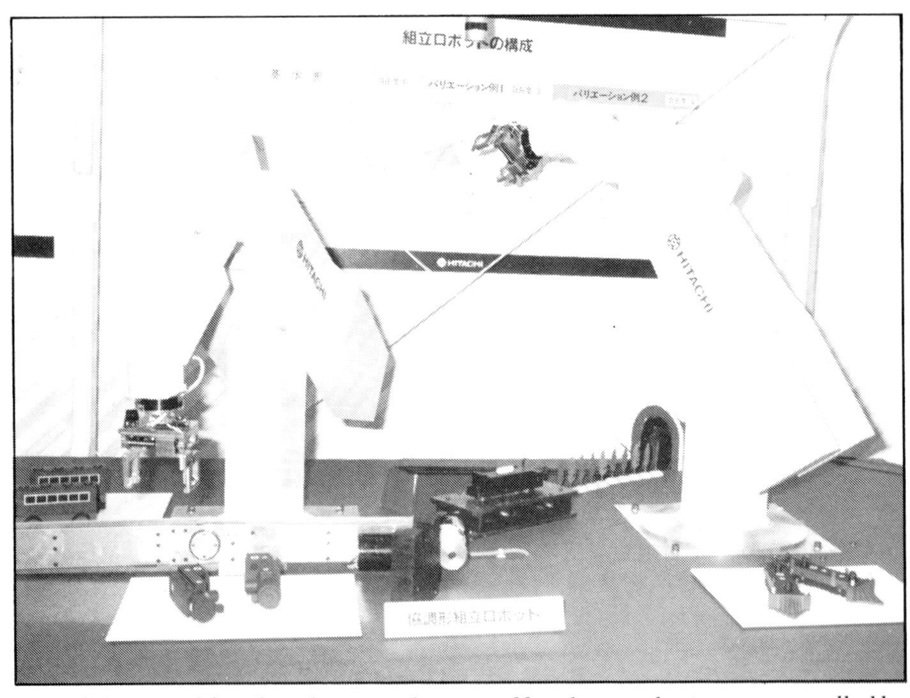

Hitachi's assembly robots have six degrees of freedom and grippers controlled by electric servo motors.

Assembly

Hitachi's new assembly robot is due to go into production within about a year at a cost of £10,000 upwards. It is of the jointed arm type, and can be built in a modular fashion with four, five or six axes of movement.

The robot consists of a column, which can rotate, driven by an electric motor beneath it. It has a harmonic drive. The main arm comprises a pair of parallel arms, one on each side of the column. The forearm pivots between the end of these arms, but has a forked end to pick up the articulating wrist. The hand can also rotate through the full 360° while the gripper takes the form of a pair of plates moved together or apart on ball screws.

An electric motor in the column actuates the main arm at the shoulder joint, while there is a motor in each of the main arms to actuate the elbow and upper wrist joints. Then, there are two motors in the wrist to articulate the lower wrist joint and to rotate the gripper around the wrist. And, as if there were not enough, there is a motor for the gripper.

With this system it is easy to build simpler versions. For example, the forearm could be omitted to produce a four-axis machine; this would leave the two motors in the main arms free to rotate the gripper and actuate the 'elbow' joint.

An important feature of the robot is that it is controlled by two Hitachi 6800 series 8-bit microprocessors. One actuates the robot while the other receives and interprets signals from external sensors – these could come from a central computer, or from visual or tactile sensors. In other words, the robot has been designed with the future in mind. In the Tokyo display the robots were shown fitted with visual sensors. Two robots were assembling a toy train at two stations with the cameras mounted remotely in the structure above and aimed at the assembly areas. Hitachi intends to use a miniature camera incorporating charge coupled devices as the visual sensor in production versions.

A special program language has been developed for the robots. This is similar to the VAL language for the Unimation Puma.

An electric motor is used to actuate the gripper since this allows the use of a potentiometer for position control. One interesting mechanical feature of the design is that the main arm can swing through some 270° vertically so that it can pick up a part in front of it, and swing it 'overhead' to a position behind it, a movement which is quicker than the alternative horizontal articulation.

Hitachi has set the control system up so that the visual sensor recognises light patches on a dark background. Thus, the components of the toy train were generally dark green, and there were about five yellow discs on the side, and the orientation was based on the position of these yellow discs. However, it is possible for the sensing system to respond to shapes since the necessary software is available.

Prosthetic

Finally, Hitachi is about to introduce a new large prosthetic robot, based on its process robot, but much, much bigger. It has a capacity of 30kg when operating at maximum reach and acceleration, but it can carry up to 70kg when not fully extended. This has been designed for handling in forges, for heavy spot welding, arc welding and the application of sealant. It has a continuous path control system.

Whereas the more interesting Hitachi developments are those yet to go into production, in the machine tool field, several robot arms are now available. These have been introduced recently to meet the need for reduced manning levels in NC machines shops.

Now that these machines can be set by semi-skilled labour, the aim is to move towards the almost-unmanned machine shop. The idea is that in addition to its robot loading, the machine tool be equipped with automatic measurement systems which detect tool wear, and then adjust the tool position accordingly. Then, there needs to be automatic detection of tool failure, or minor damage. Tool failure indicators are at the development stage only, but one or two simple systems are operational.

With this approach, and the provision of adequate stock of workpieces for the robot, the machine can operate for three or four hours unmanned. Thus, a machine shop can be virtually unattended, and can operate through the lunch break, and some overtime. In other words, the addition of this equipment allows the number of machines to be reduced, and flexibility to be increased.

Generally, the workpieces are carried on a carousel which is moved by electric motors responding to signals from the robot controller or the machine controller where the whole system is integrated. Therefore, the robot arm always picks up from the same spot, and unloads at that station or the adjacent one.

Although this approach allows the use of a very simple robot arm, which can be built into the machine, it does limit the size of the work-in-progress.

With a more complex robot, it would be possible to provide a pallet full of workpieces and let the robot get on with it, placing the finished workpieces into another pallet. The alternative approach, of course, would be to transfer to and from conveyors linking the machine tools together.

Two arms

Ikegai has developed two robot arms, one designed to be mounted near the chuck inside the guard, and the other outside the guard. Up to four axes of movement are available, in each case the shoulder and the elbow joints pivot on axes parallel with the machine tool spindle. The gripper can be rotated to pick up the workpiece, which is mounted on the carousel with its axis verrtical, and turn it so that its axis is vertical for installation in the chuck. The gripper can also slide to move the workpiece in and out of the chuck.

Okuma Machinery Works has also developed its own robot. This bears more than a superficial resemblance to the Fanuc Model 0 but it manages with one less axis of movement. Indeed, the movements required at a machine tool are highly specialised so its motions are not similar to those of a handling robot.

Ideally, a double gripper is needed so the robot can grip and withdraw the

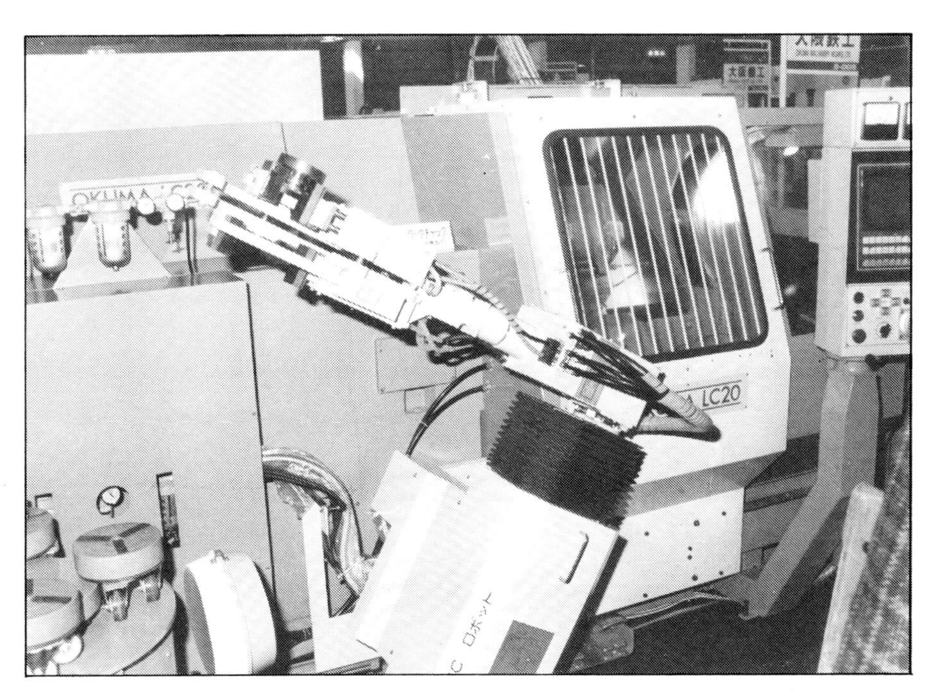

The Okuma robot operates between the machine tool and the carousel.

The Mori Seiki arm has a complicated system of linkages.

finished workpiece, and then rotate to allow the new workpiece to be installed. Then, the arm must be withdrawn from the machine and moved so that it is above the carousel.

The carousel should be immediately in front of the machine to minimise the space requirement. Of course, what constitutes the front or back is of academic interest so long as both operator and robot can reach what they need to reach.

The other requirement is that all operations to be carried out while the machine is idle should be carried out as quickly as possible, but once the guard has slid back into place, the movements can be relatively slow.

Okuma, like Fujitsu Fanuc, has designed a robot to carry out these motions only, with one axis of movement used for more than one of the required movements. The gripper is of the double type, and is carried on a long forearm that pivots on a spindle parallel with the machine bed.

The arm can also be moved linearly on the axis of the spindle. However, that spindle is carried in a jaw attached to a horizontal spindle at right angles to the machine bed. Thus, the two spindles can rotate and the arm can be moved linearly; the arm can be rotated also around its own axis.

Complex

In practice, therefore, the arm pivots on the spindle to move into the machine, moves linearly to remove the workpiece, rotates on its own axis, moves linearly to install the new workpiece, moves linearly again, and then pivots out of the machine back to the vertical position. The arm and first spindle then are rotated together through 90° on the second spindle so that the arm is horizontal and parallel with the machine bed.

The arm is then above the carousel, and the linear movement axis can be used to lower the arm to place the workpiece on the carousel. To move from one workpiece to the next, the arm pivots on the first spindle – this is the same axis as that used when the arm is moved into the machine.

The result of this approach is that the robot arm is compact and can be mounted on the front of the machine. In addition, few axes of movement are needed. On the other hand, this is a very specialised set of motions and suitable for this particular set-up.

The fact that this is intended for small batches, and not for semi-conveyor line machining, is shown by the option of an optical checking system. A lamp projects onto a screen the silhouette of the workpiece to be used next, and the image is identified by a solid state camera. Therefore, using comparison with the computer memory, it is possible to call up a different machining program to suit the component – giving random unmanned machining, so long as the sizes and tooling are similar.

Okuma quotes a capacity of two 5kg workpieces, and there are two sizes of gripper. The smaller one can cope with components of 20-85mm diameter, and the larger one of 80-150mm. A price of around £9,000 is quoted for the basic arm.

Trolley-mounted

Mori Seiki has produced a robot arm mounted on a trolley which is then locked into position at the machine. It too has a capacity of two 5kg workpieces, but instead of grippers mounted back to back, this machine has them in line, simplifying the movements. There are two sizes, one for workpieces of 30-150mm, and the other for 75-170mm diameter.

Actuated pneumatically, the device consists of a block above a column carrying an arm. There are four pneumatic cylinders: one to raise and lower the block and arm through 150mm; a second to rotate the block and arm on the column; a third to articulate the arm to move the workpieces in and out of the chuck; and the fourth to rotate the arm and gripper together.

When the arm is in the machine the two grippers are one above another and the arm is hoizontal. To move out of the machine, the arm and block are rotated 90° horizontally on the column, but at the same time a ram actuates a lever to rotate the arm. However, from this lever extends a link which is parallel with the arm and connects to another lever fixed to the spindle carrying the two grippers. Thus, as the arm rotates so the grippers are also articulated; by the time the arm arrives at the end position the grippers are horizontal.

This mechanism looks as if it will be prone to wear, and seems to lack the

Typical installation of a Fanuc Model O robot.

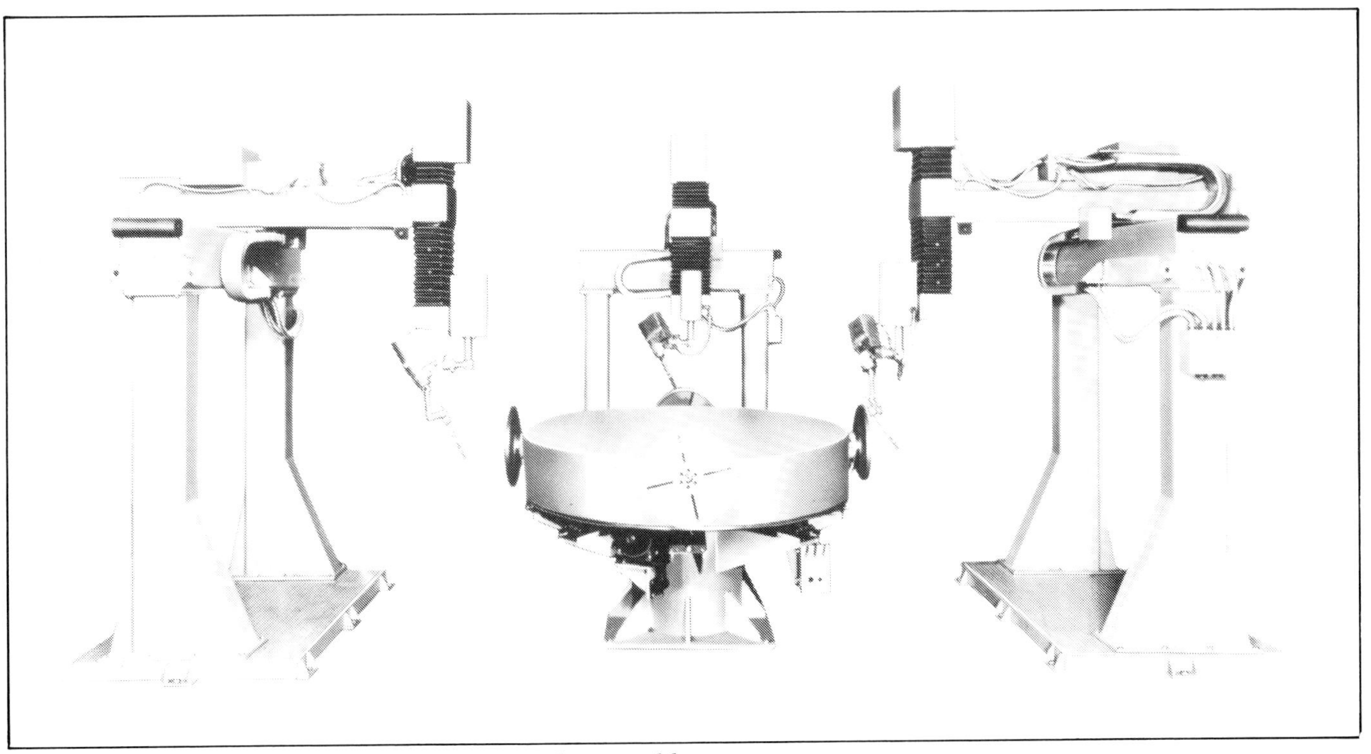

Shin Meiwa's new PW752 robot arranged around a turntable.

essential flexibility of motion that should be inherent in a robot. It is not cheap at around £10,000 in Japan.

Fujitsu Fanuc has introduced a new robot, the Model 3, to handle workpieces up to 50kg – the capacity of the Model 1 is a total of 20kg. This is a column and arm robot built on much the same lines of the Model 1. The column can rotate through 300° on its base, and there is 1,200mm vertical travel combined with an arm stroke of 1,200mm. In addition, the wrist can rotate through 270° and bend through 180°, giving five axes of movement.

To provide stability with such large workpieces the arm is supported on two vertical guides; the arm also has a pair of guides to carry the wrist. All movements are actuated by dc servo motors and an advantage claimed for this robot is that it can pick up workpieces from a low level. The hand, mounted on the lower portion of the wrist, consists of a plate from which jaws project. With the plate horizontal, the robot can pick up workpieces 500mm above the ground.

Fanuc Model 3 is designed to service up to two CNC machines, but it may also find applications in handling heavy components. Repeatability is put at ±1mm, control being on a point-to-point basis, with motion in all axes simultaneous. There is the usual teach unit, with playback, and the standard memory can hold up to 300 points. However, Fujitsu Fanuc is also producing bubble cassette memories for all its robots now, and in the case of the Model 3, these can provide an extra 1,300 or 2,700 points.

Welding

Recently, Kawasaki has developed a new version of the Unimate 2000 for arc welding while Shin Meiwa has extended its range. Called the A2850, the new Unimate has six axes of freedom, with two more being incorporated in the specially developed table. Thus, dexterity is provided without affecting the angle of the welding wire, which should be kept constant for good welding. A microcomputer is used to control the robot and table.

Kawasaki has streamlined programming to reduce the number of points needed in a series of movements. It has arranged also that during teaching the movements do not affect the inclination of the wire. The wrist is also said to be more compact than usual.

Shin Meiwa has replaced the PW751 with the new PW752. Although the two machines are similar, the new one has more axes of freedom, and is designed for incorporation into automated systems. The PW751 has three axes of movement at the table, and only vertical movement of the arm with orientation of the welding gun. The PW752 is mounted on a gantry so it has movements in the x,y, and z axes, while the wrist can be rotated, and the angle of the gun varied. It is designed to weld workpieces up to 750mm in cube.

The robot can be mounted on a trolley to provide up to 3m linear movement, and it can be used with a variety of workpiece positioners. The simplest of these merely rotates the workpiece on a faceplate – and with the trolley, the robot could move between a pair of faceplates. Other configurations include up to three machines working at a turntable, or several machines straddling a moving conveyor system.

Shin Meiwa plans to start selling the PW752 in the next few months in Japan at a price of around £18,000 for the basic robot. The company is also making available its sensorless sensing system, which allows the robot to find the start of a welding path. This costs about £3,000 in Japan.

Currently, Shin Meiwa is concentrating on the larger PW150 in export markets, this being the unit Grundy Robotic Systems is planning to sell in the UK.

Nachi-Fujikoshi, which makes handling and arc welding robots, has also introduced a painting robot of the prosthetic type. This is being sold in conjunction with a transfer system to move large workpieces into position and orientate them. The first set is being used by Komatsu to paint large buckets for shovel loaders, this being approximately 3m by 1m in size. Since it is deep, the operation takes quite a long time, and is quite strenuous for an operator.

This is one more example of the way in which the Japanese are now pushing the use of robots, not just to increase productivity but also to eliminate difficult jobs – especially in poor working environments. At the same time, the use of robots in car body welding continues with the accent being on the use of smaller robots.

More new robots help fuel Japan's new 'robot fever'

A new spray painting robot and several new assembly robots have recently joined Japan's army of robots. Japan's robot fever is, however, confined to only a few industries and its robot population is not quite as people imagine.

FOR SOME TIME NOW, 'robot fever' has gripped both the Japanese media and most of the foreign correspondents resident in Japan.

Eager to find a reason for Japan's industrial growth in recent years, they seize on robots for the simple reason that such machines are evident for all to see. What is more, a few companies like Nissan Motors make a point of taking visitors to their most robotised plants.

And indeed there are many robots operating in Japan but the figure is not quite so startling when the size of the workforce is taken into account. For example, there are some 6,500 programmable and play-back robots in use in Japan spread across a working population of 55 million. That is equal to 118 robots/million workers, against 225 robots/million workers in Sweden.

In fact, the true figure is probably even less in Japan's favour, since more are employed by proportion in the service industries in Sweden than in Japan.

In any event, statistics show that in Japan some 25% of the workforce is employed in manufacturing and mining – about 14 million people. In this sector therefore there are 460 robots/million workers. Of course, these statistics include all sales and administrative personnel, but as far as manufacturing as a whole is concerned, the use of robots would seem to have increased productivity by less than 0.1% in themselves.

Even if it is assumed that 20% of the people employed in this sector are productive workers, the figure is still very small. On that basis there is one robot for 430 workers, equivalent to an increase in productivity of some 0.25%.

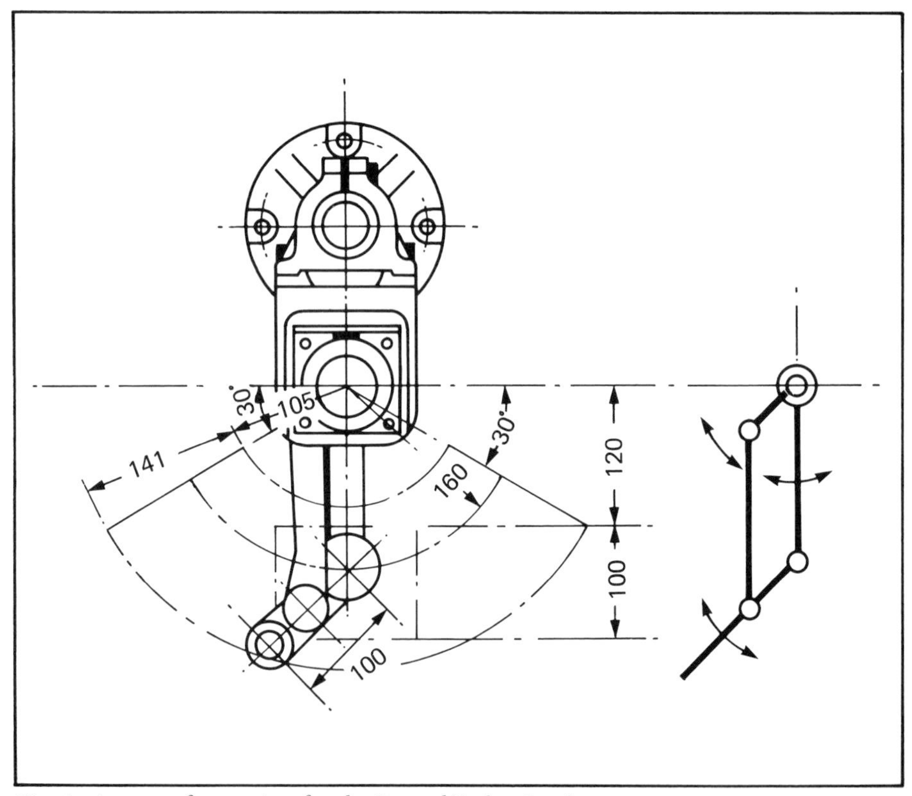

Fig. 1. Arm configuration for the Pentel Puha-1 robot.

Few robots

In practice, of course, many industrial sectors have no robots at all, whereas others have made significant gains in productivity. The comparison is valid for two reasons, though. First, it demonstrates that Japanese industry is not for practical considerations as highly robotised as the Japanese would have us believe.

Secondly, it illustrates why, combined with 'robot fever', so many companies are starting to make robots. They realise that this is just the tip of the iceberg, and that in the next few years, sales of robots will multiply enormously.

In this atmosphere, Kawasaki, NEC, Pentel, Sankyo Seiki have introduced new robots. Kawasaki has introduced the 9753-EP seven axis machine for spray painting, an application still waiting for a breakthrough, and this could be it. Until now, the use of robots in spraying has been confined to fairly simple components and underseal on car bodies.

The application of robots to the spraying of underseal to car bodies is simple, in that the area to be covered can be reached easily from one position. One of the most recent installations is not in Japan at all, but at British Leyland's Cowley plant, Oxford, where two Hall Automation CompArm robots are being used to apply underseal to the Triumph Acclaim – otherwise known as the Honda Ballade.

Currently, most car bodies are sprayed mainly by reciprocators which are capable of covering the whole exterior of the body, except the lower portion at front and rear. However, the reciprocators cannot reach inside the body, nor can they cover the pillars around the doors. But with robots it is possible to paint inside and to cover those critical 'door shut' areas.

To fit in with existing paint plants, robots will need to be able to spray the body as it moves along, and they may also need to be able to open and close doors, although it may be easier to do this job with some simple automated device.

Kawasaki's 99753-EP robot seems to meet all these requirements. It is a jointed arm robot which has six axes of movement itself; but there is a seventh in a longitudinal slide on which it is carried – thus it can move along the track, synchronised with the movement of the car.

The slide normally allows 3,000mm of travel, there being 135° rotation of the robot about its base, and 70° and 100° articulation of the two arm joints. Full 360° rotation is available at two of the wrist joints, and 220° at the third.

Overall, 16 programs can be stored and the robot can be taught in the normal way with the teach unit, or with the aid of a joystick it can be taught directly.

Assembly

Another area where robots are beginning to make their mark is in assembly. In Japan however the emphasis is on simple robots doing simple jobs, not esoteric devices looking many years ahead. Thus machines are being developed to install bearings and seals, and to tighten screws and nuts.

These jobs can be automated easily but the flexibility of a robot is increasingly worthwhile, either owing to short product life or large variety. These robots are likely to have another application in the future in the area of laying adhesives on components as more and more assemblies are built as bonded structures.

Among new assembly robots are two with similar basic characteristics – the Pentel Puha 1, and the Sankyo Seiki Skilam. Both are based on the SCARA concept developed by Professor Makino of Yamanashi University. The design centres on the fact there is a jointed arm on a vertical pillar with the spindle at the end of the arm having vertical motion.

In the Puha 1, there is actually a parallelogram linkage to actuate the 'forearm'. The upper arm – it is of course horizontal – is 160mm long, and the forearm is 100mm long. The auxiliary actuating arm pivots on the main pillar, there being a connecting link from the end of that arm to midway along the forearm. The whole arm is mounted on a pillar which is a large diameter screw giving additional vertical movement.

The Puha 1 is controlled by a Z-80 microprocessor and used to insert rubber retainers for pencil leads in the bodies of clutch pencils, a job involving a great deal of precision. Pentel also sees applications in packing, application of fluid droplets, caulking, and tapping and drilling small holes.

Sankyo Seiki developed its Skilam in conjunction with Prof. Makino. This too is a small device with a capacity of 6kg. Movement and construction is much on the lines of the original Scara robot, with a jointed arm carrying a vertical spindle at the hand.

The Skilson is a four-axis machine with the upper arm able to rotate through 200° on the pillar; the forearm can rotate through 135°, while the wrist can rotate through 270°. All of these axes are vertical. The normal linear movement of the operating spindle is 50mm. However, forearm, wrist and spindle movement can be 160°, 360° and 75mm respectively if required.

Repeatability is put at ±0.01mm and the complete robot weighs 60kg. There is a separate control cabinet.

Sankyo Seiki, of 1-17-2 Shimbashi, Minato-ku, Tokyo, has already sold a number of these robots, mainly to companies in the electrical hardware industry making such products as lamps, washing machines, refrigerators, and so on. In Japan, the cost of the robot is around £11,000, and for export, the price delivered to a Japanese port, is about £12,500. Thus, this small assembly robot has a more than competitive price.

Newcomer

Also entering robot manufacture is NEC, the giant semi-conductor and telecommunications company. It has already made a number of special-purpose machines for its own use but it has now unveiled a range of three machines – the Nebot 10, 50 and 300, which have capacities of 1, 5 and 30kg respectively.

These robots are based on a pillar and jointed arm, along the lines of the Scara robot, although they are quite large. For example, the Nebot 10 has upper arm and forearm of 300mm long each; while on the Nebot 50 the lengths are 350 and 450mm, and on the Nebot 300, they are 400 and 600mm. Vertical movement of the arm ranges from 200 to 500mm. Like the Puha and Skilam, the Nebots are driven by dc servo motors.

Fujitsu Fanuc: the rising star of Japan's robot manufacturers

With new assembly robots, new flexible manufacturing systems, Japan's most publicised robot maker is poised to live up to its potential – and the plans of its management.

Seiumon Inaba, president of Fujitsu Fanuc and his interpreter. At Fanuc, all employees from the president down wear bright yellow jackets. The buildings are painted the same colour.

IN THE PAST year or so, no robot maker has been in the news in Japan more than Fujitsu Fanuc. Early this year (1981), it started to operate a flexible manufacturing system in a new factory with virtually unmanned operation at night. Then, in the autumn, it introduced two new robots intended for assembly work. Next August (1982) it will open a factory in Luxembourg with a lot of robot assembly, while a month later, robots are due to start assembling dc servo motors in a new factory in Japan.

Clearly, those are very ambitious developments, intended to take Fujitsu Fanuc into an extremely strong position in the robot world. But these plans are to some extent the beginnings of a robotic age that was foreseen, and planned for, many years ago by Seiumon Inaba, the company's president, and his executives.

In many respects, Fanuc, as it is generally known, is unusual. It started off as a division of Fujitsu, Japan's largest computer maker, and also a leading company in telecommunications, in 1956. Between then and 1972, it concentrated exclusively on the development of NC – numerical controllers – for machine tools. In 1966, it produced its first NC based on integrated circuits, and in 1972 it developed its CNC – computer numerical control. That year, the current company was constituted, with Fujitsu taking a 52% shareholding, Fuji Electric and Siemens of Germany each 6%.

In 1974, an important step was taken with a licence being obtained from Gettys to make dc servo motors, which has since become an important part of both CNC machines and robots. In 1975, Fanuc produced its first robot for sale, the Model 2 and this was followed by the Model 1 in 1977, the Model 0 in 1979, the Model 3 in 1980, and the Model 00 and A Series Model 0 and 1 robots for assembly in 1981.

World leader

During this period, though, the CNC has remained the mainstay of Fanuc's business, although it has developed its own spindle motors and produces small CNC machines and electric discharge – EDM – wire-cut machines as well. It is without doubt the leading CNC maker in the world, having sold some 90,000 units, with monthly production now around 2,000 units. Even now, robot manufacture accounts for only 5% of the turnover, but to the Press generally, a robot is far more exciting than a CNC and hence the wide publicity in Japan. The other reason, of course, is that Fanuc has relied on its own technology in developing robots, rather than buying a licence.

A controller for a robot is basically, a CNC, so in branching out into the robot field, Fanuc was using its technology as a base. Before the energy crisis of 1973-4, Fanuc was developing a hydraulically actuated robot – an

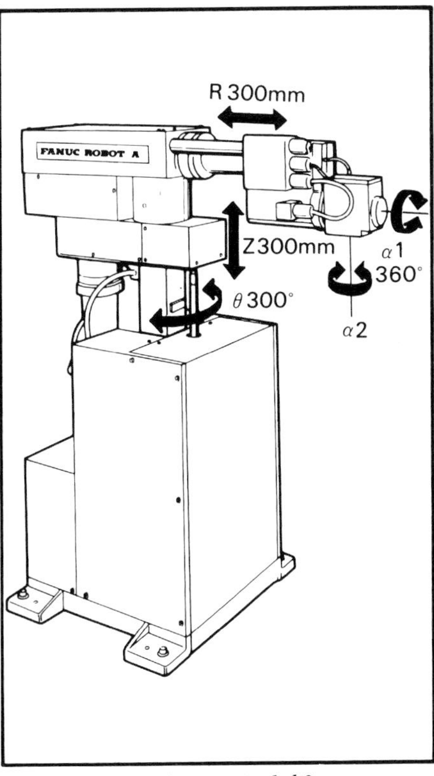

Movement of the A-Model 0 assembly robot.

indication that the company was into the business early – but subsequently concluded that electric dc servo motors were a better base – hence the deal with Gettys.

Fanuc also bought a Unimate robot which it mounted overhead on rails so that it could load and unload a number of machines. It was evidently too costly an installation, so there was no interest from customers.

In any event, it was not until 1979 that production really got under way, with Fanuc being able to offer two machines – both designed for machine tool handling, the business Fanuc knew well, and one in which it could exploit its own robots. Indeed, Fanuc has developed its robots by using them; initially this meant that it had its own test area in a real working environment, but later on it was able to gain from the increases in productivity before its competitors could.

Overtime

By the end of 1979, Fanuc was using 15 robots to load and unload components in its dc servo machine shop. These replaced 13 men, and were already working through the lunch break, and were capable of working up to four hours 'overtime' with very little supervision. The limit was in the supply of workpieces, these being stored on oval carousels that indexed around in sequence with the operation of the machine, so that an unmachined part was always ready, waiting to be picked up.

Since then, Fanuc has progressively increased the use of robots to handle workpieces at lathes and small CNC machining centres in its factories. It has also experimented gradually with assembly, until the new models were developed, and now it is ready to go into unmanned assembly – in some areas.

Fanuc is very much a systems company, so it did not just develop the robot, and then the carousel for workpieces, but also devices to monitor machining operations. Thus, by 1979 it was using devices to monitor tool wear, and adjust the tool as necessary – not during machining but between operations. But also needed was a device to monitor the operation of the machine, and to detect tool breakages or peculiarities in the workpiece.

It was not until last year that the Monitor Model A was introduced to do this. Basically, this device monitors the electric current at the main spindle of the machine, and compares this with information stored in the memory. Obviously, if a tool has been chipped, or broken, the current required will increase sharply – and in that event, the machine will be stopped immediately. In addition, as the tool wears, so the current will increase, and once the present limits are reached, the tool is changed. In addition, the Monitor will detect whether a workpiece is too hard or too soft, and again, if the result is a too-high or too-low current requirement, the machine will be stopped.

Unmanned factory

Once this Monitor had been developed, Fanuc was able to produce 'machining cells' based on CNC machine, robot, workpiece carousel and Monitor for small parts; for larger parts, an automatic pallet changer replaces the robot. In Fanuc's new factory near Mount Fuji, a number of these cells are combined with automatic warehouses, unmanned trolleys and buffer stores near the machines to provide a truly Flexible Manufacturing System.

Fanuc invested £16 million in this plant, whereas it is estimated that £12 million would have been needed for a conventional plant based on CNC machines. It produces parts for 100 mini CNC machines, 100 EDM wire cut machines and 50 robots a month. These are assembled in the same plant to produce an output valued at £3.6 million a month.

Fanuc president Inaba admits to being a bit embarrassed by reports that at the Fuji factory 'robots build robots'. Basically, the reports were made by people who did not understand what FMS was about, and they are just not true. The fact is that there are 23 machining centres and seven NC lathes. These are serviced by eight robots and about 20 automatic pallet changers, while two other robots are used for welding. The assembly is all done manually.

Nevertheless, this is a very impressive plant, producing 250 different components, with men loading workpieces from the buffers to the pallet changers. Only 15 men work in the machine shop – there are 60 people in assembly, and another 25 in the plant altogether – and they work one shift only. Before they finish at night, they load up the machines and pallets, leaving the factory to operate unmanned during the next 16 hours. However, there is a man in the computer control room all the time.

As a result of this development, Fanuc has been able to increase its productivity/investment ratio 1.5 times, and productivity/worker by five times. The result is that parts are machined more cheaply, increasing profits.

Products

Fanuc handling (M series) robots, and assembly robots (A series) are all cylindrical coordinate type machines. Thus, they are based on a column giving vertical movement, and a telescopic arm giving horizontal movement, with further degrees of movement at the wrist. The M-Model 1 and 3 – the Model 2 has been phased out – are free-standing machines designed to operate with a number of machine tools at the same time; up to five, depending on the cycle time. The Model 1 has a capacity of 20 kg – a double hand is used, so this means a workpiece of 10 kg – and the Model 3 up to 50 kg.

There are two versions of the Model 1, the cheaper having a simpler controller, which actuates only one axis at a time. In the other model, movement in three of the five axes is possible simultaneously. Fanuc recommends that the Model 1s be used with one or two CNC machines only, but in some applications more machines can be loaded/unloaded by one robot. To suit these eventualities, three different arms are available, with 500, 800 or 1 100 mm telescopic movement. All models have rotating columns with 500 mm vertical travel, while the wrist can be rotated and twisted.

The Model 3, designed for loading/unloading at machining centres and forges, has greater movement, and simultaneous control of all five axes for rapid operation. Vertical and lateral movement of the arm are

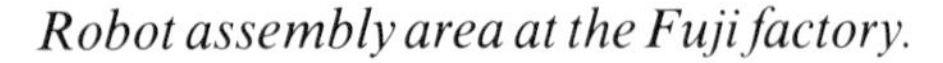

Robot assembly area at the Fuji factory.

Part of the new Fuji factory, where robots and automatic pallet changers are used to keep machines operating.

1 200 mm, the column can rotate through 300 deg, while the wrist can rotate through 300 deg, and twist through 190 deg.

Fanuc's Model 0 is designed to be fitted to the front face of a CNC lathe, and its five axes of movement have been arranged specifically to suit this installation. Only one axis can be controlled at a time, but that is all that is needed with this layout. The robot arm is carried on a trunnion, the arm being in two parts, so that the portion with the gripper can slide laterally on the other portion, which resembles a block.

Thus, the robot arm is normally vertical, and to unload/load, it rotates through 90 deg into the machine tool, and then the 'lower' arm is slid forwards to pick up the workpiece. It slides back, returns to the vertical, twists its wrist through 180 deg, and then rotates back to the horizontal position to load the new workpiece. After loading, the robot returns to the vertical, then turns through 90 deg on an axis parallel to the main spindle of the machine tool, and deposits the workpiece on the carousel. It is then able to pivot on the trunnion to move over the new workpiece and pick it up. The capacity at the wrist is 20 kg.

Pneumatic drive

Fanuc has also introduced a low-cost pneumatic robot arm, the Model 00 which is mounted on the headstock of the machine tool. It is combined with the carousel, in the normal Fanuc manner, the robot being a simple three-axis machine. The arm pivots on an axis parallel with the machine spindle – to load and unload at the chuck – and the wrist can be rotated through 270 deg to pick and place workpieces on the carousel.

To keep costs to the minimum, this device, which might not qualify for the name robot, can be set to a number of present positions only – two for linear movement and articulation, and four for wrist rotation. Nevertheless, it is ideal for many applications where the demand is for automated handling with some flexibility. Repeatability is put at ±0.3 mm, against ±0.5 mm for the Model 0, and ±1 mm for the Model 1 and Model 3.

Assembly robots

More exciting, though, are Fanuc's A-Model 0 and Model 1 robots intended for assembly operations. These are just going into production, and as usual, Fanuc has been using them for assembly work, on a pilot production basis for some time.

The differences between these machines are limited to size and capacity, these being cylindrical co-ordinate machines. The column can be rotated through 300 deg, and can raise the arm of the Model 0 300 mm, and the Model 1 by 500 mm. Arm extension is also 300 and 500 mm respectively, while the wrist can be rotated through 360 deg in one of two planes – either on a vertical or a horizontal axis. Compared with the handling robots, the assembly robots operate far more precisely – to ±0.05 mm, ten times better than the M-Model 00 – and about 20% more quickly. For example, the column moves the arm vertically at 600 mm/s, while the arm extends at 1 200 mm/s, compared with 500 and 1 000 mm/s respectively for the Model 1. Rotation is also rapid at 120 deg/s.

The M-Model 00 apart, the Fanuc robots share a number of special features, such as the magnetic bubble memory, a semi-conductor memory which provides a permanent record of programs even if power is lost. Thus, programs can be produced, and, if the number of programs is too great for one magnetic bubble cassette, another can be installed, the first one retaining its programs.

Fanuc has also developed a range of grippers, the first being obtained under a licensing deal. For the new assembly robot, new grippers have been developed, one of which grips the workpiece in three parallel pins. These pins are carried on eccentrics in the housing, so that they move in a rotary motion as they move in or out. Thus, the grasping action is progressive.

Assembly viewpoint

President Inaba's views on the use of assembly robots, which is the next stage for Fanuc, are surprisingly down-to-earth and very practical. Clearly, this is a result of his policy of using the robots before selling them.

Despite all the talk of 'unmanned manufacture' in which Fanuc's name is usually bandied about, Inaba said: "I think a factory with five men is sufficiently unmanned," and he is talking of a small factory. He thinks that men should be used at the 'interfaces' between warehouse and machines, and between the machines and automatic testing.

For example, at Fuji, it would have been possible to have devised automatic loading devices with robotic features between the robot trailers and automatic pallet changers. "We wanted more flexibility, so we made extra stations between the trailers and pallets. This handling is an easy job for a man, but a difficult job for a robot. Also, you need men for maintenance, and in our company we no longer have 'workers' and 'engineers', we have 'engineer-workers'," he said. Probably, engineer-operator is a better description in English, but the point is that a few men can do many things that very intelligent robots will not be able to do for many years yet.

In the past year, Fanuc has been gradually increasing the number of its robots for use for assembly of dc servo motors. Now, some 50 robots are being used at the Hino factory, but in September 1982, the second Fuji factory is due to open to assemble motors on a more highly automated basis. "We shall have 150 robots and 50 workers; now we have 150 workers and 50 robots," said Inaba with a chuckle.

Cell

Fanuc's policy is to use one handling robot and to arrange the work stations for an assembly cell around that in a semi-circle. In the basic cell adopted so far, an M-Model 1 robot is combined with three assembly robots – all Model 0s – simple presses, a vibratory bowl feeder, and a special feeder for long studs. There are also three carousels carrying workpieces, an indication of the systematic approach.

The operations involved are: fitting the oil seal and bearing to the rotor; insertion of the rotor assembly in the flange; and fitting the cover. These operations take place in the first two stations, the second being the press. At subsequent stations, the studs are fitted, and nuts are mounted and tightened.

The first assembly robot can reach a small table and press, as well as the main table and press. It places the oil seal and bearing on the rotor on the first table, where these are subsequently pressed in position. Then, it can take a rotor/seal/bearing assembly from the intermediate carousel, and place it on the main table. The handling robot then lowers the stator over the rotor assembly on to the end flange. A cover is subsequently added.

At the next station, an assembly robot picks up four long studs in sequence, and then tightens them in position. At the final station, an assembly robot picks up four nuts in sequence, lowers them in position, and tightens them up.

Inaba claims that this cell, which can operate for 24 h. continuously, can produce 300 assemblies in 24 hours. With normal manual assembly, the output is 30 motors/man/shift, so that one cell can replace 10 men theoretically.

Experimental

Inaba emphasises that the cell is experimental, and that by the time the new factory is built, it will have been further refined. Nevertheless, he has no intention of trying to completely automate assembly. "The men will fit the brushes, resolvers and the tachogenerators, as these jobs require very great accuracy," he said. Also, of course, delicacy is needed.

Although Fanuc is involved in a project to develop an intelligent robot, Inaba says: "I don't think the intelligent robot will be a main factor in the use of robots. We need it to supervise the machines at night, but it will be very expensive, and its application will be limited."

Nor is he enthusiastic about robots with visual/tactile sensing, even if these are simpler than those being advocated today. "For assembly by robot, parts need to be designed with care," he said. He emphasised the need for tapers, chamfers and leads to overcome minor misalignment. In other words, take an assembly robot designed to work with precision, then design your part for assembly by robot and you are in business.

This, of course, is precisely what Fanuc is doing. In its factory in Luxembourg, to be operated by a subsidiary called Mechatronics SA, Fanuc will assemble System P CNC programmers – basically a small computer with vdu. The plan is for this operation to be almost entirely done by robots, with three or four cells, and handling robots being used in between. Overall 10-12 robots are likely to be used to produce 50 sets/month initially, but with capacity for 100/month.

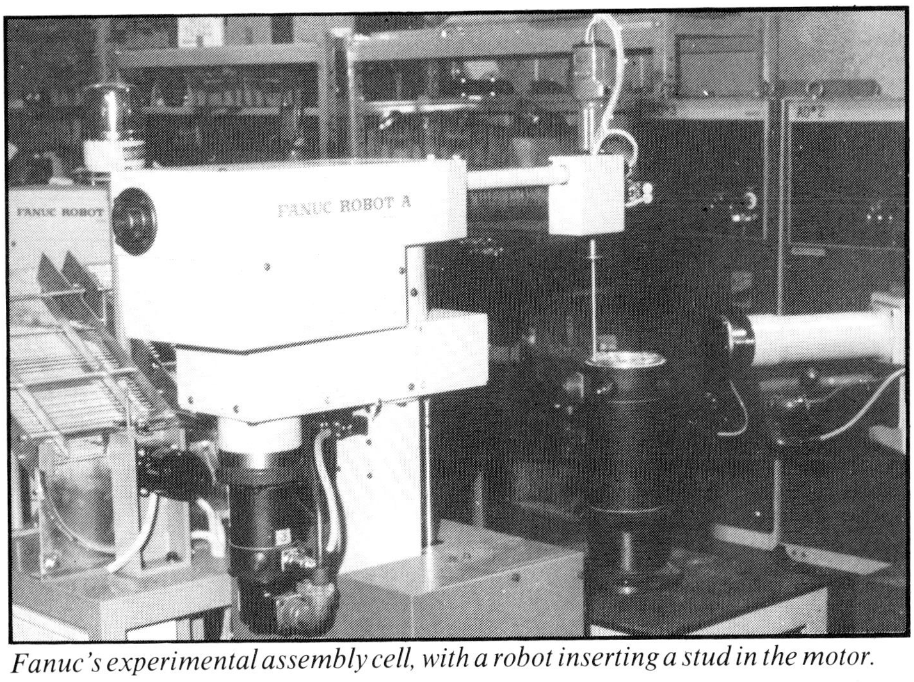

600 Group

In addition, Inaba said that he had agreed in principle for the 600 Group to make the complete range of Fanuc robots in Britain, and by the time this is published, negotiations should be complete. This move should have a significant effect on robot useage in Britain, if it is successful, not least because the price of the robots should be far lower than those currently being imported via Siemens.

Clearly, Fanuc has risen to the top level of robot manufacturers in the past three years or so. By the end of 1979 it had produced only 50 robots or so, and now it is building 50 a month. Inaba is not happy with this volume, claiming that his sales people are not doing well enough, and that he expects this to reach 100 a month by the time the new dc servo motor factory is opened.

For all that, robots account for only 5% of Fanuc's turnover, but Inaba expects a dramatic change in the next five years. The Fuji factory where robots, mini-CNC and EDM machines are assembled, can be turned over to robot assembly alone, Inaba pointed out, and this would presumably give capacity for 300 robots/month to be built. Without doubt, Fanuc is ready for the robot explosion.

Fanuc's experimental assembly cell, with a robot inserting a stud in the motor.

Applications diversify

Some robot makers look for new applications, while others seek overseas partners, but demand for more of the same robots continues.

IN THE PAST few months, there has been a flurry of commercial activity as companies with robotics technology attempt to negotiate licences with foreign companies. Fujitsu Fanuc has been active, while Mitsubishi, Komatsu and the licensees of the Scara robot, are all involved in negotiations.

Meanwhile, Dainichi Kiko is looking for applications to generate enough business to keep a big factory going, Matsushita is increasing output, and the run-of-the-mill applications continue to expand demand for robots. Against that background, though, Japan's Ministry of International Trade and Industry (MITI) has warned that the 'honeymoon period' between labour and robots is coming to an end.

Fujitsu Fanuc, still relying on robots for only about 5% of its turnover, has been talking with all sorts of companies about sales and licensing deals recently. Of course, it did conclude the deal with the 600 Group in the UK for local manufacture of its robots, but since then has been talking with both General Motors (GM) and Bendix with a view to manufacture in the USA. The talks with GM raised a few eyebrows, since it announced some time ago that it planned to make and install a large number of robots in the next decade. In addition, GM was involved in laying down the specification for the Puma robot. The question is whether GM is intent on using the Fanuc robots to help it develop its own, or whether it justs wants to take the easy way out.

Fanuc has also started talks with Tatung of Taiwan, and draft agreement has been reached for Tatung to sell robots in Taiwan only. However, Seiuemon Inaba, Fanuc president, has said that he is prepared to let Tatung build the robots for sale in Taiwan if it wishes. Tatung is Taiwan's biggest electronics engineering company, with products ranging from televisions and cookers through personal computers and telecommunications equipment to machine tools. In consumer electronics, much of its technology came from Toshiba, but it also imports some American computers and electronic medical equipment.

At present, there are no robots at all in Taiwan, and although industry is labour-intensive, the government of the Republic of China, Taiwan, has mapped out a plan for turning to automation. Thus, potential demand is high, and with its diverse products, Tatung itself is well placed to take advantage of the future use of robots.

It has also been announced that Westinghouse, in addition to developing its own robots, is negotiating to market arc welding robots produced by Mitsubishi Electric and Komatsu. More startling, perhaps, was the announcement that IBM was to sell the Sankyo Skilam robot in the USA as the IBM 7535. The Skilam is one of the Scara group of robots developed at Yamanashi University by Professor Hiroshi Makino for assembly.

In addition to Sankyo Seiki, NEC, Nitto Seiko, Yamaha Motor and Pentel provided funds for the development of the Scara. Then, they designed their own units around the basic concept of an assembly robot that has some compliance built in; basically, so long as the object being inserted in a hole strikes the chamfer of the hole, and not the flat surface, it will be inserted successfully.

Only Sankyo decided to incorporate the motion curves developed by Makino to reduce vibration at the end of the movement. These are polynomial curves, the deceleration period being twice as long as the acceleration curve. Thus, the arm decelerates smoothly, so that accuracy of positioning is enhanced.

The Sankyo Skilam is a sound design, with a stiff arm pivoting on the pillar. The basis of the Scara is that the arm has a shoulder and elbow joint, both of which pivot on vertical axes, while the normal tool can rotate and move vertically. There are two models, both of which can have three or four axes of movement. The larger one, SR-3 has upper and lower arm lengths of 400 and 250mm, while the SR-4's dimensions are 200 and 180mm.

Owing to the use of the polynomial

A Part Time robot being assembled at Dainichi Kiko's factory.

acceleration/deceleration curves, the Skilam can operate very quickly, a 3s cycle being normal. Makino claims that the Scara robots are faster than the Fanuc assembly robot or the Puma, and in Japan, at any rate, they are about half the price of the Puma – about £12,000. In the past year, almost 400 Scara robots have been built, 200 by Sankyo, over 100 by Nitto – mainly for tightening screws – and about 80 by Yamaha, which has just announced that it is to start selling its Came machine.

Pentel has so far built few, but is using them for a demanding job. The robot picks a conical rubber moulding, approximately 2mm in height from the sheet in which it is moulded, and inserts it in the tip of a clutch-type pencil. The cycle is 2s.

Dainichi Kiko robots on test.

Patents

Hirata has also started to build Scara-type robots, but these are made under licence, and other companies in Japan are evidently following suit. The patents situation is complicated, in that the basic Scara patents are held by a small company called NAIS, but each manufacturer of the original Scara group has free use of them. Thus, any foreign company wishing to build this type of robot could approach one of the existing group, or could approach NAIS directly, in which case, its design would need not to infringe any patents filed by the Scara group companies.

It is interesting that IBM should have approached Sankyo, as did CGMS of France last autumn for a similar deal. One reason for IBM's interest no doubt was that the Sankyo Skilam is suitable for assembling small electronic components, so the US company will be able to use the robots in its own factories. On the strength of this deal, Sankyo is forecasting that it will build over 600 robots this year, and probably more than 50% will be sold through IBM.

With all the growth that is projected for robots in Japan, it is hardly surprising that new companies are coming into the business rapidly. What is surprising is that so many European companies seem to think that they need to come to Japan for a design, instead of doing the work themselves. Obviously, the situation regarding patents is a minefield, but there is little magic in Japanese robots – after all, most of them were based on foreign designs. The exception, of course, is the Scara.

In any event, faced with the fact that robots are going to be used far more than in the past, MITI has decided to start a full-scale investigation into the effects of their use. In fact, an advisory council is being set up in the Ministry of Labour, largely in response to pressure from the Trades Union Council. The study will last two years, and will tackle such subjects as unemployment and redeployment of workers.

At the same time, the Japan External Trade Organisation (JETRO) has published a booklet entitled 'Industrial Robots' (publication No. 32/1981). It is based on the usual misconception that all Japanese seem to have about robots, namely that if something can move a workpiece it is a robot. On the subject of labour relations, this publication suggests that in Japan many workers may have to shift from manufacture to the service industries. This seems highly unlikely, since this sector is in need of rationalisation which will inevitably mean employing less people. So another solution must be found.

What is particularly interesting about the JETRO publication is that among the case studies, none are of large-scale use of robots. One or two of the stories are not about robots at all, but handling devices. It seems that JETRO had as much difficulty in tracking down robots as do foreign visitors when they come to Japan.

One application was at Nihon Press Kogyo, a small company involved in

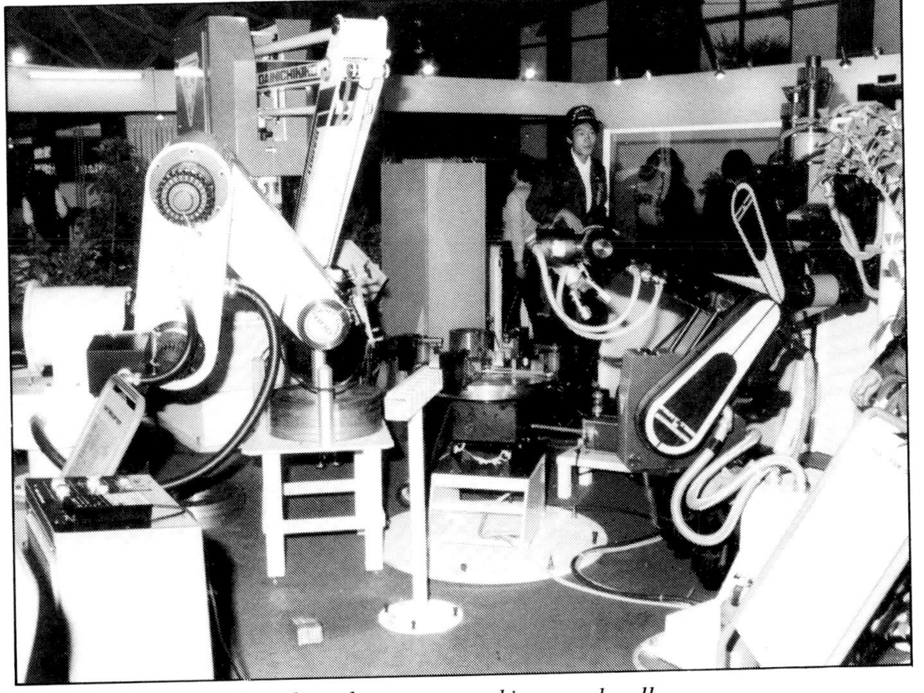

A number of Dainichi Kiko robots arranged in a work cell.

making small fabrications. It decided to install a robot when it received a large order for damper brackets from an automotive company, and knew it would need extra staff. In the end, it chose a Yaskawa Motoman to handle nine different jobs. Two robots were needed, but the company decided to install one owing to the high cost of robots and the limited space. Instead, therefore, of buying two robots, it leased one, and hired four young welders who could be paid low wages.

The company pays 230,000 Yen a month in leasing charges, and another 20,000 Yen in running costs, so the total costs are 250,000 Yen (£600/month), which is a little less than the labour cost of one man. The robot welds up batches of six workpieces which are assembled on a jig manually. Working this way, the robot is faster than a man.

Another company that has installed a few robots for arc welding is Toyota Tekko, one of the suppliers to the Toyota car group. It has installed six robots, and has a target of 40 by 1985. Previously it was using arc welding machines, so much of the welding is simple. Nevertheless, each robot is producing 2400m weld/month, against 1500m for the machines. The robots are being used mainly for batches of up to 5000/month, and it is claimed that they have increased profits by almost £50,000 a year, implying a payback period of three years.

With demand for robots in Japan, and probably in Europe as well, set to increase at the rate of around 50% a year, there are clearly plenty of opportunities for robot manufacturers. One company that has decided to exploit this situation is Dainichi Kiko. It used to be a small company making special handling equipment, such as small trucks designed to handle large bars or plate glass. Thus, it had experience in the manufacture of grippers, and now plans to become a fully-fledged robot maker, offering a complete range of robots. It claims to be a completely independent company, and is privately owned.

Dainichi Kiko

Dainichi Kiko still makes some handling devices, but its main products are the Babot and Part Time robots. The Babot is a large jointed arm robot available in three sizes with total capacities of 50, 100 and 350kg. The Part Time is a much smaller jointed arm robot. The 500 is mounted on a pillar, so that it hangs down like a human arm, whereas the 300V and 800 are mounted on bases in the normal way. The 300H is mounted sideways on a pillar so that it moves in the same way as the Scara.

There are three controllers: the R-510, for sequential control of the movements, one at a time; the A-200 which can control three axes simultaneously, and the A-300 for control of up to six axes simultaneously by interpolation. Misao Fujiwaki, head of one of the engineering sections at Dainichi Kiko, says that so far they have concentrated on robots for handling, sealing and welding. There is a plan to develop a robot for spray painting. He says that both the car and electrical companies are now spending most on robots in Japan, but most of Dainichi Kiko's customers have bought one to three robots for trials, and that big follow-up orders are expected this year.

Thus, most of the applications involve a few robots only. One of the most interesting is the use of five Part Time 500s to handle wire at a spring maker. There are five processes from the cropping of the bar to the finishing of the coil spring, and the robots work at two lines alternately. The cycle is

The Yamaha Came robot, one of the Scara family, being used to tighten screws on Yamaha motor-cycle engines.

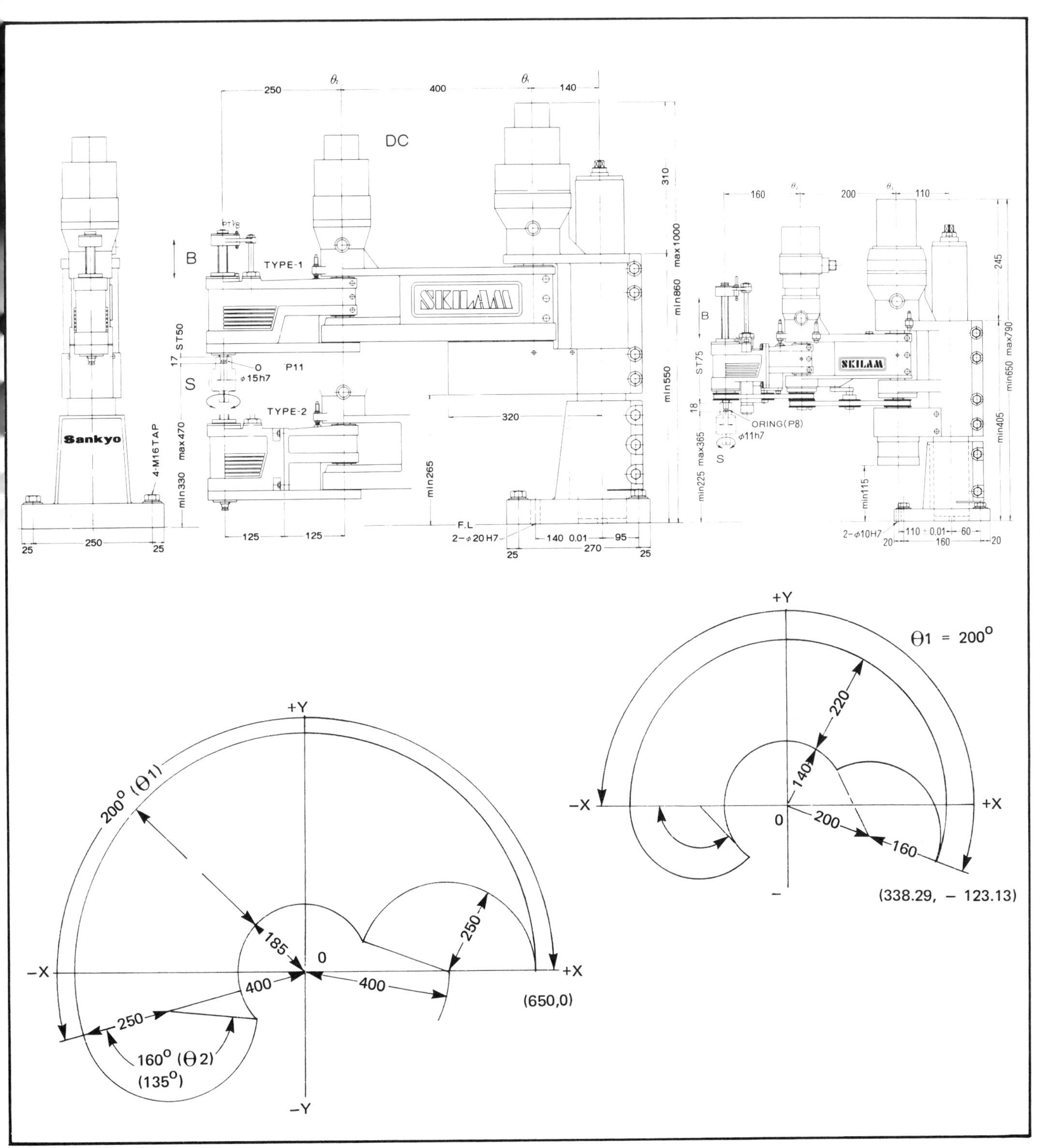

Configuration of the Sankyo Skilam robot.

under 20s, and the robots operate completely unmanned.

One of the Babot 1440 robots is being used to handle large bars 400mm long by 300mm in diameter at a slot cutter. The cycle takes 40min, and the robot picks workpieces from a pallet, and unloads them into another. The machine can operate unmanned for 14 of the 22-h working day.

A shoemaker is using a robot to remove synthetic rubber moulded shoe inners from the dies. The die resembles a foot and ankle, several being arranged on a rotary table. The Part Time robot removes the moulding from the die while the moulding is still warm, so a delicate grip is needed. In addition, the robot needs to be able to push the moulding off the die as it grasps it. Despite the fact that Dainichi Kiko claims to have a lot of experience in the design of grippers, in this case the gripper was made by the user of the robot.

Currently, Dainichi Kiko claims to be making about 30 robots a month in its new factory. However, it is an assembler only, all the machining being subcontracted and all the servo motors, controllers and so on being supplied by specialists. There is a large hall for assembly and testing, and at the end of February there were about 30 robots on test awaiting delivery. In the assembly area, there were many partly-assembled robots, but nearly all of these consisted of a base and a few parts only. Meanwhile, a group of 4-5 workers were finishing off a Part Time robot; other people were working in the test area, but the impression was that far less than one robot/day was being produced. In fact, Dainichi Kiko now employs about 140 people, including 60 in engineering. It has a team of 20 involved in applications development, and apparently, when the production shop is busy, the engineers help assemble robots.

Nevertheless, Fujiwaki said that turnover had reached 2 billion Yen (£4.7 million) last year, and should reach 6 billion Yen (£14 billion) this year. Whether that turnover is reached depends on the success of the new factory, due to be completed in the summer – rather later than initally planned.

In this plant, the company aims to machine parts with partially unmanned operation, and also to assemble robots, partly with the aid of robots. Fujiwaki said that the aim is to use vision sensing provided by Autovision of the USA to identify and orientate parts fed to robots. These robots will then pick up the parts and assemble them. He also said that the company planned to have some tactile sensing devices in operation by the summer. 'We are working with a specialist maker of sensors now,' he said, 'and will develop a complete range, including one that senses temperature limits, and another weight.'

Dainichi Kiko is investing 4.2 billion Yen (£10 million) in the new factory, which should have a capacity of 250 robots/month. That equals the combined production of Fanuc, Kawasaki, Yaskawa, Hitachi, Matsushita, and Sankyo at present. 'We hope to be producing 100 robots a month by the summer,' said Fujiwaki. This is certainly an ambitious target, since Fujitsu Fanuc is only hoping to be able to produce 100/month by the autumn. In view of the fact that the current production level seems to be overstated, the target looks rather optimistic. One expert on the industry suggested that 30-40 a month was a more realistic figure. It remains to be seen.

In any event, Dainichi Kiko, despite its small size, has established a number of overseas agents – in the USA, Sweden, Australia, Korea, Taiwan and the UK. In the UK, it has set up a company with Sykes of Huddersfield – Dainichi Sykes Robotics – which is to start making robots. So far, Sykes has bought 'a few robots' and is to start off making 20-30/month, according to Fujiwaki. 'It is not economic to produce 5-10/month,' he said. Initially, servo motors and controllers will be supplied from Japan, but the aim is for local sources to be used, in due course. The new company will be able to sell in almost the whole of Europe except for Sweden.

In Japan, Dainichi Kiko is looking to new industries to give growth. 'We make a wide range of robots, so we will be able to provide people with complete systems,' said Fujiwaki. Often, customers want robots for handling, assembly and welding, for example, and Dainichi Kiko can supply all these. In addition, Fujiwaki said that there are many untapped applications in the food industry. In Japan, there are thousands of small bakeries making cakes and biscuits. Since the baking time is often short, there is a potential application for a robot to open the oven door, remove the tray of biscuits and insert a new tray.

Although these tiny firms, often employing only one or two people outside the family, may not have the capital to invest in robots, once 'robot fever' grips Japan, they will easily be convinced that they should get up to date and buy a robot. Therefore, this is a big potential market, and one that could help Dainichi Kiko grow quickly. In any case, this company is all set for the robot boom.

Spot welding

Japan's automotive industry continues to progress in the use of robots for body welding. The newest factory in Japan is the Tahara plant of Toyota, where it produces Corollas and light trucks in the No. 1 plant, and Celicas and Soarers in the newer No. 2 plant. Some 280 people work in the body-in-white shop in the No. 2 plant, where 90% of the welding is automated thanks largely to the use of 90 Unimate robots. Many of these are the small 6060 arms, designed specifically for body welding.

There is one underbody line with 14 Unimate 6060 and eight of the large Unimate 4000s. Unusually, some of the small robots carry large jaws so that they can reach the middle of the floor. Of course, the 4000s are used where the longest reach is needed.

Generally, the 6060 arms are mounted separately on short pillars, and are installed so that the arm pivots on a vertical axis. Thus, the arm uses its main articulation to reach into the body. In some cases, the pillar is mounted on a slide so that the robot can move right into the underbody. With this line, Toyota can weld up underbodies of seven different types, although in some cases the differences are confined to bracketry. There are two basic front ends and two basic rear ends for the underbody, to provide models with different wheelbases, different widths and engine bays.

Each robot is preprogrammed, and as each underbody starts on the line so the production control computer inputs the necessary data. The side assemblies too are welded by robot, with a similar control system.

In the framing area, the machinery must deal with three different bodies, although two of these, the Celica Supra and Liftback are the same from the windscreen pillars backwards; the Soarer is completely different, apart from the underbody, which is basically similar to that of the Supra.

At this stage, Toyota uses a mixture of multiwelders and 6060 arms, the multiwelders being equipped with multiple guns and clamps to cope with the variation. At the first framing station, though, there is one 6060 arm mounted on a gantry, and at subsequent stations, the robots are used to make welds where the differences between bodies are large. Respot welding is done by Unimate 6060 and 2000 robots. Throughout, there is one line working to a cycle of 1min 50s to produce 500 cars a day.

Recently, the Japanese car companies have introduced many new models, and it is because models are renewed so frequently that many robots are used in body production. Thus, Nissan now uses over 700 robots, Toyota 550, and Mitsubishi, Toyo Kogyo and Honda around 150 each. Most of these are Unimates.

It is also the automotive industry that is providing the demand for robots that can spray car bodies – a potential market for several hundred more robots. But all the signs are that the growth coming now is in many small orders, often for two or three robots only. In addition, the applications are diverse and the costs of development may prove to be high, owing to the need for special grippers.

Emphasis on cheaper robots

Several manufacturers are introducing cheaper robots, others are concentrating on handling machines, while Toshiba is tackling the remote inspection problem, and Toyo Kogyo is putting some robots into its paint plant.

WITH the Machine Tool Fair at Osaka, it was not surprising that a host of new robots intended specifically to handle workpieces at machine tools should have been introduced in Japan. Most of these are designed to be mounted on the machine tool – in most cases an NC lathe – and are broadly similar to the Fanuc M-Model O.

Less expected, though, were some really inexpensive assembly robots from Hitachi, with prices starting at about £2,500. Toshiba, which has seemed to be rather dormant on the robot front recently, came back with a robot arm for inspection in remote areas, and also some robots for assembly. Then, Toyo Kogyo demonstrated the state-of-the-art in robot use in automotive manufacture when it opened its Hofu plant in western Japan. Included in this plant is robot painting, and some simple vision sensing.

At the Osaka Machine Tool Fair the accent was on FMS, so many companies were exhibiting their own handling robots. Among these were Yamazaki, Okuma, Mori Seiki, Ikegai and Hitachi Seiki. Of the robots designed to be mounted on the front of the machine tool, Hitachi Seiki's was one of the smallest and neatest. Daikin Industries was one of the independent companies showing a robot arm that could be mounted on the front of any machine. The arm articulates on a horizontal spindle, and can also move longitudinally in its housing, and the wrist can rotate.

Ikegai, which started off with a small robot arm mounted on the headstock, has now developed a larger, but simpler, arm that is mounted in front of the machine. Speed is of the essence with this type of robot, and the L-Robo swings out of the machine at 90°/s. It has a double gripper, which can be rotated through 360°, and has three other axes of movement. It is priced at just over £10,000 in Japan.

Toyama introduced the RP-1, a free-standing cylindrical co-ordinate robot, not unlike the Fanuc M-Model 1. It has a capacity of 50kg, including the wrist, and has up to six degrees of freedom. The arm can extend at 1,000mm/s, and can move vertically at 700mm/s. Similar in concept is the Kitamura Spaceman, although it is smaller and neater, with a capacity of 10kg.

Yaskawa

Yaskawa showed the first results of its liaison with MIC – a Yaskawa robot operating with a vision system and an x-y table. The camera first plotted the position and orientation of a pressed plate which was fed to the table. Then the table centralised the plate, and the camera changed to the close-up mode to inspect it for defects – in this case, either bolt holes that were missing, or unwanted semi-circular holes in the side. The robot then placed the part in either a 'Go' or 'NoGo' chute. The vision system was MIC's standard unit, of which about 45 have been sold,

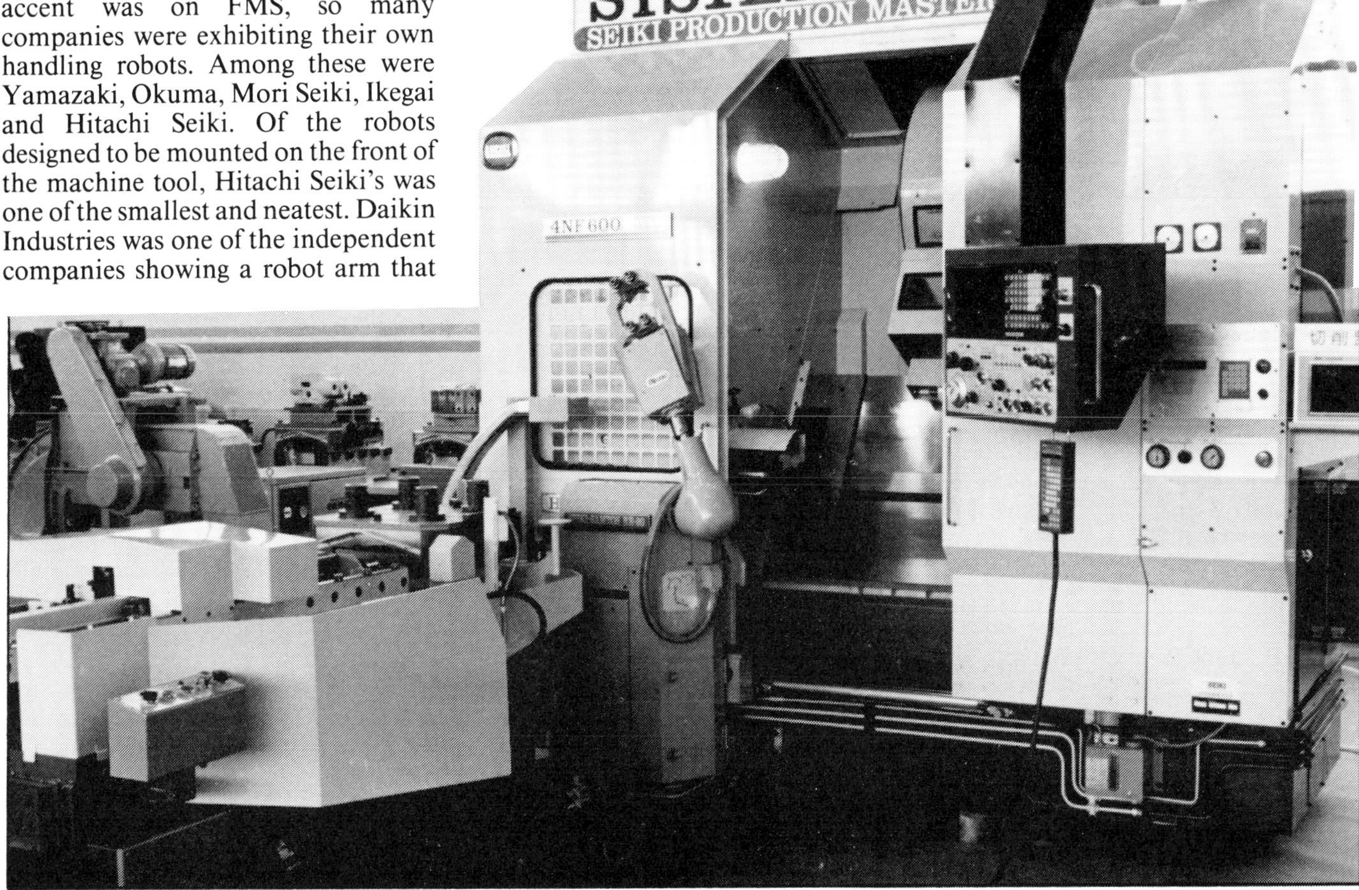

Fig. 1. Hitachi Seiko's robot is mounted neatly on the front of the CNC lathe.

although at the time of the show none had been sold in conjunction with a Yaskawa robot. Of course, this is coarse inspection.

Incidentally, Yaskawa has finished a deal with Bendix, under which the American company will sell the Motoman L3 and L10 robots on an OEM basis in the USA and Latin America. The expected volume is 50-100 units pa, at first.

Fanuc

Fanuc was also well to the fore at Osaka, demonstrating the S-Model 1 and 3 jointed-arm robots. There are models with gripper capacities of 10, 20 and 30kg in the range – Models 1, 2 and 3 – and these are intended for spot and arc welding, spraying and laying sealant. These all have ac servo motor drives, which are now being introduced to the Fanuc range of robots to reduce weight and size. The robot can rotate on its base through 300° at up to 300°/s, or can be mounted on a linear slide with 1,000 or 2,000mm travel. The shoulder joint provides 75° rotation, at up to 120°/s, the maximum speed of the elbow and wrist joints.

The largest S-Model, the Model 3, has a reach equivalent to the large KUKA machine, whereas the small Model 1 is only 1.31m tall – with the arm horizontal. The arm, from elbow to wrist joint is 700mm long. Repeatability of the S-Model 1 is ±0.2mm, and ±1mm for the two larger models.

Fanuc has joined forces with Osaka Denki to market the new S-Model robots for arc welding, and they are marketed as Daiden-Fanuc units. Osaka Denki is a welding specialist, and it has devised a tracking system similar to that of Yaskawa, in which the gun weaves as it welds. The current at the ends of the weave are compared, if they differ, the controller adjusts the welding path.

Of course, Fanuc has also been busy signing deals with The 600 Group and Kolon to set up 50:50 joint ventures to market and possibly build robots in the UK and Korea, respectively. Now that Fanuc is involved in Britain, its robots are likely to be sold much more aggressively, with better service back up than previously. It is also likely that prices will be lower than before.

However, Seiumon Inaba, president of Fanuc, is evidently not keen on local manufacture, principally because Fanuc now has three big factories in Japan to produce controllers, robot mechanical parts and servo motors. None of these is working at anything like full capacity – the robot factory, for example, is producing 100 robots/month compared with a capacity of 300; no big increase is expected in the near future – so only assembly and minor applications engineering is likely to be carried out in the UK.

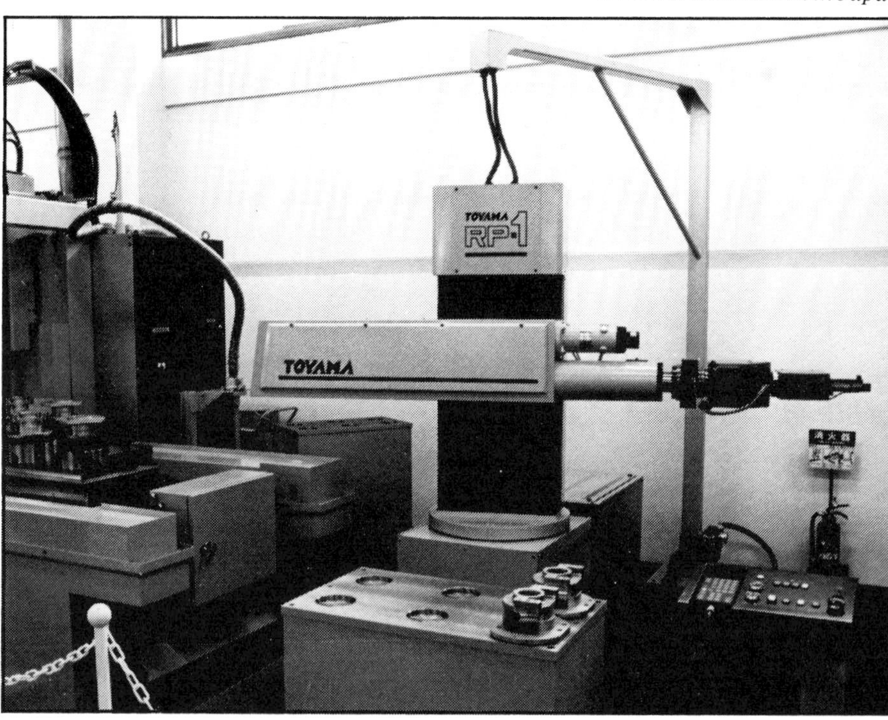

Fig. 2. Like Fanuc, Toyama adopted a cylindrical co-ordinate concept for its handling robot.

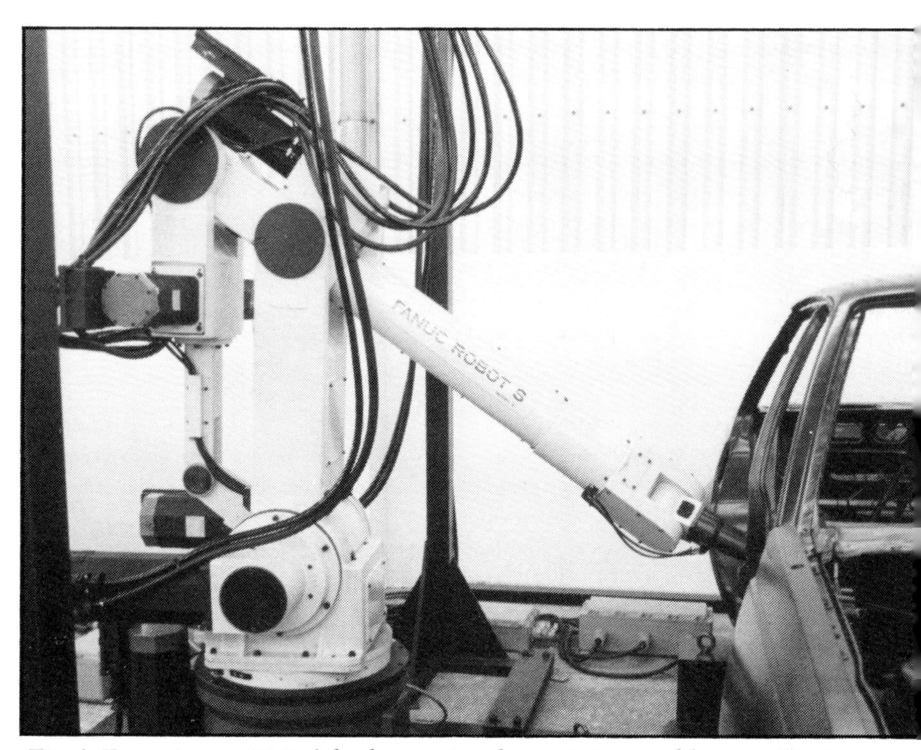

Fig. 3. Fanuc's new S Model robot is aimed at painting, welding and laying sealant.

Toshiba

Toshiba, a massive engineering organisation, is now expanding robot work in a number of different directions. First, a robot intended for inspection in nuclear reactors and underwater sites; also under development are some inspection robots for factories and pipes. Then, there are some newly developed assembly robots.

The remote inspection robot, now ready for production, consists of a multi-jointed horizontal arm mounted on a column, which travels along rails. There are eight joints in the arm, each consisting of a universal joint with two axes of freedom, to give a total of 16 axes. Each joint can articulate through 100° conically. The arm is controlled by a 16-bit computer in which a number of programs are stored, and it inspects with the aid of the camera at the end.

Normally, the robot is pre-progammed to wind its way around obstacles to the object to be inspected.

If it encounters an obstacle, it stops and awaits further instructions from the operator – no interactive system is built-in to help it find its way around the object.

The main innovation is in the universal joint and the sensors. The universal joint consists of the normal spider, a semi-circular gear being mounted on each leg of the spider, the teeth facing away from each other at 90°. Small motors, whose spindles are axially in-line with the arm, drive through a worm gear and idler to the semi-circular gear to move the joint. Thus, each joint has two axes of freedom.

Since the arm is long the weight had to be kept to the minimum so that the motors and other equipment need not be large. Therefore, to save weight, the gears are of titanium, the casing of aluminium, and the covers for the arm CFRP. The 2.25m long arm weighs 23kg.

There are a total of 56 touch sensors in the arm, which are actuated when the cover of the arm is subjected to a load of 100 or 200g. The cover of each section of the arm is cylindrical, and collars at its ends are connected to the universal joint housings by leaf springs with low rates. The collar at the end of the cover fits inside the universal joint housing, and between the collar and the housing are four equally-spaced pressure sensitive pads. These are plastic or rubber and respond to small loads. There are also 34 limit sensors, and 17 positional sensors.

A Micro 7 micro-computer with a 16-bit processor and 128 kilobytes of main memory is used to control the robot; the programs, which are written in Assembler, are stored in 64 kilobytes. There are various modes of control, the most complex being one in which a graphic display of the route is shown on the video screen, and then the arm is programmed to follow the required route. However, that system requires a 32-bit mini-computer, so it is costly.

Normally, the arm is equipped with a pair of cameras or rangefinders for inspection purposes. The rangefinders can be either optical or ultrasonic. The level of precision of movement is not high, however, as accuracy is not important with inspection robots.

Toshiba expects to be able to sell the robot arm to nuclear power companies from £65,000 upwards. Since this robot can find its way into radio-active chambers, thus increasing safety and reducing the currently uneconomic downtime in these facilities.

Toshiba exhibited three much smaller inspection robots at a show it held in Tokyo in December. These were based on similar units, but are intended for different environments. One runs along a monorail; another finds its way along pipes with bores of at least 0.5m; and the third is a little tracked vehicle for traversing rough ground.

All of these are equipped with cameras for inspection purposes, the one designed to go inside pipes has two cameras, and three arms, each with a wheel, guide the robot along the pipe. The tracked robot is interesting in that it has a telescopic body, so that the wheelbase can be shortened for travelling up stairs or around corners, and half-moon shaped wheels carrying tracks. To some extent, these look like devices seeking jobs, although there is clearly a case for an inspection robot for gas pipes.

Closer to industrial reality are Toshiba's assembly robots. These have six axes of freedom and a ccd camera acting as a visual sensor. Two of these robots are to be used to solder wiring terminals to components in the base of electric fans – a product made in huge volumes Japan. The robot is similar in concept to the Puma.

The tiny camera lens is mounted in the middle of the gripper on one robot, and it first identifies the position of the wire and terminal. Then, the robot arm is positioned to pick up the wire, which has been identified by colour, and place it at the correct terminal. The other robot moves across and solders the wire in position. Toshiba engineers say these robots are about a year away from production.

These machines are developed from Toshiba's assembly robots, which are in use experimentally in one of its factories. These earlier designs have five basic axes of movement, and are similar to the Puma except that the wrist incorporated an extra vertical joint. Thus, the wrist and insertion tool is always kept vertical.

Vision sensing for assembly is still thought by most researchers in Japan, to be some way from everyday use on the factory floor. Hitachi and Mitsubishi Electric have made it seem even further away by introducing cheap robots.

Fig. 4. Toshiba's robot intended for inspection of remote areas has a long multi-jointed arm.

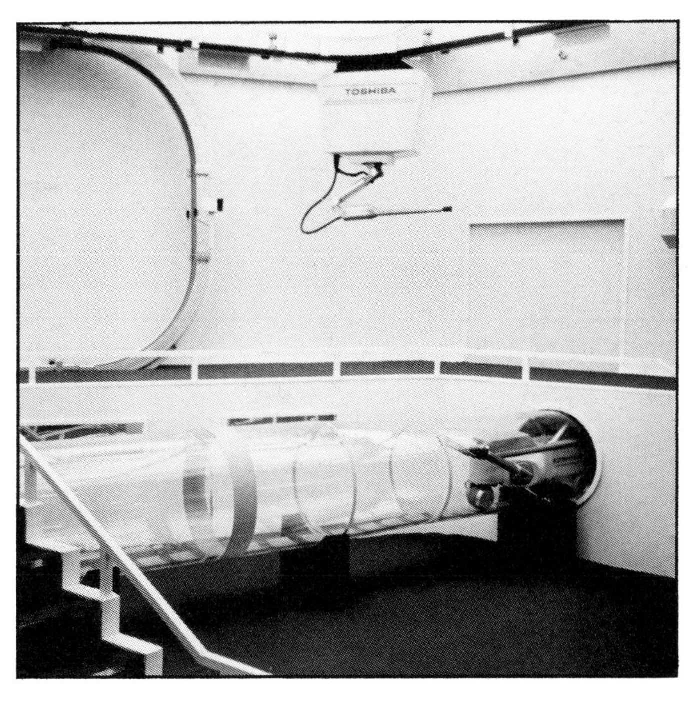

Fig. 5. The Toshiba inspection robots intended for working in pipes and on monorails.

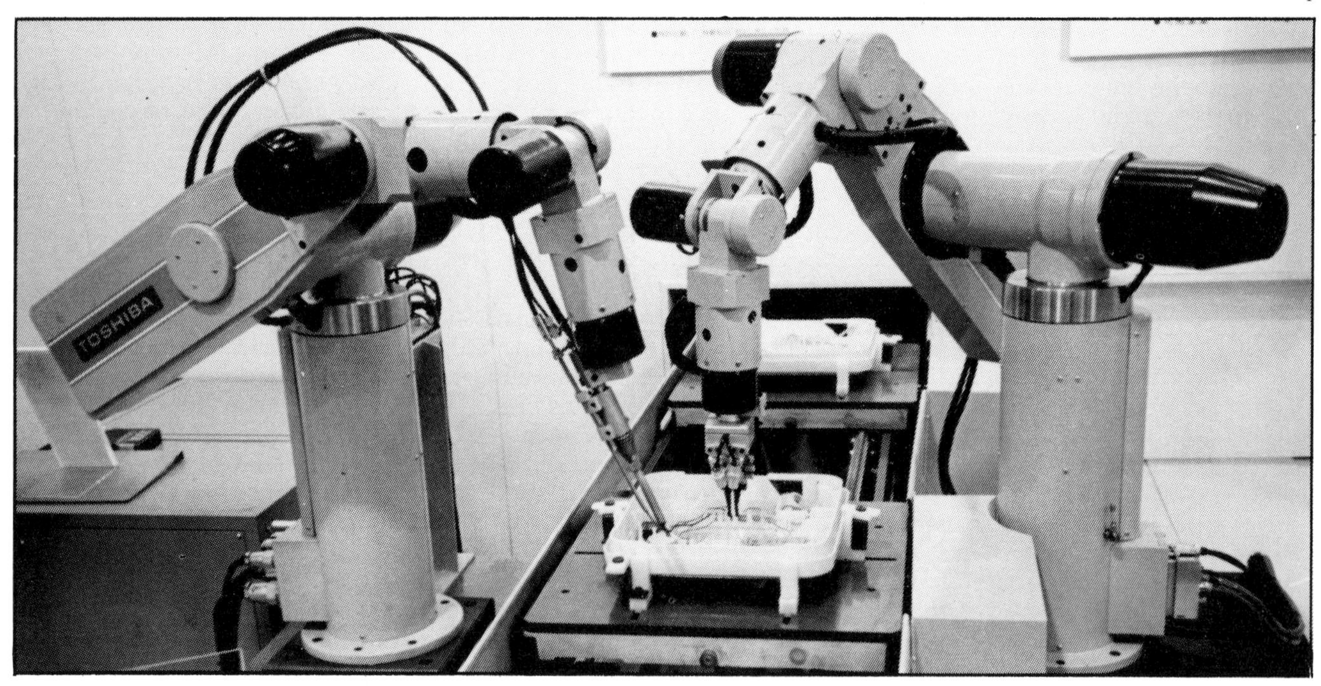

Fig. 6. Toshiba has produced some neat assembly robots with vision sensing.

Hitachi

The cheapest and simplest of the new Hitachi assembly robots, the A4010, costs around £2,500. This simple four-axis assembly robot is therefore about one-third of the price of most competitive machines. Intended for use in assembly of electronic consumer products and automotive assemblies, this machine is certain to increase dramatically the number of robots used for assembly work.

The A4010 is similar in concept to the Scara machine, but the arm is mounted on top of a console which contains the controller. The arm can articulate on a vertical pivot on the console, and it also has a vertical elbow joint. At the end of the wrist there is a vertical telescopic column giving vertical travel. It can handle parts weighing up to 1kg, and the complete assembly weighs only 30kg. Repeatability is put at ±0.2mm, while maximum speed of operation is 730mm/s. Upper and lower arm sections are respectively 300 and 200mm long, and they can both articulate through 120°.

One reason for the low price of the A4010 is that it has a small cmos (complementary metal-oxide-silicon) random access memory with enough capacity for only five points and 25 steps. It has a point-to-point controller, and can hold only one program at a time.

The A4010 is the cheapest of a range of three assembly robots, the others being the A3020 and A6030, which cost around £12,000 and £20,000, respectively. The A3020 is similar in principle to the A4010, but it is twice as fast, with a maximum speed of 1,500mm/s. In addition, the level of repeatability – ±0.05mm – is clearly aimed at assembly rather than simple pick-and-place operations. Upper and lower arm lengths are 400 and 300mm respectively, while articulation angles are 120° and 155°. Capacity is 2kg.

The A6030 operates at the same speed, but is a six-axis jointed arm machine with a capacity of 3kg. Upper arm and forearm are both 400mm long, and they can articulate through 160° and 140° respectively. The robot can rotate through 300° on its base, and there are three axes of freedom at the wrist.

These two larger machines have magnetic bubble memories with a capacity of one megabit to store 500 steps and 2,000 instructions. They are due to become available in April and Hitachi plans to build up production to a staggering 400/month.

Mitsubishi

Mitsubishi Electric intent of getting into the pick-and-place business has produced a stronger version of its MoveMaster educational robot – the Movemaster II. This tiny articulated-arm robot has a robust plate construction and upper and lower arm lengths of only 220 and 160mm respectively.

There are five axes of freedom, with 300° rotation at the base, 130° at the shoulder, and 90° at the elbow. The wrist can also roll and pitch, while there is a simple two-finger gripper. Maximum speed of movement is 400mm/s, little more than half that of the Hitachi A4010. However, repeatability is ±0.5mm, and the robot can handle parts weighing up to 500g.

However, the strange feature of this robot is the controller. The actual motions are programmed and stored in the drive unit, and Mitsubishi Electric is offering a personal computer for programming. However, the computer is not suitable for the factory environment, so instruction for programs can either be stored in eprom (erasable programmable read-only memory), or a sequence controller can be used. So long as a device is available to program the eproms, then the system is cheap, with

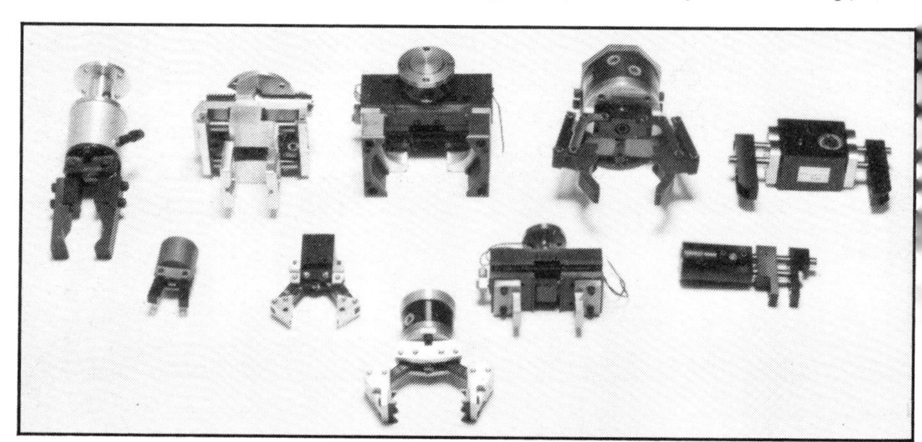

Fig. 7. Hitachi's A6030 and A3020 assembly robots have a large range of hands

Fig. 8. Mitsubishi Electric's Movemaster II is designed for simple pick-and-place operations.

the arm costing only £2,800, and additional eproms about £7.50 each. A sequence controller boosts the price of the system to about £3,500. Mitsubishi Electric is hoping to sell 100 units a month, and is planning to use some in its own factory to transfer small spindles between a parts feeder and a grinder.

Shin Meiwa

More man-size, but still smaller than its existing robots, is Shin Meiwa's RJ45, which is similar in concept to the larger RJ65. The arm of the RJ45 is 900mm long compared to 1,300mm for the RJ65. There are five axes of freedom. The teach unit is a neat integrated system with a number of digital displays, indicating the program, step, and the menu. There is the option of a small, separate display to give the programmer further information. Overall, it is a neat design suitable for mounting on the floor, on a slide, or on a gantry. Sales are due to start in Japan by February, with exports starting in the summer.

The RJ45 is one more example of a smaller and cheaper robot being introduced; user price in Japan is set at about £17,000 as opposed to £21,000 for the RJ65.

Toyo Kogyo

Toyo Kogyo has installed 155 robots in its new car factory, which is some 100km west of Hiroshima. The plant is designed to produce 20,000 Mazda 626 cars a month, to a cycle of 54-56s. There are 1,800 employees, and the main interest as far as the use of robots is concerned is the paint plant, where 20 Tokico robots are installed, two with a simple form of vision sensing.

The vision sensing is used to adjust the path of two robots that apply weld sealer. The robots are pre-programmed in the normal way, but minor adjustments are needed to compensate for movement of the body on the conveyor. A lamp near the gun directs a beam at right-angles to the panel, directly at the seam. A photo-sensitive cell near the lamp detects the seam as a joint between dark and light areas, and signals are transferred to the controller to adjust movement of the robot as necessary.

In addition to these robots, two more without sensing are used to apply weld sealer inside the car. Two more robots apply pvc, and four more apply the two types of underseal used. Then, there are six applying primer/surfacer, and four applying colour coats.

In the primer/surfacer booth four robots, working in conjunction with two 'Minibell' electrostatic systems, apply all the paint. There are a pair of simple robot arms at one end of the booth to open the car doors, two pairs of robots to do the spraying, and finally a pair of door-closing robots. Thus, the robots can spray all the inside of the body, including the tricky areas around the hinge pillars, while the Minibell units spray the exterior.

There is a third pair of robots further down the booth to spray matt black paint, wet-on-wet, on the door surrounds on some models. In the

Fig. 9. To adjust the tracking of the Tokico robot used to apply weld sealer to the underside of the Mazda 626 body at the Hofu plant, photo-electric sensing is used.

Fig. 10. The body side assemblies are carried on trolleys while they are welded at Toyo Kogyo's Hofu factory.

Fig. 11. A Fanuc A Model O robot is used to handle the flywheel spigot bearing when it is pressed in at the new Toyo Kogyo engine plant.

colour booths – the conveyor splits into two lines to pass through two parallel booths – there is first, a set of guns on reciprocators, then there are two robots and five men. Finally, there are the Minibell sprayers.

The robots are used only to spray the front and rear of the body, and the cowl forwards of the wind-screen and the rear pillars – the parts that the reciprocators miss. To minimise wastage, the paint selectors are mounted on the robot arms, the controller switching among any of 27 colours as necessary. The men are left to do the tricky areas, such as inside the door pillars.

In the welding shop, 88% of the welding is automated, 76% by 130 robots, and 12% by multiwelder. Although three versions of one model only are being produced at present, the plant has been designed to produce three versions of different models. To make this approach practical, robots are of course used widely, while at the framing station, the side assemblies are lowered on to fixtures which rotate through 90° before sliding into the main fixture. These side fixtures are three sided, one side for each body type.

TK

The main innovation in robotics in the body shop is a 'tunnel' robot, developed jointly by TK and Kawasaki Heavy Industries. It actually consists of three pairs of robots, each pair hanging down from a common slide on which they can traverse horizontally. The arms are grouped together to form a tunnel, each arm being able to extend vertically, and articulate at the wrist. The robots are used to make 33 or 35 welds between the front, centre and rear floor panels. The 18 axes of these arms are controlled by one console unit. Overall, 62 functions can be controlled, compared with 51 on a normal robot.

Robots for direct glazing and wheels

Two other points of interest in the body shop are that, firstly, 32 robots were transferred from Hiroshima to Hofu – evidence that the life of robots is long – and secondly, a camera is used to identify the body approaching the respot line. The camera, which is mounted on a slide, reads a number on the body, and this is compared with the order number. If it matches, then the robots are programmed accordingly, and if not the line is stopped.

In assembly, TK uses one Tokico robot to apply the adhesive to the windshield prior to direct glazing, and a big Nachi jointed-arm robot picks the spare wheel from a conveyor in a special gripper, and then puts it into the boot.

TK has also installed two robots in its new engine assembly plant at Hiroshima. One Fanuc A Model O is used to pick up the flywheel pilot bearing from the end of a chute and position it for pressing into the flywheel. The robot holds the bearing in position for the press to push it into the flywheel. Thus, the robot not only replaces a man, but eliminates the need for a fixture. The other robot, a Taiyo Toffky applies a bead of sealant to the sump to form a fixed-in-place-gasket. The path is complex, and of course, the robot lays a bead of precise thickness on the designated path, increasing quality control.

New robots all round in Japan

The Robot Exhibition in Tokyo featured many new robots and demonstrated several vision systems.

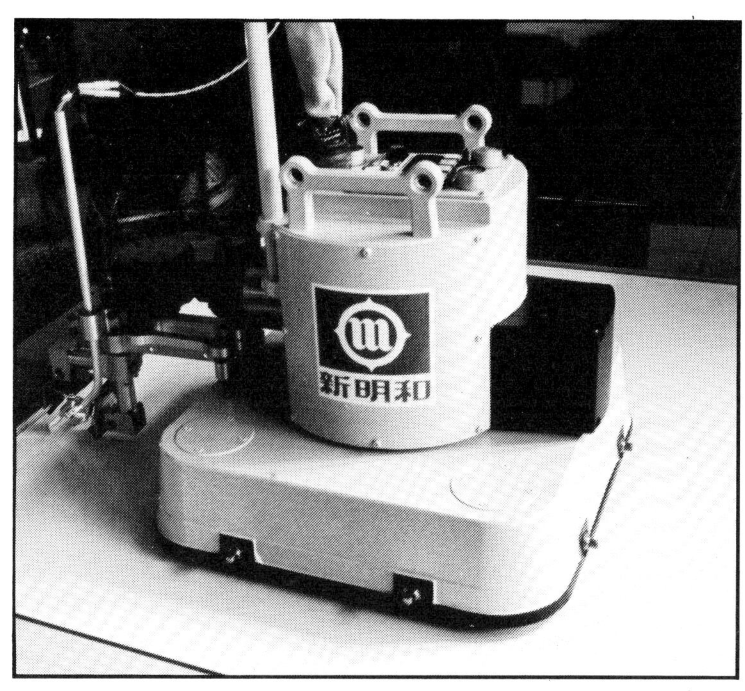

Shin Meiwa RV110 portable arc welder.

WITH over 40 companies exhibiting robots, and something like 20 new models on display, it was the sheer numbers that impressed most at the International Industrial Robots Exhibition at Tokyo.

That does not imply that the robots were not worth seeing, because there was another feature of some importance: for the first time at a robot show in Japan most exhibitors gave the impression that their machines were actually intended to perform useful work. In the past, there has been a tendency to show robots either doing nothing, or doing something that has no bearing on manufacture. But this time, several robots were shown in work cells. Moreover, practical vision systems were in evidence.

Among the trends to be seen, apart from the vision systems, were a number of new machines capable of lifting workpieces of 60kg and upwards. In this class were new articulated robots from companies such as Kawasaki, Toshiba, Dainichi Kiko, Meiden and Yaskawa. Then, there were several new Scara type robots, including one or two large units from companies such as Fuji Electric, Nachi, Toshiba, Shin Meiwa, Silver Reed, and Diesel Kiki.

Also significant were a number of robots carried on gantries or four-post structures. Dainichi Kiko, Yaskawa, Mitsubishi Heavy Industries, which has joined the move to electric actuation, and Sankyo were all showing robots of this type.

On a business note, it was also significant that several big companies were either making a stronger showing, or taking their first step into robotics. Included were: Toshiba, which despite being an early starter, has been producing only a couple of models for several years; NEC, which has moved into the heavier end of the business, IHI, and Komatsu. Meanwhile, Nachi, Mitsubishi Electric, Shin Meiwa and Yaskawa were all showing extended ranges.

Finally, it was good to see both ASEA and Cincinnati Milacron, who are setting up assembly operations in Japan, making a showing. Until now, the Japanese have faced no foreign competition on their home market, except from the Unimates built and sold by Kawasaki. Admittedly, Kawasaki is strong, especially in spot welding, but that is a different matter from a fully-fledged foreign company selling in Japan. It will be interesting to see how much progress they make.

On the sensing front, Matsushita was most impressive, with a variety of positioining and orientation systems. It was also showing its range of robots, as well as some mock assembly lines. One of its interesting sensors was an ultrasonic device used to find the position of a hole in a plate, for insertion of a pin in 0.1-0.3s. It can centralise the pin so long as the hole is within a tolerance of $\pm$3mm, which means that a robot can be used to drop the plate roughly into a fixture which is larger than the plate. The ultrasonic sensor scans the hole in two directions, so that the position of the centre can be calculated. This device seemed to be working very well.

There were also 'eye sensors' being used to find the centre of the circles and rectangles. The image has 256 × 240 pixels, and positioning takes less than 0.2s. One system can also detect the angle of orientation of the rectangle. These systems cost Yen two million (about £5,700) in Japan. The shape recognition system is similar, but takes up to 0.8s.

Mitsubishi Heavy Industries has also adopted vision, with a system for its painting robot. It also costs Yen two million, and is designed to recognise patterns, and compare them with

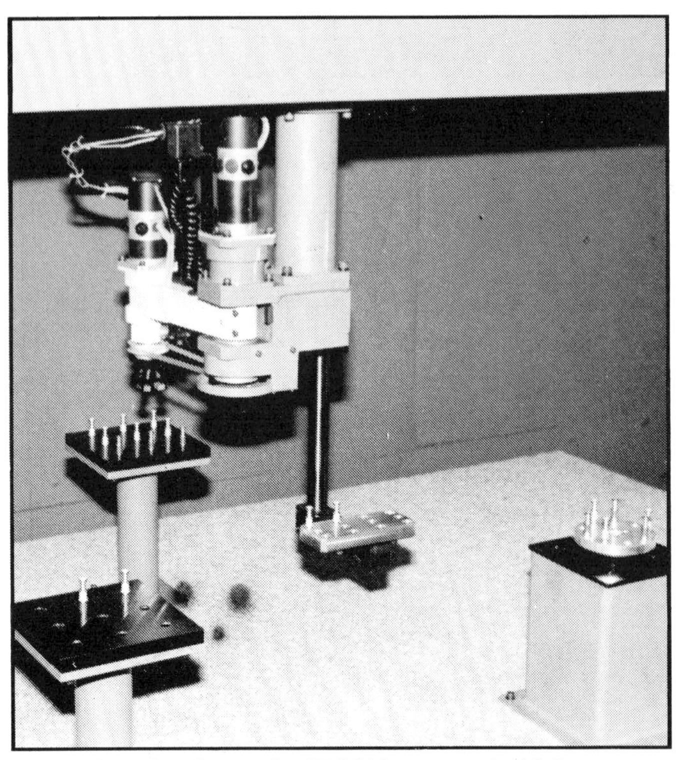

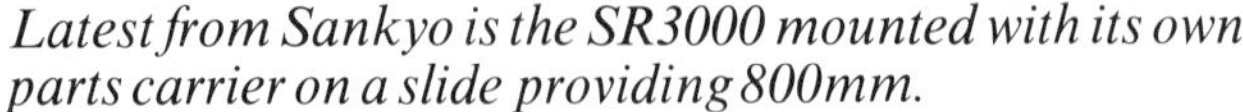

Latest from Sankyo is the SR3000 mounted with its own parts carrier on a slide providing 800mm.

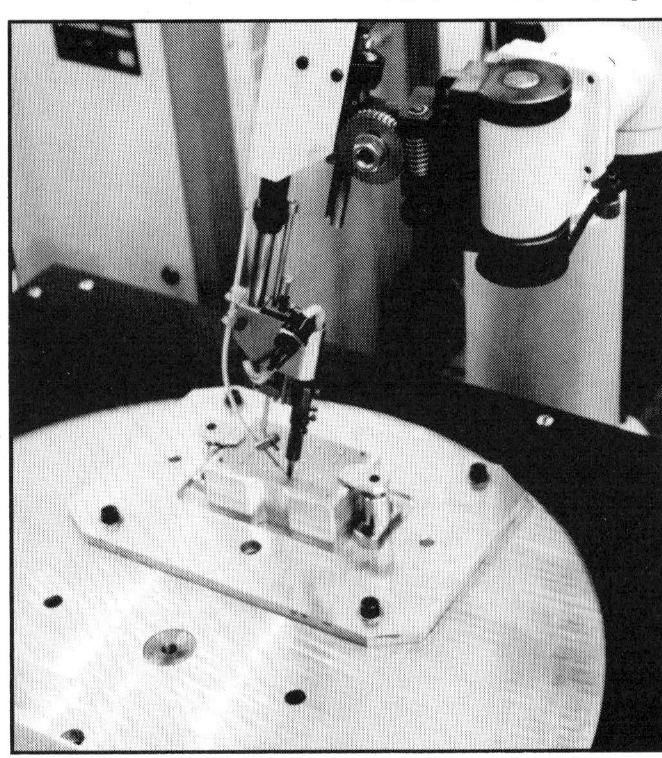

Silver Reed's new ARX3 set up as a soldering unit.

templates stored in the memory. Up to 128 shapes can be stored, and even if parts are the same shape, so long as the basic dimensions differ by 5mm the camera is claimed to be able to distinguish the parts. In addition a 5mm square cut-out in the corner is claimed to be enough for the controller to detect the difference.

The Mitsubishi consists of the Multi 16 16-bit microcomputer made by Mitsubishi Electric, a ccd camera, and a large memory. Images of the shapes of the workpieces are first recorded through the camera into the memory as reference templates, and subsequently, the controller compares the images transmitted from the camera with the reference templates. Whereas the robot costs Yen 10.4 million, the vision system costs Yen two million, to produce a system costing Yen 12.4 million (around £35,000). Whether this is a practical application is a moot point, because the robot and controller still need to be programmed beforehand, so it only eliminates the need for a device to indicate which type of component is coming next.

Module vision system

Fuji Electric, which has been producing various types of optical systems for some time, has built some prototype robots, and was showing these operating with vision systems. Some were assembling six components, supplied in random positions on a belt conveyor, while another was palletising rotors. In both cases, the vision system is a module, but for the large rotors two cameras were used. One is used to survey the table of workpieces, while the other is used to find the centre of the bore of the rotor. The controller then positions the robot correctly.

Little information was available about the Fuji Electric robots, but there were three prototypes, all of the Scara type. The largest has primary and secondary arm lengths of 560 and 440mm respectively and a capacity of 50kg. There is 600mm vertical motion, provided by a ball screw, while harmonic drives are used at the arm joints. To give sufficient rigidity, there are two vertical slides to carry the substantial arm.

One of the small Scara robots is carried on a horizontal slide. The arm is about 250mm long, and it has a capacity of 2kg. Repeatability is put at ±0.05mm, the machine being aimed at assembly, as are the slightly larger column-mounted units.

One of the most interesting displays was a practical cell on the Mitsubishi Electric stand. The cell was built around two Scara type robots, an RH-211 and an RH-212 Scara, to assemble magnetic contactors. Such a system has been operating in the factory for six months. The contactor consists basically of a laminated U-shape core which is inserted in a housing. However, there is a bridge piece with two pins which locates the core in position. To complicate assembly, between the bridge piece and pins are rubber bushes.

The two robots work at a table, the first having three grippers to choose from. There is a fixture between them, and a fixture adjacent to each. Rubber bushes are presented in pallets, in front of the first robot.

The first robot picks up a pin, and with the aid of a camera, finds the centre of a rubber bush on the pallet, and guides the pin into position. It mounts the pin/bush in a cup in a bridgepiece, and repeats the process for a second pin. Then, it changes grippers, and performs an operation that seems unlikely. First, it holds one pin, and steadies it as the other robot starts to lower the laminated core into position. Then, it releases that pin, and holds the other one while the laminated core is pushed home.

Assembly

At the next robot station, the core/bridge piece assembly is lowered into the housing, and then a spring clip is lowered over the core, and again the method of assembly is unusual. The clip, which is shaped like a wide inverted 'U', has to be sprung over the corners of the core into recesses, but the robot lowers it so that one end only is registered. Hanging down vertically from the side of the arm is a pushrod, with a plastics knob on the end. To secure the clip, the robot lowers the pushrod, so that the knob touches the clip, and then the arm moves along, so that the knob presses the clip down over the corner of the core.

In another display, a pair of small Mitsubishi RV-133P articulated robots were loading small brackets into a pair of pedestal spot welders.

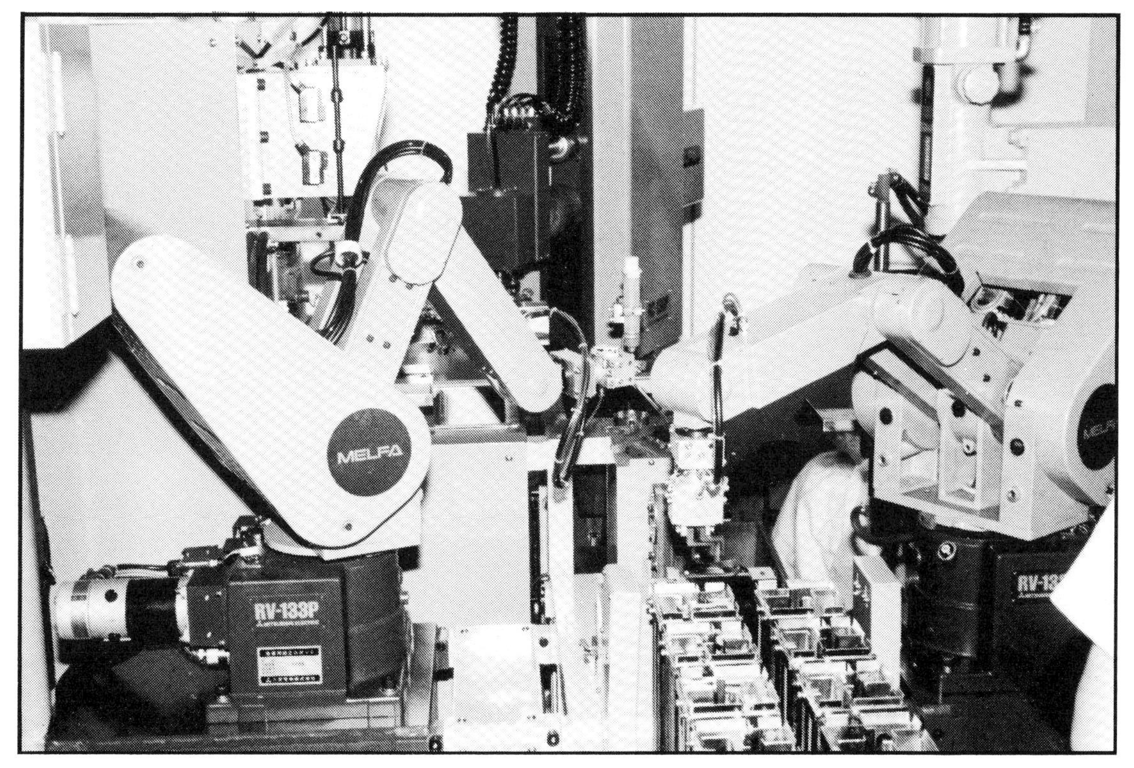

Mitsubishi Electric set up three robots in a welding and palletising station.

Then, an RC-136P cylindrical co-ordinate robot was palletising them.

Diesel Kiki, a manufacturer of diesel injection equipment, developed the Flex Arm Scara type robot for use in its own plants and is now selling them – Yamaha has made a similar move, although a rather longer time has elapsed between initial use and sales than with Diesel Kiki, which is already using 40 of its Flex Arms in its factory. This is a typical Scara intended for assembly, and costs only Yen three million (about £8,500).

Diesel Kiki demonstrated one robot loading injection pump cam rings into a jig, and then inserting the four roller assemblies in a cycle of 24s. It works in conjunction with the assembly machine for the roller/spindle/thrust washer assemblies. The robot picks the roller assembly from the last station of the assembly machine, and transfers it to the cam ring which it has placed on a jig. After the four roller assemblies have been inserted, the robot moves the sub-assembly to an inspection device where the height of the rollers is checked. Then, the robot palletises the cam ring assemblies.

Pentel and Silver Reed were demonstrating their Scara robots in soldering systems. Silver Reed's ARX3 has primary/secondary arm lengths of 300/200mm, and is a three-axis machine, with 75mm vertical motion at the toolholder. The machine costs from Yen three million (£8,500) upwards, although with soldering machine and a turntable, the system costs Yen 6.98 million (£20,000).

The ARX3 is actuated by dc servo motors, the two for the arm joints being mounted in line with the shoulder joint. It is a typical Scara design, the maximum speed of the arm at the tool holder being 1,375mm/s, compared with 2,700mm/s for the Yamaha Came, and 1,500mm/s for the Diesel Kiki Flex Arm.

Toshiba has also produced a new Scara type robot, the SR653/654H models, the primary/secondary arm lengths being 400/250mm. It is actuated by ac servo motors, one at each of the two main joints, and can move the gripper at up to 2,500mm/s. It is controlled by a 16 bit micro-processor – eight bit devices are more usual on these fairly simple robots – and 999 points can be programmed.

Capacities

Nachi is now producing three sizes of Scara robot, the 200-AL, 400-AL and 600-AL, with primary/secondary arm lengths of 200/160mm, 400/250mm and 630/400mm respectively. Capacities are two, six and 30kg. Speeds are 1,270mm/s for the smallest, and 1,100mm/s for the others.

Now, however, that Nachi 600-AL is not the largest Scara type robot; NEC, Daido Steel, and Shin Meiwa have all produced larger units. NEC's new NR-231, 232 and 233 models have arm lengths of 700/700mm, 875/875mm and 1,050/1,050mm respectively, and capacities of 30 to 80kg. The arm is mounted on a column with 700-950mm vertical travel. NEC quotes speeds of 60°/s for each section of the arm, which is fairly slow. Daido Steel's ATR-HA100 has three joints in the 1,970mm long arm, and the wrist can bend through 195°, and rotate continuously. It is powered by ac servo motors, and the shoulder joint can articulate at up to 90°/s, while the other two arm joints can articulate at 85°/s, which is fairly fast for a machine that can carry up to 150kg.

Shin Meiwa's two latest arc welding robots are variations on this theme, the RJH45 incorporating a parallelogram linkage in the first arm to give vertical travel. The primary/secondary arms are 450/450mm long, but the parallelogram arm is 495mm long, and it is supported on a beam 565mm long from a column or the gantry. Of course, this machine can weld over a large area, but is overshadowed by the new concept in the second robot, which has one arm to carry the work-piece, and another one to do the welding.

It has a Scara structure, with an arm length of 1,000/800mm. The arm can be carried on a horizontal slide giving up to 6m travel. The main robot arm, called the B arm, consists of the Scara-type arm and the gripper. The primary/secondary arm joints can articulate through 180° and 240° respectively, while the joint carrying the gripper can also rotate through 400°. The lower section of the column carrying the gripper can move 400mm vertically, while 90° bending is also provided.

In addition, there is the arm for welding. It is an articulated arm hanging down from the column carrying the gripper. It can swivel through 280° around the column, while the

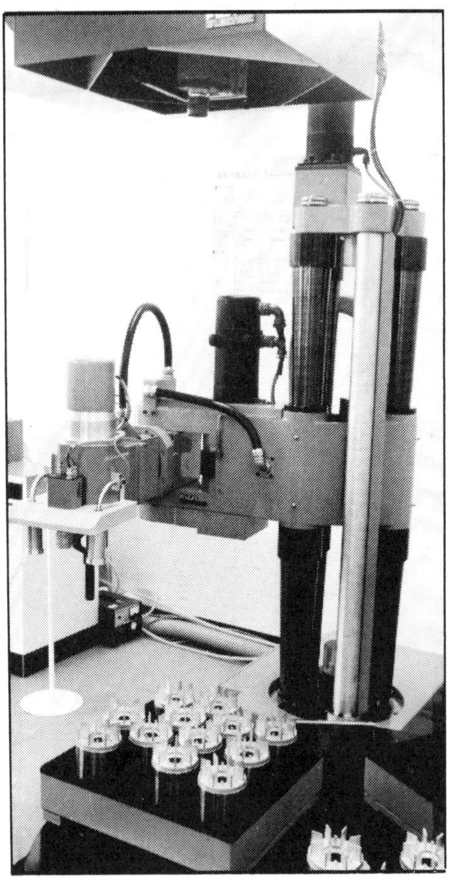

Fuji Electric Scara type robot with 50kg capacity and vision sensing.

shoulder joint can articulate through 90°, and the elbow joint through 75°. The welding gun can also be articulated through 360°.

All this means that the B robot can pick up a plate or beam, lay it in position in the assembly, and then hold it while the welding gun is manipulated to tack weld the seams. Then, the gripper can release the workpiece so that the gun can be moved along the seam to complete welding. This is an interesting concept, currently at the development stage, but it shows one way to go; with the aid of vision tracking it would make a self-contained welding cell.

Magnetic gripper

A magnetic gripper can also be used, in which case the load capacity is 150kg. If used as a handling robot, though, the capacity is increased to 350kg.

Toyoda Machine Works, which introduced a range of articulated, Scara and gantry type robots in the Spring, was showing the latest addition, the RA6-2 six-axis articulated assembly machine working with a pair of CNC nut runners. A RC4-2 robot fitted bolts loosely, while the RA6-2 then tightened them. Also to be seen was a video presentation of the new Toyoda assembly systems based on robots.

Among the new articulated arms, the Yaskawa V6, a six-axis unit with dc servos and harmonic drives stands out. It is a compact design, the servos being built in. There are three axes of movement at the wrist. It can be floor-mounted or suspended from a gantry. Speed of movement of the arm is up to 1.5m/s, and it has a capacity of 6kg. Primary and secondary arm lengths are 360 and 374.5mm, and the joints can articulate through 180° and 300° respectively. The wrist can yaw through 380°, bend through 380°, and rotate on the axis of the arm 330°.

It is taught on a PTP basis, with linear and circular interpolation, and there is capacity for 1,900 positions and 900 instructions in the semiconductor memory. Incorporated in the control console is a 9in crt. Clearly a machine that will enhance Yaskawa's range of business outside its traditional area of arc welding. The company is also starting to import the French Scemi robot, a small articulated arm device intended for precise assembly.

Meanwhile, Kawasaki has started to produce the tiny Puma 280, which will be sold at the high price of Yen six million (£17,000), Toshiba has come up with a Puma lookalike in the form of the SR-606V, a five- or six-axis machine with a pedestal 630mm high carrying a 600mm long arm.

Among the large articulated arm robots were the Yaskawa L60, which as its name implies has a capacity of 60kg. This is a six-axis machine, yawing being included in the wrist motions. Toshiba's SR-206V is another 60kg machine. It has a parallelogram linkage, and can move at up to 2,600mm/s. ON the other hand, the latest big Dainichi Kiko machines, PT-1200 and PT-2000, are six-axis machines with capacities of 50 and 120kg respectively. The PT-1200 can move at up to 2,000m/s.

Meidensha's MeiRobo GHR-200 is a 200kg machine. It was being demonstrated with a rather ponderous vision sensing system. Not to be outdone by all this was Kawasaki, showing a new 65kg machine, the EA65, which is designed principally for spot welding. The transformer is mounted on the tail of the arm as a counterbalance. Kawasaki quotes a speed of 1,892mm/s at the hand, whereas the fastest Unimates, can move at up to 4,000mm/s.

Diesel Kiki Flex Arm being demonstrated with an assembly machine for injection pump can ring assembly.

The Yaskawa V6 articulated arm robot.

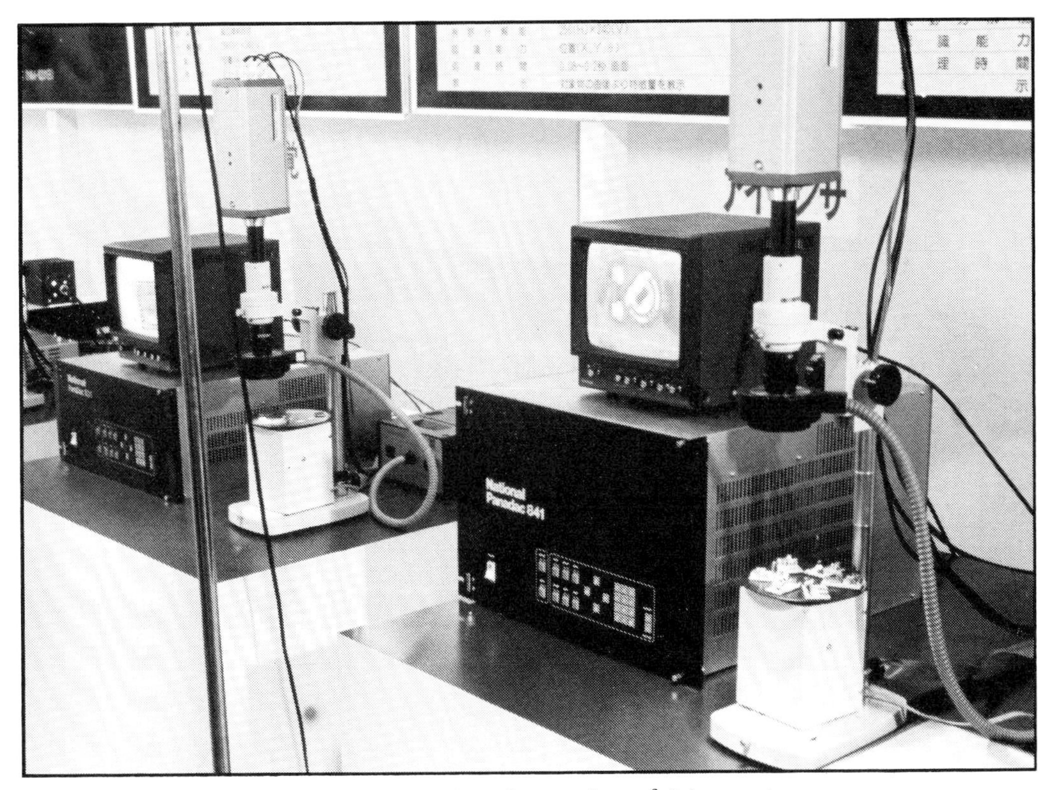

Matsushita has developed a number of vision systems.

Arc welding

On the arc welding front, there was an important newcomer in IHI, which is involved in shipbuilding and heavy engineering. More important, though, was the fact that virtually all the Japanese companies seem to be sensing the welding current for tracking on fairly thick plates. First developed by Yaskawa, the gun is arranged to weave, and the current at the ends of the weave are continually compared. The controller adjusts the path to equalise the current, so the robot can weld around curves without prior teaching.

In fact, IHI will not be offering its equipment till 1984, but most others including Komatsu and OTC have recently introduced such tracking systems. IHI has chosen dc servo motors as the power source for its larger articulated machine, but the smaller FH robot, intended to be mounted on the ceiling or wall, is actuated hydraulically.

The FE floor-mounted unit is similar to the Yaskawa L10, the FH gantry-mounted model is more compact, but has limited movements in five axes, so the working envelope is fairly small.

Shin Meiwa's new RV110 portable robot caused quite a stir. A tiny machine looking rather like a vacuum cleaner, it is intended for welding inside ship hulls. It is placed inside a tack-welded assembly with bulkheads on four sides, and can then drive around on wheels to weld all the joints between bulkheads and floor without being taught. The main body is 410mm square and 360mm high, the welding gun projecting from one side. The gun can be moved linearly and vertically, and it can be rotated through 360° in two planes.

There are a pair of touch sensors adjacent to the wire, one to contact the vertical plate, and the other the floor. On contact, the ceramic probes are deflected so that they hinge, thus actuating linear variable transducers to stop further movement. Therefore, the robot moves up to the plate, and starts to weld along the seam, and when it comes to the corner, the gun is manipulated around it, again with the aid of the touch sensors.

The machine can traverse a length of 30m, and since it weighs only 30kg, it is truly portable. However, there is the inevitable controller, which is 800 × 570 × 530mm connected by cables. In a factory, it is anticipated that this would be carried on an overhead conveyor, allowing the robot to cover quite an area.

In fact, this machine, and a number of other small robots of more conventional types, were developed in response to a plea from the Japanese shipbuilders, who are all anxious to raise productivity to keep ahead of the newly industrialised nations.

Matsushita was showing yet another small articulated arc welding robot, which is still under development. The AW 4010 is a six-axis machine, with primary/secondary arm lengths of 450/450mm. It is mounted on a short base, which rotates, while the wrist can also be rotated.

With so many new machines, one is bound to wonder where they are all going, although it is clear that assembly is now seen as a major application.

At the same time, other established companies have recently expanded their overseas networks, but the newcomers will need to fight all the way. Significantly, though, many of them, such as Silver Reed, Diesel Kiki, Komatsu, Matsushita, and IHI have opportunities to install many robots in their own plants – and that is the way to start off.

3. Assembly

Although flexibility is now the aim, several of the installations in this section are not flexible. Nevertheless, they represent good examples of automation in areas where the high volume output justifies a dedicated approach. In some cases, the installations give an indication of the flexibility to come, with a robot being used for palletising, or some other flexible equipment being installed.

An excellent example of high throughput and flexibility is given by Nippondenso's fuel gauge line, where, with some well-disciplined human intervention, changes between batches can be accomplished in one cycle – in other words one dummy pallet is pushed down the line to switch the stations over to the new assembly. The company's relay line, with an output of 50,000 units a day, is also a fine example of flexible automation.

On the other hand, Suwa Seikosha relies on hard automation for most of its assembly lines, but is now gradually introducing its own robots where appropriate to do so – generally for complex operations where the watch case is assembled.

Hitachi and Toshiba have also introduced some good examples of robot assembly, which are covered in this section, and as Brian Rooks describes, Hirata is moving ahead with integrated assembly systems. This is certainly one trend in Japan, with companies such as Fanuc, Matushita, Mitsubishi Electric and Seiko producing systems based on robots, conveyors and parts feeders to speed up installation of new lines. Therefore, the use of robots for assembly is now increasing rapidly.

3.1 Assembly – Automotive products

How the Japanese put fuel gauges together in one second

Fast automatic assembly of fuel gauges is just one of the many advances the Japanese are making in automated assembly.

THE IMPORTANCE of automatic assembly to the future of manufacturing industry proved to be a topic on which wide agreement was made by speakers at the 4th International Conference on Production Engineering, held in Tokyo. And while there were less papers on automatic assembly than on some other subjects, they created a lot of interest.

Prof. Warnecke of Stuttgart University summed up the matter when he said: 'In the 1980s, flexible automation in assembly will be the main topic for production engineers. However, solutions will only crop up at the end of the decade.'

It was not surprising therefore that some of the papers were concerned with research aimed at making more automatic assembly possible while others described new applications or equipment.

Without doubt the main interest was to be found in the first speech of the conference presented by Katsuo Aoki, executive vice president of Nippondenso, the automotive electrical company in the Toyota car group. He attempted to give an insight into why Japan has adopted automation so successfully, and in some respects his reasons were the opposite of what people in Europe expect.

The main reasons are the high level of education in Japan – over 94% of the population stay at school until they are 18 – and the lifetime employment system adopted by big companies. The lifetime employment system means firms are prepared to train people continuously to improve their performance, or to suit their aptitudes, so that people can be persuaded easily to adapt to change.

This means, for example, that a man can be taken off a manual assembly line and trained to look after an automatic machine. That is, not just to watch it, but to maintain it and keep its performance up to standard. It also means employees are suitable for training.

Another feature of the Japanese system is that employees' wages increase with age as much as they do with skill, so that as long as the company continues to prosper, employees have 'something to look forward to', as Aoki said. And in this situation, all employees are treated alike – be they shopfloor workers, engineers or accountants.

In fact, engineers are something of a natural elite in Japanese companies because they have special skills. Almost all the other company managers are 'salarymen' whose skills have been taught by the company. So the head of accounts could easily have studied chemistry at university, but would have been retrained by his company as an accountant – and as a salesman or production controller on the way. In this way the flexibility of the workforce is not confined to the shopfloor; it is throughout the company, but generally with the exception of the engineering department.

Cycle times

Aoki said his company had found that to obtain high productivity and good line balance on manual assembly lines, a cycle of 50-60s was needed. With longer cycles there is little improvement. However if the cycle is reduced to 30s, the balance efficiency is cut from 80 to 70%; and with a 20s cycle, the efficiency falls to 60%. Thus, wherever a cycle time of less than 30s in needed, Aoki sees little alternative to automation.

An indication of the way the Japanese refine their automation – and they don't just install it and accept the reduced manning levels – came from a case of a robot used to remove castings from a high pressure diecasting machine. When the robot was installed, the removal cycle took 45s; following improvements to the programming over a five month period, the cycle was reduced to 30s.

In most companies in Europe that would have been considered acceptable; but not in Japan. For a further nine months programs were refined and, following some investigations and suggestions from the workforce, the mechanical operation of the robot was improved as well. The result was a cycle of 15s. This shows what perseverance can do.

Aoki defined the criteria for automation at Nippondenso as:

□ Equipment must have an 'uptime' of at least 80%;
□ Failure rate should be less than 0.3%;
□ Payback period should be no more than 3.5 years.

He defined these as the criteria for 'investment efficiency'. Nippondenso has installed three fully automated lines, and, with the aid of a film Aoki described an extremely impressive one designed to produce fuel gauges for cars. This particular machine has an output of about 500,000 units/month, although in some cases, Aoki said, the target was 600,000-700,000 units.

The fuel gauge in a car consists basically of two sub-assemblies, the dial and needle, and the back casing with the shaft, voltage regulator, bimetal strip, base and terminals. The line is divided into two sections: in the first straight section of seven stations, the components are assembled to the casing. In the second section, which is a rotary machine, the dial and needle are assembled, while the voltage regulator and needle are adjusted – all automatically.

Automatic assembly is not as flexible as manual assembly, as was shown by the modifications needed in the case of the line at Nippondenso. Men can cope with terminals of several different lengths with no extra tooling – just extra instructions. But for each extra piece of variety in automatic assembly, an extra piece of machinery is needed. Thus, before designing the new machine, Nippondenso rationalised the range; otherwise the machine would have been too costly and complex.

The variety was reduced in the following way: Three different types of casings were retained but 13 types of terminals were reduced to four; eight bimetals to four; 20 voltage regulators to three; two bases to one; and the two shafts were retained.

Previously, about 110 types of inner structure assemblies were produced and this was reduced to 60. However, it is mathematically possible to produce 288 varieties.

But the variety of 60 models is still considerable, especially when it is desirable to produce each model every day in a single eight hour shift. In fact, to suit the Toyota 'Kanban' system of manufacture, very small batches are needed, so that the line is usually producing 200 batches a day, each being of 40-100 units, and sometimes even smaller.

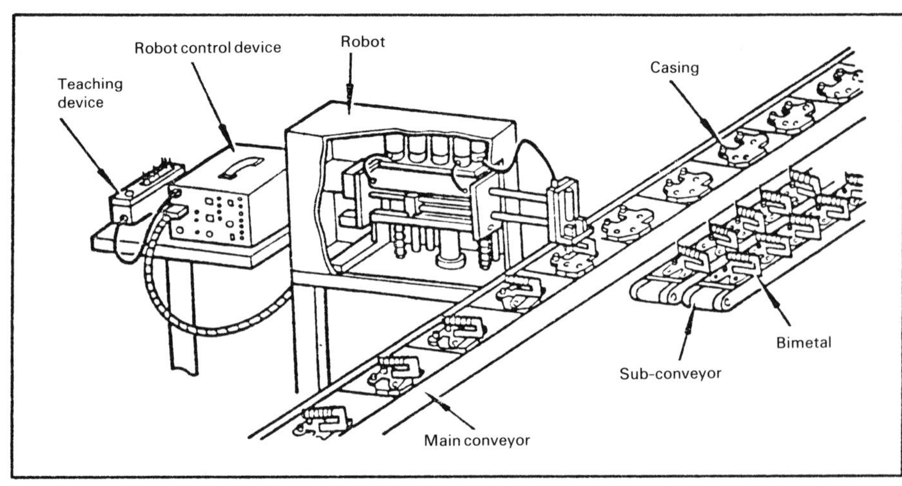

High-speed pick and place robot developed by Nippondenso for automatic assembly.

Microcomputer

The complete system is controlled by a microcomputer, called a flow controller. The foreman simply programs the controller to produce given quantities of certain models and the microcomputer controls the line accordingly.

The key to the system is use of 'dummy casings' which pass down the line. Each one can be 'read' electrically by a sensor to indicate the models to be produced. The sensor puts out a signal to the flow controller which signals back to the automatic assembly machines to select the correct parts and carry out the right operations. A sensor is placed before each station so that the station's equipment is reset immediately before the batch of new models arrives. This avoids any delays.

Resetting takes the same time as one cycle – in this case 1s – so the minimum of time is lost. In a day, with as many as 200 resettings, only 0.7% of the working day is lost in resetting.

Most of the equipment on the line is simple with arms and clamps actuated pneumatically. But there are some simple robots too. For example, at the beginning of the line, there are three lines of buffer stock of the different casings. They are parallel to the main line.

A small robot, set in accordance with the 'dummy' that has just passed, reaches across the line, picks up the casing, and deposits it on the line. This is a programmable robot which can be used in several different stations. It has a PROM (programmable read-only memory) and mechanical stoppers which combine with pneumatic power to give it very rapid operations. It completes its cycle in 0.7-1.0s.

It was the development of this high-speed operation which made a production cycle time of 1s possible. This short cycle also reduced the cost of the installation since with a five second cycle five machines would have been needed. This would have raised the £600,000 investment cost to well over £2.5 million.

In the first seven stations, the components are assembled by these robots using simple operation. Then on the rotary machine the voltage regulator is adjusted and some more components are added. The position of the needle is also adjusted. In the manual assembly line, adjustment accounted for about 30% of the work involved.

Machines

Some particularly interesting machines were developed to automate these jobs. To adjust the voltage regulator for example, allowing for manufacturing tolerances, a microcomputer and pulse motor are used. Contact is made with the terminals so that output voltage can be measured; the necessary adjustment is calculated and a plate bent accordingly. The algorithm for adjustment is modified since statistical data for the previous 36 adjustments are taken into account. Therefore, if the mean value of

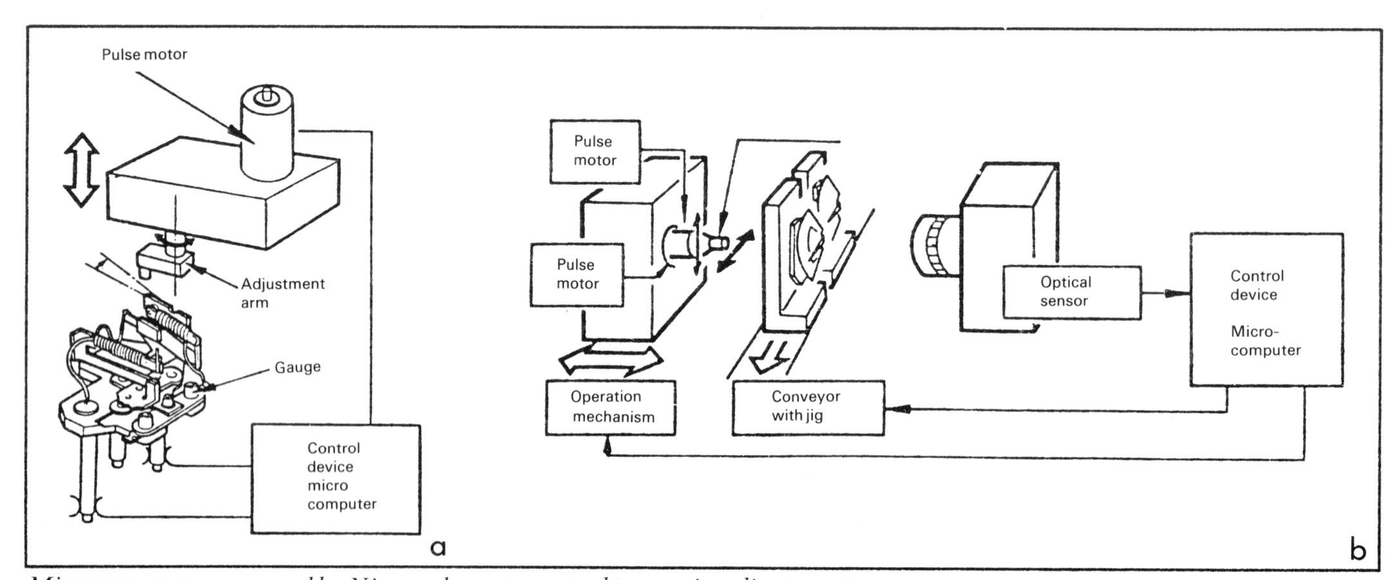

Microcomputers are used by Nippondenso to control two main adjustments: (a) the voltage regulator, where electrical signals from the terminals provide data; and (b) the needle position, where a line image sensor is the key product.

distribution of adjustment results deviates when a new batch is started, allowance is made.

In principle, the device that adjusts the position of the needle is similar to that for voltage regulation, except that a visual sensor replaces the electrical sensor. To cut costs, a line image sensor is used instead of the more normal two-dimensional image sensor.

The line sensor recognises the position of the needle in 0.005s, within an accuracy of ±0.12 deg, and a further 1s is needed to adjust the needle to zero within a tolerance of ±2 deg. Incidentally, the voltage regulator is adjusted to a tolerance of ±0.15V. In fact, each of these adjustments takes 2s, so the machines are duplicated, but even so, that is half the speed of the manual operation – and the scatter is reduced.

Nippondenso has devised therefore an automatic assembly system capable of producing 500,000 fuel gauges in the 190h available each month, or 22,000 units/day. There is a buffer stock in the line – housed on the belt used to transfer between stations – of just 40 units, and according to the data, the line is operating at about 75-80% efficiency, which is just about acceptable.

Disappointing

The conference keynote paper was extremely disappointing since the speaker, Daniel Whitney of Charles Stark Draper Laboratory, USA, concentrated on a very specialised subject. He mentioned in passing there were some new robots on the market aimed at assembly.

The bulk of his speech was devoted to the compliant assembly device, developed at his laboratory, for attachment to robots to ease assembly of parts with tight tolerances. This device was described adequately at the International Robots Conference in 1977.

This time he discussed the mathematical quantification of the limits within which the device will operate successfully. This device is designed not just to be compliant, but to reduce the angle of misalignment as it moves in response to mismatch. So it certainly has a place in automatic assembly by robot.

In addition, the Laboratory has developed an instrumented version of the device which may help overcome some of the insensitivity of robots in assembly, For example, it should make it possible for a robot to be able to distinguish between different types of errors.

The new robot under discussion at Tokyo was the Scara assembly robot developed by Prof. Hiroshi Makino and his colleagues at the department of precision engineering at Yamanashi University, Japan. The robot has four axes of freedom, and is based on a vertical pillar with a horizontal arm with two joints, both pivoting on vertical spindles. In addition, a pneumatic actuator gives vertical movements of the hand. Movement of the other axes is controlled by electric motors – dc servo motors for the arm, and a stepping motor driving through belts to rotate the hand, or tool.

The first machine was compact with a height of 722mm, and a total arm length of 500mm. The second machine, intended for bolt-tightening, has a longer arm. A Z8080 microcomputer is used to control movement; the researchers have developed special software to obtain reasonable acceleration and deceleration in rapid assembly motions. Makino claims that a cycle time of 3s is the average for pick-and-place operations.

Among the merits of the Scara robot are its simple design, large working area, a load capacity of 30kg – very large for such a small machine – and some built-in compliance.

Makino claims a 20mm diameter shaft can be inserted in a hole with 0.01mm clearance with a mismatch of 1mm (so long as there is a 1mm chamfer) within 0.06s.

'The time taken is almost always less than 0.1s,' he said.

Makino said he expects this robot to be commercialised by a Japanese manufacturer in the near future.

Debugging

Frank Riley, vice president of Bodine Corporation, USA, talked of the way his company had adopted microprocessor control for special purpose assembly machines. These controllers can be made in regular production volumes and programmed to suit the application.

Moreover, they can be debugged without changing hardware; and when changes are made to the products or production system they can be reprogrammed again. Standard controllers have rack capacity for 265 inputs and outputs and are combined with the Gould-Modicon 484 programmable controller with 4,000 words memory capacity. To overcome inherent problems with the rotary cam switches the 484 controller has been modified so camshaft rotation can be measured optically or by magnetic resolvers.

Riley claimed the decision to use this system, in preference to 'hardwire' systems, had reduced wiring time by 40%, and debugging time by 30-50%.

An essential aspect of automatic assembly is bolt tightening. This is one area where use of a machine should increase reliability. Tests by Nippon Electric Co. showed that with manual tightening, the scatter in tightening torque can be ±30-60%, and of course, it is well known that torque is only a very rough guide to tightness.

Kazuo Maruyama and his colleagues at the Research Laboratory of Precision Machinery and Electronics, Tokyo Institute of Technology, have tested various methods of measuring bolt tightness. They advocate use of the 'gradient' method, in which the change in torque against rotational angle, $dT/d\theta$, is measured continuously.

The aimed value is the point at which this differential suddenly decreases. Since gradient control involves the use of expensive equipment, it is suggested that in less critical applications, a mean tightening torque should be obtained by the gradient control method on a number of the fasteners in that application. Then, a setting of about 80% of the mean torque value should be taken.

Helter-skelter

Vibration feeders of various designs have long been used in automatic assembly systems, but Rin-ei Precision Instruments and Yamaha Motor Co have built balanced vibration feeders to suit new applications.

In the Yamaha design, conducted in conjunction with Kanazawa Uni-

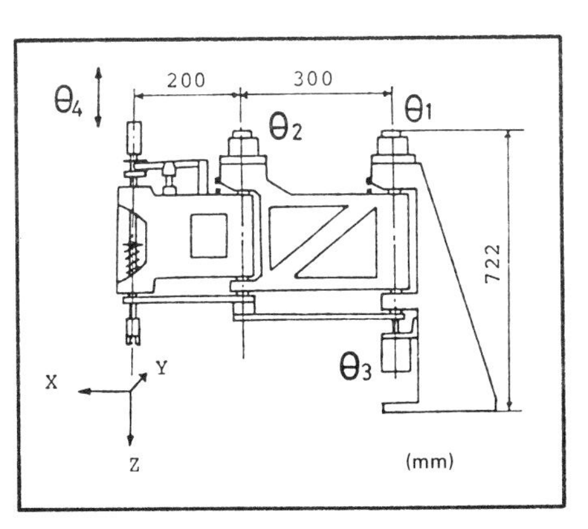

Although the Scara robot looks simple, it seems to be well suited to assembly.

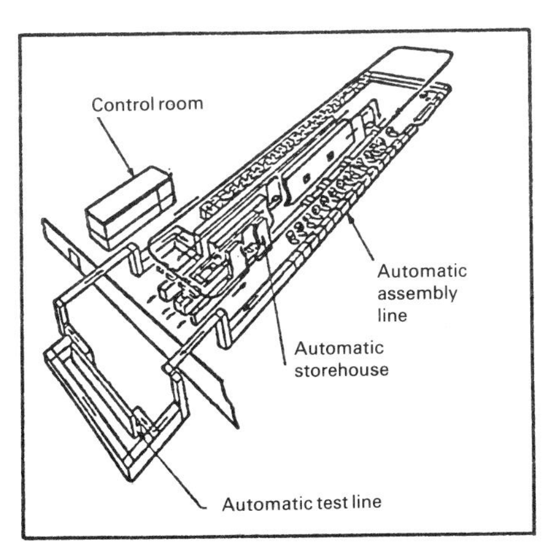

General layout of the Hitachi tape recorder assembly line.

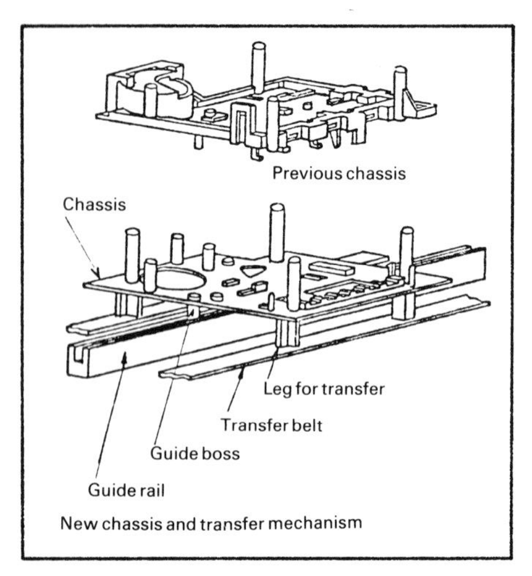

The chassis was redesigned from the upper form to the lower one to facilitate automatic assembly.

versity, a number of circular tracks are combined in one tower, like a helter-skelter, although the tracks are independent of one another. There are six tracks and the tower is divided into an upper and lower half. The result is the top three feeders vibrate independently of the lower ones – and they can be out of phase feeding small parts to automatic assembly machines. The Rin-ei feeders are linear, with two tracks moving parts in opposite directions.

In both cases, the mechanism is simple. A pair of electro-magnets are excited to attract one another; then they are de-activated. Since the troughs are carried on leaf springs, the combination of electro-magnetism and the spring force keeps them vibrating.

Japan's growth as a supplier of electronic equipment results largely from its ability to automate manufacture. At the Conference Naoki Takahashi of Hitachi explained how the company manufactures cassette recorder mechanisms automatically.

Hitachi has evolved a system for evaluating automatic assembly, taking into account the ease with which parts dropped into position can be assembled automatically. As a result of this programme, a new design of tape mechanism was designed in which: all parts could be fitted easily from above the chassis; parts were fewer than in previous designs; and nearly all parts are used on many different models – in other words, the designs were standardised.

About 70 components/sub assemblies are involved, including electric parts such as the motor and magnetic heads, pressed parts such as the chassis and control levers, plastic components, and soft components such as the drive belts.

A U-shape track, with an automatic stores between the lines, was devised to assemble the mechanisms. There was also a separate automatic test line.

To suit automatic assembly, some changes to the design were made. For example, the chassis was redesigned from a moulding to a pressed plate with a number of legs projecting upwards or downwards. There is also a skid to rest on the guide rail. The legs rest on the belt used to transfer the chassis from station to station. There are 52 stations on the line and each is actuated independently so that if one machine breaks down, the line can continue to operate. In that event an operator takes over from the broken down machine.

Special machines

One operation which presented many difficulties was assembly of the rubber belts. This required a special machine to be developed to fit an 80mm diameter rubber belt, 4mm wide and 0.5mm thick, around the drive pulley and flywheel.

The machine consists of a drum feeder, a screw feeder, a feed track, a fitting unit with a chuck, and a work transfer unit. The principle of operation is: belts are stored in the drum and picked out by a screw-feeder transporting them to the feed track. They then move forward to be picked up by a pick-and-place robot carrying the chuck. The chuck claws lift the belt and place it on the pulley and flywheel of the workpiece.

The process begins with a hopper feed to the drum. A large number of pins are located on the inner surface of the drum so that as the drum rotates belts are picked up on the pins. The belts hang down on the drum and subsequently fall on to the screw feeder to find their way into a groove.

If two belts fall into the same groove they are forced out by the friction between them as they rotate. Since the screw is inclined slightly uphill from the drum to the track, a belt that cannot find a groove moves back to the drum end, where it will eventually find a groove.

It is impractical to use normal bearings at the ends to support the shaft because they would prevent belts feeding through. The alternative is sets of three rollers which bear on the periphery of the shaft to give support. This allows the belts to pass below the rollers unimpeded.

When the belts have been wound up on the screw they fall on to a track feeding them to the pick-up position. This acts as a buffer stock.

The chuck on the robot has four claws in two pairs. These stretch the belt sufficiently for it to pass over the pulley and flywheel, The robot releases the belt and the assembly operation is complete.

Hitachi is patenting this device in the USA, Japan, Britain and Germany. The robot device is especially useful since it helps keep the overall cycle time on the line down to 4s.

Integrated circuits

Automation is indispensible in semi-conductors and integrated circuits where there are enormous production volumes and a need for greater accuracy. There are generally two processes involved – diffusion bonding of the pattern on a silicon wafer and assembly of the electrodes to external leads. Whereas the diffusion process – or patterning – is rapid and automated, the assembly process is inherently longer, especially as circuit packing densities on the same wafer are increasing.

To overcome this problem NEC, and other Japanese makers, have produced automatic wire bonding machines. Since these incorporate visual sensing, they are in many respects the trendsetters in automatic assembly equipment. The headaches associated with designing such systems are that distances are very small, great accuracy is needed, and high speed assembly presents problems of its own in acceleration and vibration.

In the NEC design, a ccd camera is combined with a number of wire bonders. The ccd camera can not only recognise the position of the pellet or wafer but also the positions of the electrodes. The data is used to control the position of the x-y table carrying the pellets, and the bonders.

One ccd camera controls five

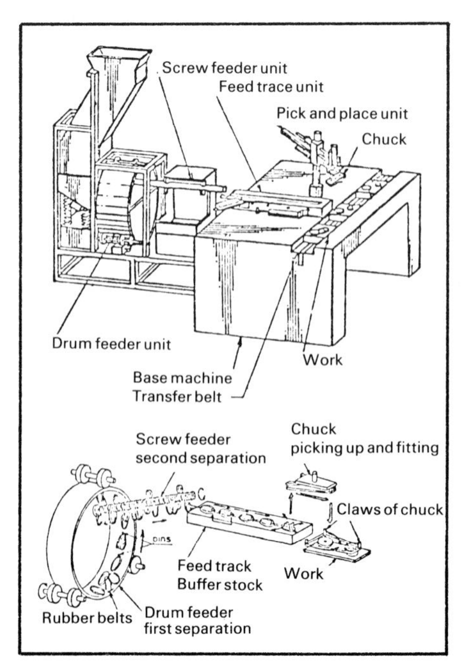

The automatic rubber belt machine developed by Hitachi, with the drum and screw feeder shown separately.

bonders and feeds its data through an image processor to an 8080 microprocessor. The data is then sent to the microprocessor controlling the bonder and x-y table.

On each pellet there are two points, near the electrodes, and diagonally opposite one another. The image processor relays this information to the microprocessor where the deviation from the standard position is calculated. Then, the image processor detects the positions of the electrodes in the image.

For this purpose, two techniques are used to prevent spurious noises or images from being mistaken as electrodes. First, square masks, covering about 80% of the area of the electrode, but leaving the central portion of the electrode clear, are superimposed on the image. Secondly, the 'black' level around the electrode image must be above a certain value for the light to be accepted as a real electrode.

To align the electrode with the bonders, the x-y table is moved by a dc servo motor driving a ball screw. Position and velocity encoders on the shaft of the motor are used to generate signals used to control acceleration and positional accuracy of the table. The table is actuated in two modes: first, the motor is actuated so that it follows a predetermined velocity curve; and secondly, it follows a positional program.

NEC claims this machine can detect 99.8% of electrodes, and the operation takes 0.5s for each pellet – and this includes camera movement. Then, the bonder can operate at the rate of 0.2 s/wire, or 5 wires/s. The bonding accuracy is put at $\pm 15\mu$m. The x-y table has an overall average speed of 40mm/s, although the actual cycle is about 2mm in 0.05s.

NEC is also using a gang bonding technique where there are many electrodes on standard pellets, but the wire bonding technique has been found to be more effective where there is a lot of variety. In the next step, the company plans to control the bonding head by computer, so that the bonding pressure, time, and other parameters can be controlled.

Clearly, this system will not be considered the last word but there will be continuous developments which the semi-conductor manufacturers can and must make owing to the dramatic growth in demand for these products. Inevitably, many of these developments will spin off into other areas of industry – which is one reason why the Japanese are anxious to have a strong electronics industry.

Simplicity is the key to automatic assembly

Much of the automation in Japanese assembly lines is simple and relies heavily on co-operation from the operators to keep it going.

VISUAL and tactile sensing robots are not needed for automatic assembly. Instead, simple robots, or hard automation with sufficient flexibility to cope with a small variety of products is adequate.

This statement is clear from some examples of automation currently in operation in Japan. Nevertheless, the trend is towards even greater flexibility in the future.

Moreover, although the lives of products seems to be shortening all the time, many components change only in detail, or in size; so the operations required and the form of the components remain the same.

There is, for example, a host of assemblies that are basically circular housings with components built-up on shafts, including electric motors and generators. Although automotive engine development is moving ahead quite quickly, the basic form of the engine seems unlikely to change much in the next decade or so. This is especially true in the small single-cylinder units of mopeds, small motor cycles, snowmobiles, generating sets, lawnmowers and other devices. Thus, with some flexibility in the equipment, it should be possible to automate assembly of this type of product.

That such automation is practical is shown by the factories of Nippondenso, the Japanese automotive electrical company, and Yamaha Motor Co, the motor cycle manufacturer.

Job cut-back

It is interesting that Nippondenso, having enjoyed rapid growth over the past decade, is now looking to a future with limited growth but increased productivity.

Fig. 1. The Nippondenso assembly line for alternators relies on manual parts supply.

The result, according to Katsuo Aoki, executive vice president, is that it will no longer be possible to transfer people displaced by automation to expanding plants elsewhere in the company. Thus, there may be some reduction in total workforce.

Aoki said that currently 30% of the workforce are women, mainly young, and that in Japan, once they marry they tend to stop work immediately. Therefore, he reasons, there will be an easy way of reducing manning levels. It will just be a case of not recruiting so many people to fill the gaps.

In fact, although Japanese women are often persuaded quite strongly to stop work after marriage, there are growing signs that not all are happy to do so any more. So, the future for Nippondenso may not seem to be as simple as it seems at present.

But when it comes to the decision to install an automated line, Aoki says uptime must be 80%. In this case, uptime is defined as the proportion of the eight-hour shift the line is operating; preventative maintenance being carried out when the shift has finished. Small maintenance jobs and minor breakdowns are looked after by the operators.

Nippondenso's alternator plant, although almost ten years old, is interesting in the way it is integrated and in the level of automatic assembly. There is a high-pressure diecasting shop and cold forming plant in the factory producing components exclusively for alternators. There are some plastic injection moulding machines as well; these are alongside the assembly lines at the places where the mouldings are required.

Kanban

As part of the Toyota group, Nippondenso operates the Kanban system, in which stock and work in progress is kept to the minimum. The basic principle is that the department or section can produce components only at the rate at which they are being used by the section ahead. There are buffer stocks for a couple of hours or so, but the men are not allowed to fill extra pallets.

In any event, assembly of alternators is carried out with a cycle time of 5s on two lines to give a nominal capacity if 500,000 units/month. However, the preliminary lines operate at a cycle of about 10s, so four are needed. On these lines some 36 different models are produced, ranging from a unit for a medium-size motor cycle to a large truck.

The automotive alternator, of course, has a rotor built on to a shaft, a laminated stator carrying the windings, and aluminium end housings. There is also a drive pulley, and some ball bearings.

At Nippondenso, automatic assembly starts with the laminated stator, which is formed from strip, then rolled into rings. The rings are then stacked automatically and placed on a small square pallet to pass along a conveyor system. There are six stations in the rotary projection welding machine where the stacks are welded together. Then they are sized, powder coated, and the windings assembled in three stages. The conveyor throughout this section, and subsequent sections, is synchronised.

Although winding wires onto the stator is largely automated, it was necessary to split this operation into three to make automation practicable and the cycle time not too long. Even so, stators emerge from these machines with some wires projecting and these are subsequently fixed in place manually.

Linked machines

Rotor assembly is also fully automated with a Koyo Fab Tec conveyor system taking parts from station to station, since in the first section this is a question of linked machines rather than a simple conveyor line. This section starts with some centreless grinding machines, so the conveyor is an overhead system using automatic loading and unloading.

The main section is another synchonrised line, after which there are four inspection machines, each made by Citizen Watch Co. Here, electrical resistance, as well as height and other critical dimensions are checked. In the case of the height, this involves a tolerance of 0.002-0.003%.

When assembled, the rotor passes to an Akashi automatic balancing machine before being fed to the main assembly line. The rotor is mounted on a pallet, and passes down a line in which 32 operations are carried out automatically. Meanwhile, stators come down a line in the opposite direction to join the main line midway along its length.

In the rotor assembly line, bearing and end cover are assembled first. Then, the stator is fitted, followed by the fan, retaining nut, and ancillary items.

Throughout this area, there are small stocks of parts in small Perspex cages alongside the line, and simple pick-and-place units transfer the parts to the assembly.

The fans, for example, are stacked on spindles in small pallets in two rows of four. An arm moves across, picks up three or four fans – one from each stack – and drops them on a conveyor belt. Then, one of these is lifted by a pick-and-place unit and put onto the alternator. All these machines are actuated pneumatically, but one feature of the fan feeder is that there must be an empty space on which the fan can be dropped. The presence of a fan is detected magnetically.

Four stations are needed to install the nut to retain the fan to the spindle. At the first station, the nut is spun down; at the second, the body is rotated so that it is in the correct position to be gripped; then, with the body held in a gripper, the nut is tightened to the correct torque; and finally, the torque value is checked.

Flexibility

With so many different models, and some five basic frame sizes to cater for, quite a lot of flexibility is needed. Nippondenso achieves its flexibility in the simplest way possible. It merely installed extra stations, one being used for two frame sizes only, for example.

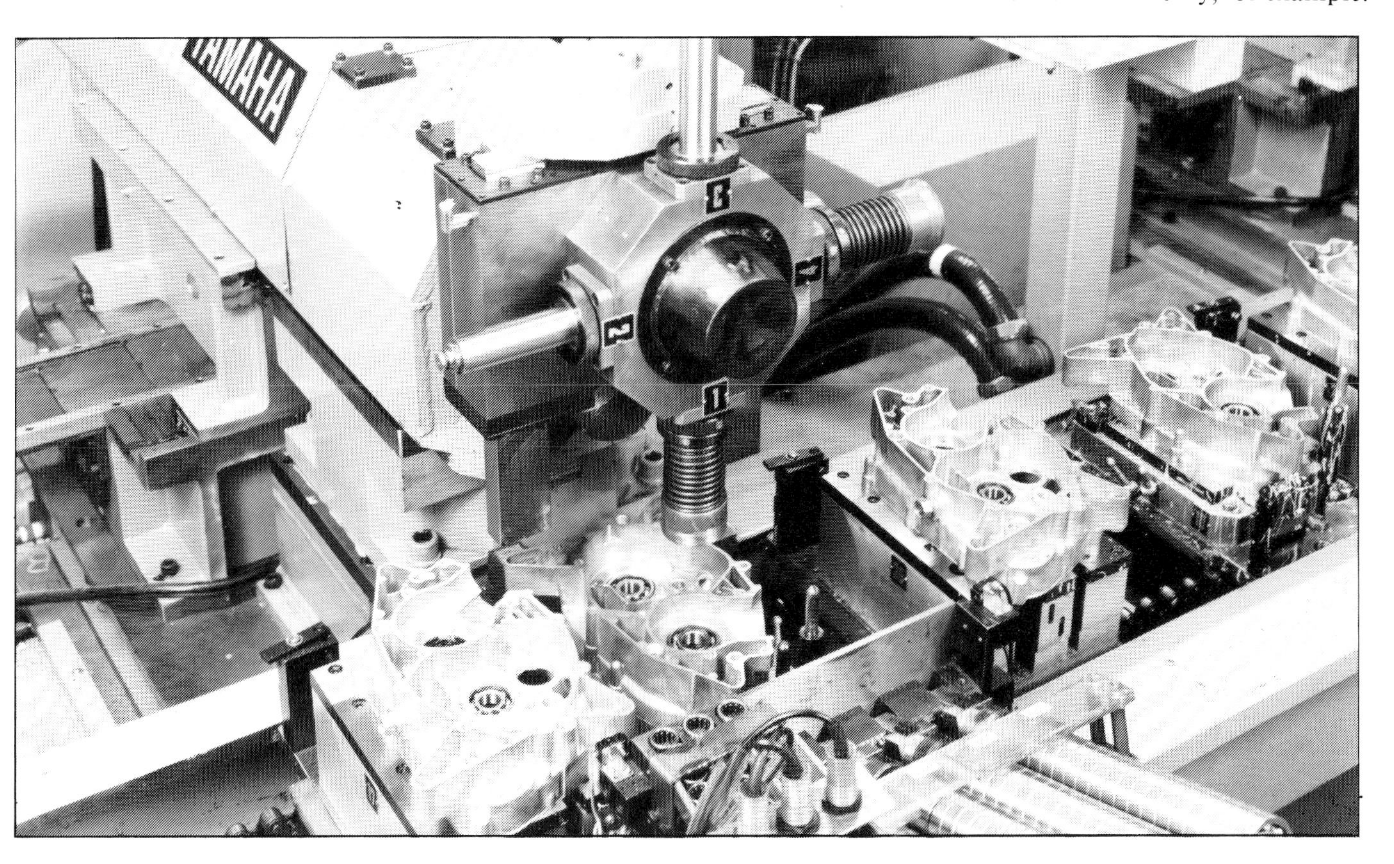

Fig. 2. The Yamaha robot can move horizontally into two directions and has a turret to suit different parts.

An NC controller is used to select the correct station.

As Kazuhide Naruki, general manager, production engineering said: "The stations are cheap, so it is easy to build in idle stations." On the other hand, the operators are responsible for ensuring that the appropriate parts are supplied for each type of alternator.

Compared with that line, the fuel gauge line is more advanced, principally because there are few idle stations, and because switching from model to model is controlled by microcomputers.

In fact, there are four sections to that line, starting with one fully automatic section for the assembly of the mechanism below the dial. At the second stage, the regulator wires are wound around the terminal and fixed in position, with some ten operators involved. Then, two operators fit the tiny needles and washers, and the final section is automatic.

By installing this line, Nippondenso has reduced the manning level from 125 to 25 people – for an investment of £600,000. At the same time output has been increased to 700,000 units a month, 480,000 of which are the same model. On the other hand, several models are produced in quite low volumes.

Fig. 3. A special purpose machine, rather than a robot, is used to insert and tighten setscrews in the Yamaha line.

By contrast, there are 124 people working in the alternator shop for an output of 500,000 units a month, this being operated for two shifts. Naruki says that the alternator line is the most highly automated in the world, but adds that Nippondenso was able to invest heavily in the lines because it was assured of a growing output.

When the line was installed demand was 200,000 units a month, but the company took a long view and automated for a higher output. Of course, that kind of philosophy is more difficult to adopt in these troubled times than before the energy crisis.

An interesting feature of the installations is that although the level of automation is high, there are a number of people around the line. Their job is not just to load parts but to troubleshoot. In fact, the lines appear to stop quite frequently. Then, the operator either flicks a switch, or presses a couple of buttons before pushing a component in place. Then assembly continues. Since the gauge line is non-synchronised, these stoppages cause the minimum reduction in output.

Engine assembly

Automatic assembly of engines in the automotive industry is usually limited to relatively simple operations carried out by hard automation. For example, the tightening of fasteners with multiple nutrunners is one application, and the orientation of the engine to suit assembly another. Then, in some plants, the crankshaft is lowered into position automatically.

One reason for this limitation is that for the installation of many of the parts considerable dexterity is required. Such operations are costly to automate, especially since many intermediate inspection operations are needed.

At first sight, it is surprising therefore that Yamaha Motor has installed four lines for the automatic assembly of engines. In fact, the automatic operations cover the first half of each line, where the bottom end is assembled; however some manual operations are still required. These engines are all single-cylinder two-strokes, which is the simplest form of internal combustion engine.

Equally surprising, though, is that the company installed four lines, all identical, which are operated for one shift only, instead of two lines operating on two shifts. The choice of this layout gives the company greater flexibility in mixing the various models without the need for very small batches.

Engines of five sizes, from 50 to 450 cm^3, are produced, but there are many varieties of each. The biggest volume is for the 50 cm^3 units, but there are quite a number of differences between models. However, the differences are either in the later, manual stage of assembly in internal shape which does not affect the assembly process.

In the sense that the variety is in the later stages of assembly, and that this variety is handled manually, the Yamaha installation represents a good model. However, the way in which the lines are automated differs from the normally accepted method. The volumes do lend themselves to automation, however. Daily output is 1,100 to 1,500 engines with cycle times varying from 15 to 20s, implying an efficiency of about 80%.

The lines are divided into halves. In the first half, the crankcase assembly and gearbox is built up, and these operations are largely automated. In the second half, the piston, cylinder barrel and head, carburettor, electrical equipment and so on are added manually.

All the engines are designed to produce similar assemblies. Thus, the crankcases are split vertically, and there are extra side covers – one for the primary drive to the gearbox, and the other for the alternator. Each crankcase half carries a number of bearings and seals for the crankshaft and gearbox. The gear cluster is built-up as a sub-assembly on a separate line.

Simple robots

The interesting features of the automated lines are the use of simple robots and balanced bowl vibratory feeders. These bowls are needed to supply parts for the full range of engines. Generally, five or six bowls supply parts with appropriate components for the engine being assembled. Only the required parts are discharged from the shutes to the work area. These shutes, like the robots, are controlled by NC equipment.

These robots, which were built by Yamaha, have a substantial main arm which can move across the line, and along it, or in a combination of both motions at once. On the end of the arm is a turret, generally with four equally spaced arms that can be rotated about the axis of the arm. At the end of each arm is a special tool to pick up a particular type of component.

The robots are used to install bearings, oil seals, plugs, a lock-pin, and to apply sealant to a joint face. They also tighten setscrews. A variety of grippers or tools is therefore needed.

To install a bearing, a split rubber spigot is used. Inside is a steel pin with a bulb at the end, and this is pushed through the rubber spigot after insertion in the bearing to hold it in position. A steel shoulder takes the force when the bearing is pressed in; afterwards the steel pin is withdrawn and the press-fit holds the bearing in position while the tool is withdrawn.

For oil seals, a simple spigot is used, the rubber lip gripping the spigot until the seal has been pressed in. The plugs are picked up in cups, to provide precise location.

There is one operator at the beginning of each line loading bearings Subsequent operations are:

First station: Installation of oil seal, ball bearing and slim shaft seal in one crankcase half;

Second: Installation of a second ball bearing and cup needle bearing;

Third: Insertion of lockpin and large needle roller bearing for selector drum;

Fourth: Application of sealant to dealing face of crankcase half. This involves a complex path, since the sealant must be applied to internal and external sealing faces, and all these are very narrow, being ends of walls rather than flanges;

Fifth and sixth: These stations involve manual operations, the first man installing the crankshaft and gear cluster, and the second man fitting the two crankcases halves together;

Seventh: Ten setscrews, fed pneumatically to the machine, are inserted in the holes;

Eighth: The setscrews are tightened, and their torques checked;

Ninth: The two shaft seals, for the crankshaft and main drive, are inserted;

Tenth: A small metal plug, a rubber plug, and a tiny shaft seal, with an outside diameter of about 10 mm, are inserted.

The basic operations of the robots are that they move across the line, pick up a component, move back across the line to install the component.

Precision

Since there is no compliance built in to the robots, great precision is needed to ensure reliability. Each crankcase half is dowelled to a pallet on which it is carried during assembly. Then, at each robot station, the pallet is picked up and located on dowels for assembly.

The tolerance on positioning is claimed to be ± 0.05 mm. It is claimed that the tolerance on the position of the robot arm is the same, giving a total tolerance between component and housing of ± 0.1 mm. Of course, this is a much tighter tolerance than is normal for general purpose robots, but it is not so difficult for a simple device with only two axes of freedom in positioning.

Although it is possible to switch the shutes by NC, it is still necessary to warn the operators that a new engine type is coming down the line. To do this a dummy pallet is sent down the line, enabling men to load the vibratory bowls and be ready to work on the new parts.

When the automatic lines were installed, Yamaha was able to reduce the manning level from 14 to 4 people per line – a total reduction in the shop of 40 people. The minimum batch is put at 100 units, but they are normally larger.

However, Yamaha is not standing still, and for its new engine assembly lines, it is building robots of a different type. These resemble the Scara robot devised by Prof. Makino of Yamanashi University. This arm has two joints with vertical axes, and some inherent compliance in the form of electric braking torque. This can be overcome by the insertion force.

Yamaha is working hard to increase its level of automation, and is just installing an automatic assembly machine for spring/damper units on small mopeds. Intended for a cycle time of some 5 s, the new machine was supplied by Shinko Engineering Research Corp, of Tokyo. It is in three sections, two rotary tables, and the final section a linear track, and is tended by two operators.

In the first section, the seal, piston and top cap for the cylinder are fitted to the damper rod. In the second stage, the cylinder, with its integral bottom mount, is fitted into a gripper vertically. At the first station, hydraulic fluid is poured into the cylinder, and at the next, the piston rod assembly is inserted. Subsequently, the top of the cylinder is swaged around the top cap to form the assembly. In the third section, the spring and shroud are added.

Finally, Prof. Makino and his team at Yamahashi University are in the final stages of developing an interesting cam mechanism for indexing rotary tables. The cam mechanism is in fact a development of one designed initially for a linear track, but it is mounted inside the table to provide indexing motion for twenty stations. As with Makino's other designs, it is arranged so that deceleration reduces before the motion stops, and the stationary position is held positively. Thus, it is ideal for very short cycle times. In fact, in the demonstration unit, a variable-speed motor is used to drive the table at speeds around 80 cycle/min. Clearly, an ideal device for assembling very small components.

Developing assembly robots on the shop floor

Simpler machines, built in-house, perform simple operations on a develop-as-you-go basis at Yamaha Motors.

AS PART of its plan to automate 50% of engine/transmission assembly, Yamaha Motors has introduced a new type of robot. Called the Came robot, the new machine is a simple jointed arm machine, with the shoulder and elbow joint pivoting on vertical axes. So far, these robots are being used to tighten screws, insert bearings, lay sealant, and insert rubber blocks between the fins of the cylinder heads.

These robots are being used on two assembly lines – one for engine/transmission units for 50 cm^3 mopeds, and the other for all four-cylinder engine/transmission units of 400 to 1,100 cm^3. What is particularly interesting about this installation is that the machines work very close to the operators, and that Yamaha has installed machines that are not fully developed. Some machines have a Yaskawa NC controller, others a newer microprocessor based controller, and in both cases, the equipment is being developed with the robots in use. Thus, problems are being encountered and solved on a day-to-day basis.

In many respects, the operations are similar to those at the Hamakita plant.

On the moped line, which operates on a cycle of 18 s, there are Came robots. The engine crankcase is carried on a pallet along the line, and at the first station, the operator places three bearings in recesses in the pallet. At the same time, the operator places the housings – the crankcase/gearcasing is split vertically – on the pallet. It was decided that it was much simpler for the operator to place the bearings on the pallet as well.

The first and second robots pick up the bearings and install them in the housings. At the third station, a robot lays a bead of sealant around the face of the crankcase – a very complex path being followed very quickly. Whereas the machines on this line are generally two-axis robots, to apply the sealant to the face needs a third axis. This takes the form of rotation of the sealant gun on a vertical axis.

Layout of the first stage of robot assembly of moped engine/transmissions at Yamaha Motors. The first two robots install bearings, and the third lays in a bead of sealant.

Nut runners

In subsequent operations, nuts are aded and tightened. It was also lanned to install a spring washer and ut to the alternator spindle by robot. owever, the spring washer is almost uare in cross-section, and it was und that washers had a tendency to t tangled up in the vibratory bowl eder. Therefore, the washers are aded by hand, and the nut only is cked up, moved across and spun own on to the spindle.

However, it proved practical to use ne robot to fit the washer and nut to he other end of the crankshaft. Again, three-axis machine is used, with an xtra gripper being carried on the ousing for the nut runner/carrier. hus, the robot is manipulated so that he nut runner picks up the nut, then he gripper picks up the washer. ubsequently, the gripper is posi-ioned over the spindle, and releases he washer. Then, the nut runner is noved across, and tightens the nut.

On the large engine line, there are even robots on crankcase assembly, nd eight on final engine assembly. The robots are used mainly in place of ut runners, their flexibility allowing hem to work on engines of five different sizes. For example, two robots working on opposite sides of the track tighten the eight connecting rod bolts at one station. Other robots tighten the setscrews to retain the oil pump strainer, and two robots tighten the eight bolts retaining the sump to the crankcase.

In the final assembly line, one three-axis robot tightens 12 setscrews to retain the camshaft bearing caps. Then, another machine installs small rubber blocks between the fins of the cylinder head – these damp vibrations.

There is a vibratory bowl feeder supplying the rectangular rubber blocks – each is approximately 10 mm square – to a channel. In sequence, an arm picks up the rubber block, and turns through 180 deg. so that the gripper is holding the block facing upwards. Then the robot picks the rubber block from the gripper, and pushes it down between the fins. This process is repeated four times, the robot placing the blocks in four different places.

Worker resistance

Most of these operations are relatively simple, and Yamaha has installed the robots, which are quite small, simply in place of a person on the line. Thus, it is able to gradually introduce more machines as the time goes on. However, the robots have not universal approval from the workers, especially those who have been cut off from their workmates by the robots. This situation is aggravated by the fact that pneumatic nut runners are used, and these are very noisy, especially since they tend to operate in unison.

From time to time, the robots do stop working owing to a malfunction, and in typical Japanese style, the whole group of workers gather round to solve the problem – the operator, the section leader, foreman and the engineers. Thus, the workforce is gaining confidence in the robots, and is learning to maintain them – an essential piece of on-the-job training, since eventually it is planned for the robots to do most of the assembly operations, leaving the workers to operate the control units and look after maintenance.

Clearly, these robots are in the early stages of development, while the applications are not fully proven. But that is the time to start automatic assembly – when you can afford to have some people to look after the machines, and to step in when a failure occurs.

A washer and nut are assembled by the same robot. Here, the gripper has just released the washer, which has settled around the threaded spindle, and the arm is about to move over to lower and tighten the nut.

Automation specialists go for car sub-assembly work

Taiyo has a small group building special purpose machines for Japan's vehicle and electronics industries.

THE GROWTH of Japanese industry has had a number of side-effects, and one of these has been the growth of companies specialising in automation equipment. Despite the fact that more manufacturing companies in Japan than in most other countries build their own jigs and fixtures, there is still plenty of scope for specialists.

One company that has grown through the manufacture of equipment for automation assembly is Taiyo Ltd of Osaka. Its main business is in pneumatic and hydraulic cylinders and valves, but for a number of years it has been combining the pneumatic units with jigs, fixtures and conveyor systems to produce automatic assembly devices. Principal customers include the car and electronics companies. Now the company employs 600 people, with a small proportion of its staff being involved in the manufacture of the automation equipment. Last autumn, the company also introduced its first real robot – the Toffky, which is designed for the application of sealant and adhesive. Like the automation equipment, this is being made in response to orders only, and so far 10 have been built.

Also on an order basis, handling devices are built. These include special grippers to carry heavy objects – one has a capacity of 800 kg, and another 4000 kg. But these are not programmable, and are derived from hard automation concepts, like the machines built for the car industry.

Typical applications of machines produced in recent years are:

- assembly of plugs in engine cylinder block;
- assembly of main bearing caps and ring dowels in cylinder block;
- connecting rod assembly;
- piston ring/piston assembly;
- assembly of sump and oil filter to cylinder block;

Fig. 1. Two connecting rod assembly machines in Taiyo's assembly shop.

— universal joint assembly.

Recently, two machines for assembling connecting rods, another for assembling pistons to gudgeon pins, and a machine to handle the drums of plain paper copiers were being built.

Fig. 2. First station of the connecting rod assembly machine.

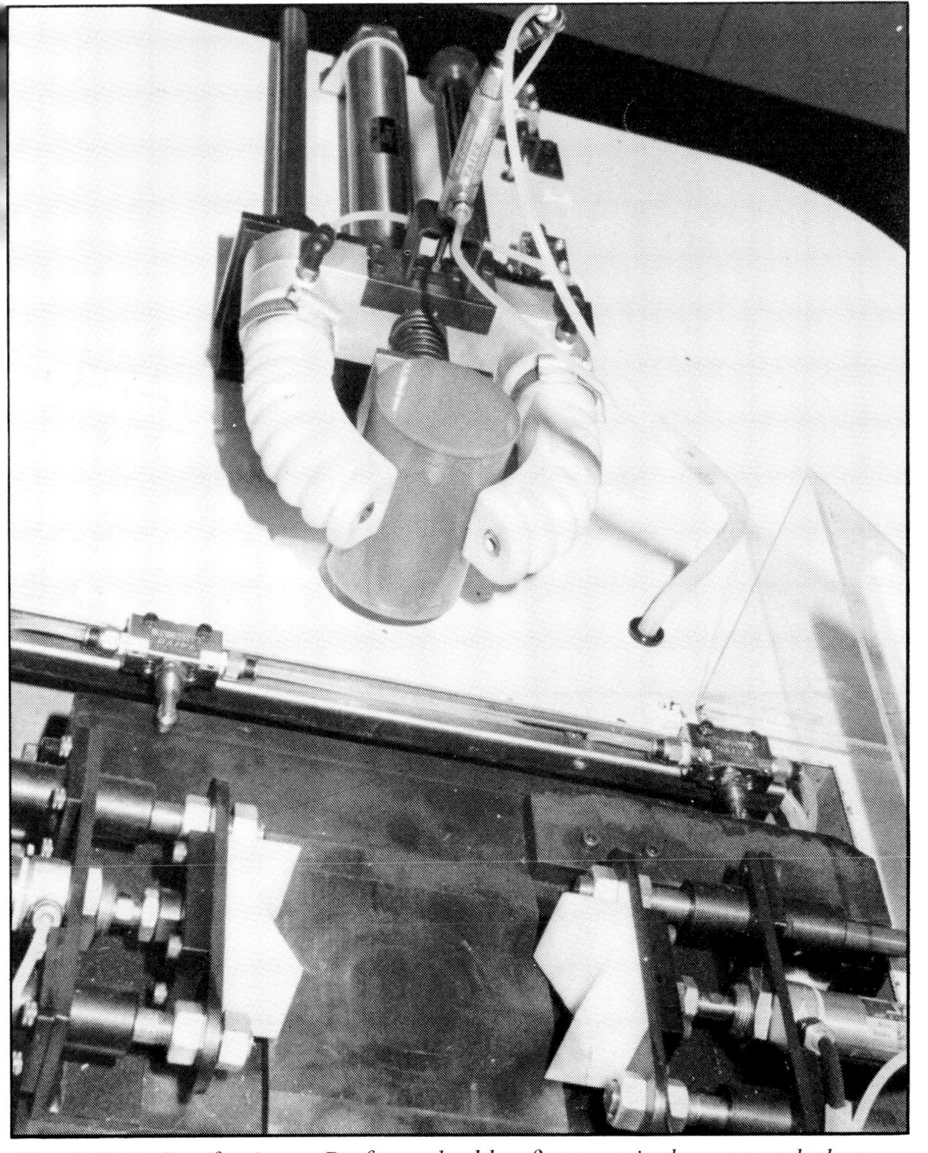

Fig. 3. Taiyo's soft gripper. Preformed rubber fingers grip the part, and when inflated straighten out to release the part.

Since the connecting rod assembly machines are to be used for diesel engines, the cycle time is 9 s, against the normal 6 s for a petrol engine – that equalling a 24 s cycle for the complete engine set.

The machines are designed to locate the connecting rod and cap together, insert the bolts, fit the screws and tighten them, and then press in the small end bush. The machine operates with linear motion, the connecting rod and cap being supplied along small parallel slat conveyors to the first station where they are moved in to mate up. At the second station, the big and small end bores are located on pins, and the bolts are inserted. The bolts have rectangular heads which locate against shoulders on the connecting rod. Next, the nut is spun on to the bolt and tightened – this is not a critical operation, since the cap and rod are subsequently dismantled. However, prior to assembly the small end bush must be orientated so that the lubrication hole aligns with that in the rod.

For piston assembly, a rotary machine with nine stations has been adopted by Taiyo. It is designed to handle pistons of two sizes, and both the pistons and connecting rods are loaded into the machine manually. In the first operation, the piston and rod are fitted in the jig, and at the second station the gudgeon pin is pressed in. Then, the position of the oil hole in the small end is checked, and numbers are engraved on the connecting rod and cap.

The next operation is an interesting refinement; the connecting rod is articulated about the small end to check the torque required. A strain gauge is used to deliver 'go' and 'nogo' signals, a buzzer indicating that the fit is outside tolerance. Finally, the nuts on the connecting rod are unscrewed so that the assembly is ready for installation in the engine.

The machines are typical of those produced by Taiyo, whose machine design manager, Fukutaro Kashiwa, says that the company has no plans to build playback robots other than the Toffky. Instead, it intends to stick to manufacturing complete systems to order, although some will include robots. For example, it is about to build a line for handling parts in a plating shop, and another for automatic assembly of the wheel and tyre to a car. In each case, one Fanuc Model 1 robot is being used.

Productivity increases ninefold

Design for production and full automation gives Nippondenso a tremendous boost in reduced costs in the manufacture of relays, while its compressor line mixes man and machine.

'WITH OUR new relay line, we have increased our productivity by nine times,' said Kazuhiro Ohta, director of production engineering at Nippondenso, Kariya, Japan. 'But one reason that we were able to do this was that the product was completely redesigned – this is an example of close co-operation between designers and production engineers,' he added.

As the largest manufacturer of electrical equipment and car parts in Japan, Nippondenso is a big volume producer of a mass of equipment, from tiny relays to starters, alternators and air conditioning equipment. Although its main customer is Toyota, currently producing over 3 million vehicles a year Nippondenso also supplies some of the other Japanese companies – but not Nissan, such is the group system operating in Japan. Its largest factory is at Nishio, some 60km from Nagoya, near the south coast of Japan. Main products are fuel injection equipment, air conditioning compressors, and the relays. A total of 7,000 people are employed there.

The fully automated relay line is Nippondenso's latest attempt at flexible automation, although of course, the flexibility is limited to small cylindrical relays. A few years ago it installed a new line for the assembly of car air conditioning compressors, and this too is interesting in the way it mixes manual assembly for the jobs requiring dexterity, and automatic assembly for the majority of simple jobs and handling. However, it is a line that seems to be a good application for robots, simply because the hard automation operates so slowly.

Relays are used in many systems in vehicles, so many different types are needed. Nippondenso's new range consists of small units, typically

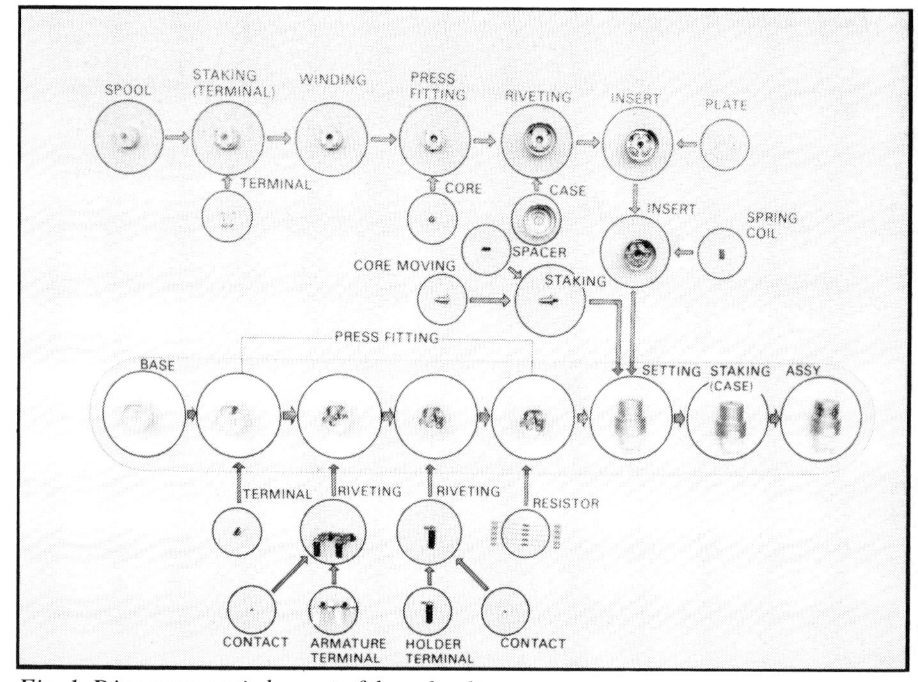

Fig. 1. Diagrammatic layout of the relay line.

Fig. 2. View of the first section of the relay line, showing the camshaft drive above the assembly line.

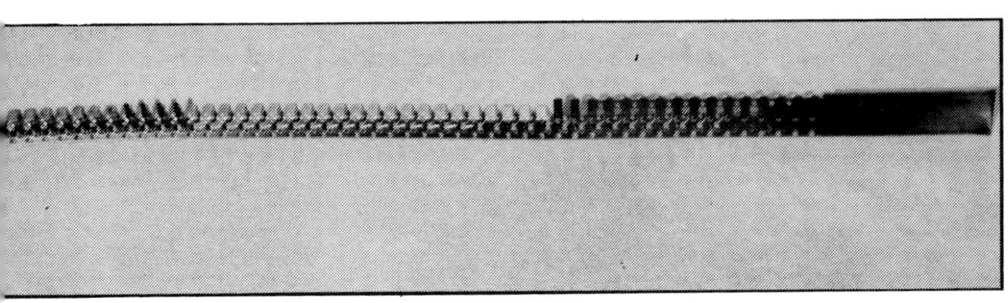

Fig. 3. A strip of copper showing the steps performed in the progression presses.

Fig. 4. Assembly of the terminals in the relay base.

Fig. 5. Adaptive control is used to set the contact clearance on these machines.

30mm tall by 28.5mm in diameter. One million are produced monthly on a cycle time of 0.9s in the new assembly shop, giving a nominal capacity of 64,000 units/day on the two shifts. Previously, about the same number were produced on 10 assembly lines with largely manual assembly, and 200 assembly workers were needed.

'We have no direct workers on the new line,' said Ohta, 'but we need 23 people to run it – that is 11 people to supply parts, load finished parts, and look after the line on each shift, with one supervisor for the department.'

Total investment in the line was £1.2 million, all the equipment having been made by Nippondenso. However, had the equipment been purchased outside, it would have cost about £1.8 million. The saving in labour costs is about £750,000 per annum, and the fact that the equipment was built in-house had a dramatic effect on the length of the payback period. Generally, young girls work on these assembly lines in Japan, and they stop work when they marry. Thus, the introduction of automation does not involve redundancies. It is necessary merely to adjust the recruitment rate accordingly. Of course, this situation hides the fact that automation does mean less jobs for unskilled workers.

Rectangular

Previously, Nippondenso was making 114 different types of relay. These were generally of rectangular form, and parts had to be assembled from the sides as well as from the top and bottom. Sizes were generally larger than now, several being rectangular, a typical model being 37 × 55 × 32mm in size, against 28.5mm diameter by 30mm tall for the new one. As new models were introduced, new assembly lines have been built.

Eventually, the decision to build a new line was taken, and the product design department was asked to redesign the part for ease of assembly. The result was a cylindrical design, with an injection moulded base to carry the terminals and contacts, and a steel casing to carry the coil assembly. Normally, the contacts in the base are closed, and when the relay is actuated, the coil moves the plunger to open the contact points. Critical features of the relay are the clearance between the contact and the plunger, and the pressure at the contact points. The clearance must be maintained within 0.03 and 0.1mm, while the acceptable range of contact pressure is 140-200g. 'But everything can be assembled vertically – that is very important,' said Ohta.

With variations in the winding of the coil, the use of two types of terminals, and other minor differences, Nippondenso engineers found that they were able to produce eight different units within the same basic design, and that these would replace the 114 types previously in use.

There are a number of stages in the process:

- ○ Complex pressing and assembly of terminals;
- ○ Coil winding;

○ Assembly of the terminals into the base;
○ Bending and setting of the contact by adaptive control;
○ Assembly of the coil and plunger in the case;
○ Final assembly and test.

The whole system is controlled by microcomputers.

In the assembly shop, there is one straight line, fed by a number of sub-lines. The injection moulded base is assembled on this line, while the metal casing sub-assembly is built up on a separate parallel line. Then, there are sub-lines for contact/terminals and for the coils. An important feature of the line is that the contacts are adjusted and set automatically by an adaptive control system, and that the complete assemblies are tested functionally before leaving the line.

All the machinery for the process is housed in an air conditioned room – but injection mouldings and the pressed steel parts are brought in – and within that room is a small sound-proofed cell housing a pair of small Aida progression presses. Two are needed to produce two different types of terminal. These are remarkable machines which carry out some assembly operations as well as pressing. Copper strip approximately 75mm wide by 1mm thick is fed to the press which performs 42 operations, leaving the finished contact still part of a strip – the metal remains continuous on one side of the strip only.

Fig. 6. Testing of the relays.

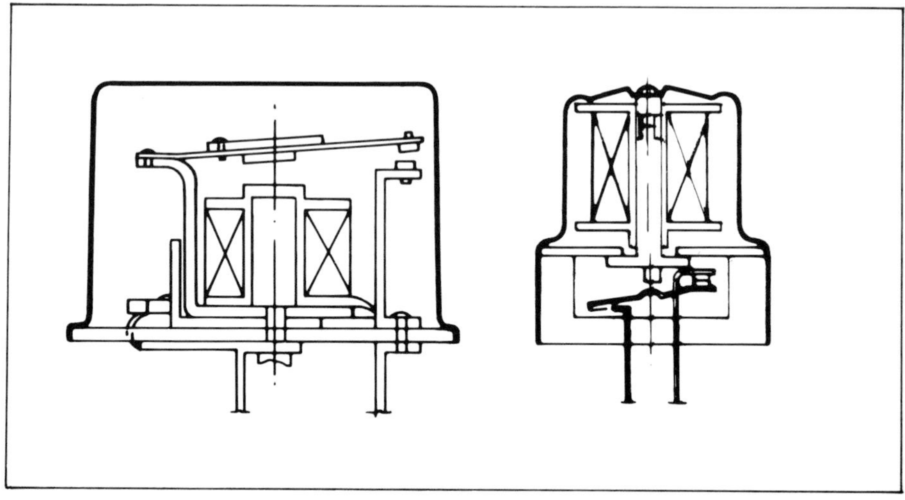

Fig. 7. The relay assembly, compared with the previous design.

Coil winding

First, the press blanks out the material so that fingers project from one side. Then, it adds various holes, and puts a joggle (or crank) in the finger, subsequently turning the finger through 90° about halfway along its length. In the joggled portion, two small holes are punched, and then wire is fed through the holes and cropped off to project about 1mm each side. At the next stage, a smaller finger is fed in from the side of the press – on a continuous strip – and is croped off, and assembled to the main finger, the two pieces of wire serving as rivets. The rivets are then peened over. Afterwards, the completed terminal is fed to the main assembly line as a strip. To act as a buffer, the strip passes over some rollers so that it bends up and down between the press and the line.

There are four coil winding machines, each machine being of the rotary type, with about 20 stations. At each station are four spools, one above the other. In addition to winding the wire, the machine solders the wires to the contacts, and applies a blob of hot melt adhesive to the coil.

In the main line, Nippondenso has adopted what is claimed to be a new concept of actuation. 'It is like an overhead camshaft engine,' said Kazuo Matsumoto, manager of production engineering. 'The camshaft for the pick-and-place arms, and for the shuttle mechanism is mounted above the assembly line – usually it is below and operates through a number of linkages. Therefore it is more direct, with less lost motion.'

Each device is operated by an overhead camshaft, usually one shaft for several stations. At most stations there are simple cam-operated pick-and-place devices, while the shuttle mechanism transferring the parts along the line is operated by a face cam.

The shuttle takes the form of an inverted L-shape plate along the line. In the top face are a number of semi-circular cutouts from the edge, and each fits the shape of the injection moulded base. The shuttle moves along one station, then is turned upwards through 30° so that it is clear of the workpiece. Then, it moves back one step. With such a short cycle time, the motion is continuous.

Adjustment

The first few stations of the line are duplicated, each fed from one press to insert one type of contacts into the plastics base. Only one section operates at one time, and after every operation, a sensor checks that the part has been inserted.

At the end of this section, the two

ig. 8. Part of the compressor line, where a seal and front cover are assembled.

ines are running parallel, so that the ne out of operation serves as a buffer. witching from one type of base to the ther is automatic, and indeed, the vhole shop is controlled by micro-omputers.

Subsequently, the other parts needed for the contacts are assembled o the base, including the tape-fed esistors. The performance of the esistor is checked before the com-leted base passes to the adaptive ontrol section.

Since 3.5s are needed for adjust-nent, four stations do the same job. Two stations are needed for adjust-nent and setting, and each consists of a ulse motor actuating a vertical rod hrough a reduction gear and worm ind wheel mechanism. The opera-ions are:

At the first station, the pulse motor presses the contact down; then, it measures the force needed to press it down; at the second station, it pushes the contact down again to bend it; finally the rod pushes down once more, and allows the contact to spring back to the correct clearance position.

'The control system stores the results of the previous batch of contacts processed, and reacts to those as well as to the figures from the contact being processed at the time,' explained Matsumoto.

Meanwhile, the coil has been assembled into the metal case, and it joins the main line for the plungers to be selectively inserted. 'We select the plungers according to length,' said Matsumoto, 'so that we can control the clearance closely.'

Afterwards, the case is fitted to the base, and its edges are crimped over to form an assembly. When complete, the assembly passes along the line to the test stations, where six are tested simultaneously. A multiple gripper, hanging down from an overhead shuttle, moves the relays in groups from station to station. Each relay is vibrated, operated at light load, tested for resistance, and operating voltage. 'Rejects are discharged into these drawers,' said Matsumoto, pointing to some empty compartments.

At the end of the line there is an incongruity; a man is needed to take the relays, which are fed to him at tremendous speed along a conveyor, and place them in compartments in boxes, four to a compartment. This is one of those jobs which sums up the tyranny of the production line – with the need to pick and place four relays into a compartment every 3.6s, the man had to work very fast. 'We are now working on a robot to do that job,' explained Ohta.

Ohta claims that the reject rate is only 10 parts in a daily production volume of 50,000. 'We produce ten to twenty batches a day to suit Toyota's requirements,' he said. He also claimed that downtime was no more than 15%, although the actual capacity and production suggest a rather higher figure.

The gross production rate is 64,000/day, against an actual output of 50,000/day, implying a downtime of 22%. Since the machines operate automatically, of course, they do not need to stop for breaks or lunch, so the actual production day can be rather longer than for a manual line – and that is one reason for using automatic assembly.

But as this line shows, it is necessary to automate inspection as well as production, and quite a number of workers are still needed to feed the line, and make sure that the machines keep working.

Compressor assembly

Nippondenso decided the compressor was too complex for completely automated assembly, so there are nine people working on a largely automated line which has 50 stations. The line operates to a 25s cycle, to produce 40,000 units a month, implying an efficiency of around 88%. It is operated on two shifts, so a total of 18 people are employed in the shop.

The swash plate compressors lend themselves to automatic assembly in that everything except connectors can be installed vertically from one end or the other. The heart of the machine is the shaft and swash plate, which is carried in a pair of needle roller bearings. The swash plate actuates the cylindrical aluminium pistons through hemispherical seats and balls, and these must be assembled so that the clearance, or axial play of the piston to the swash plate is very small. There is a two-piece aluminium crankcase, split diametrically midway along its length to simplify assembly. At each end of the crankcase is a valve plate, and end housing. Of course, sealing is very important with compressors, so there are a number of special seals and O-rings as well. The complete assembly weighs 4.5kg, and the housing is 115mm diameter by 149mm long.

There are three sections to the line. First is a sub-assembly area, where shaft/swash plate assemblies, and then crankcase assemblies are built up; then there is rectangular line carrying pallets where the main assembly is carried out. Finally, there is a Z-shape line for fitting and testing.

Shafts and swash plates are delivered by conveyor to an automatic assembly machine. At the first station, the shaft is removed from the conveyor, turned through 90° so that it is vertical, and inserted in a temporary holding fixture. Next, a gripper comes across, picks the shaft up and pushes it down into the swash plate.

At the next station, two men assemble the crankcase sub-assemblies. First, balls and seats are selected from chutes – there are 20 grades of seats – and these are placed with the shaft assembly and crankcase halves on a pallet. At the following station, the operator fits the cup/ball assemblies to the pistons, and slides the swash plate into position. Then, he places the piston/swash plate assembly into one crankcase half, and fits the second crankcase half. A job requiring a great deal of dexterity on the part of an operator, let alone a robot!

The completed sub-assemblies are then placed on a square pallet, circulating on a rectangular track. At the point where the compressor leaves the pallet, the track turns through two 90° corners, and it is near the second of these that the operators work. There are four men working in this area – shaft sub-assembly to placing the pallets on the track – and one places the end housings on the pallet, which has four locating positions. He also transfers the crankcase assembly to the pallet, which then runs back to the start of the line.

At the first station on the automated line, oil seals are fitted. About 10 stacks, each with the capacity for about 30 seals, are carried on a circular tree. The tree is rotated to align the correct seal for the model of compressor coming down the line at the delivery area. A pusher pushes the seal out from the stack, so that it can be picked up, dipped in oil, and then pressed into a crankcase. This arm, like most on the line, is a simple device, with motion in y-z axes. Therefore, it first raises the seal upwards, then moves across, and then downwards – all rather slowly – before inserting the seal.

Turnover

At the next station is a turnover device, which lowers a gripper to grasp the crankcase, lifts it up, rotates it through 180° and then lowers it back on the pallet. Then, the pallet moves along to a station where an O-ring of about 30mm diameter is inserted automatically, in much the same way as the oil seal. This is followed by another turnover station.

At the next station, is one of two men on the main line. He fits one plate, a very flexible O-ring of about 100mm diameter, and the tail-end cover. Following another turnover station, an operator inserts an O-ring and fits the valve plate.

Then, a dowel is inserted in the as-yet unassembled end cover, and a Freon seal is fitted to the main crankcase by being rotated down the shaft. At this station, the crankcase is located by two pairs of pneumatically actuated clamps. While the seal is being fitted, the height of the pin is checked.

To fit the end cover assembly, which contains an oil seal and ball bearing, another y-z pick-and-place device is used. It picks up the cover, moves it across, and presses it down. Of course, it is not practical to fit the key in the shaft until these jobs are done, so the key has to be slid downwards along the groove into position. Rectangular keys are fed from a vibratory bowl feeder through a chute to a small press. While the assembly is held in position laterally, the key is pressed straight down the groove, the key being pressed down between the shaft and a plate to ensure full engagement.

A complex device is then used to insert the six bolts and washers through the housing. The long bolts are fed from a hopper down an inclined chute to a head above the workpiece. Meanwhile, washers from a vibratory bowl feeder come in at a lower level and are fed directly into recesses equally spaced on the pitch circle of the bolts. The bolts and washers are fed in turn to the head as it rotates. Then, the head assembly is moved across to the line, and plungers push down through chucks to insert the bolts and washers.

Subsequently, the bolts are spun down, and the torques are checke 'Then we check the bolt height,' sai Matsumoto, 'as the torque and heigh together ensure that the bolts ar tightened correctly.

'It is important that the compress is filled with exactly the right amoun of oil, so we fill and check that twice, he went on. Thus, the completed compressor is turned on its side, and it i weighed empty. Then, it is filled t give the correct amount of oil. At th next three stations, the compressor i operated and tested. Owing to the tim needed, these three stations ar running in parallel. After running, th weight of the compressor is checke again, and the oil topped up as necessary.

Noise

Noise is measured in this area, s nearby is a small terminal showing th measured ball/seat/swash plate clearance for the last 50 assemblies. Wit the noise results these figures are fe back so that tolerances can be adjuste as necessary.

At the last station on the first leg o the main line is an operator fittin connectors. These are arranged in bin

Fig. 9. Insertion of the mechanical seal.

Fig. 10. The bowl feeder and machine insertion of bolts and washers.

in front of him, and as the compressor comes to his station, so lights indicate from which bins he should take the connectors. At each bin is a photo-electric cell, and should he not choose the right part, a lamp at that bin will be illuminated. Once this job has been completed, the pumps are subjected to two leak tests; first there is a 30kg/cm^2 air test, and then a helium leak test afterwards.

Although this line is not brand new, it does represent current thinking at Nippondenso. Ohta said: ‘At present, transfer lines like this are cheaper than robots. But they are not so flexible, so we are now looking into the use of robots on assembly. We are currently expanding our air conditioner assembly line to use 10 robots – we were using six.’

In that line, evaporators and printed circuit boards are inserted in plastic casings, the covers are fitted, and clips attached. The robots insert the evaporators and printed circuit boards – some are horizontal and some vertical – in the housing, and then check that the clips are correctly positioned. These are cylindrical co-ordinate robots, which are unusual in that each has two grippers at 90° to one another, and at 45° to the axis of the arm. They were made by Nippondenso, and are similar in basic construction to those used in its aluminium diecasting shop.

An automatic line has also been installed to build radiators of different sizes. No special jigs are needed for soldering, and the tools that clinch the plates over the plastic tanks are designed to handle tanks of different shapes and sizes. They consist of a number of square rods, which are moved in and out to produce the shape required.

‘We are aiming to make more different parts on each line,’ said Ohta. ‘For example, next we want to extend the radiator line to take radiators from 300 to 700mm in width. It can already handle 64 different types.’ That line, like the one for air conditioners, operates to a cycle of 9s.

Of course, one hidden advantage of the use of flexible lines for assembly is that the output/line increases. Therefore, it is economic to tool up for automated assembly, which would not be the case with several different lines. Nippondenso’s relay line is a classic example of the advantages – output up from about 100,000 to one million/month.

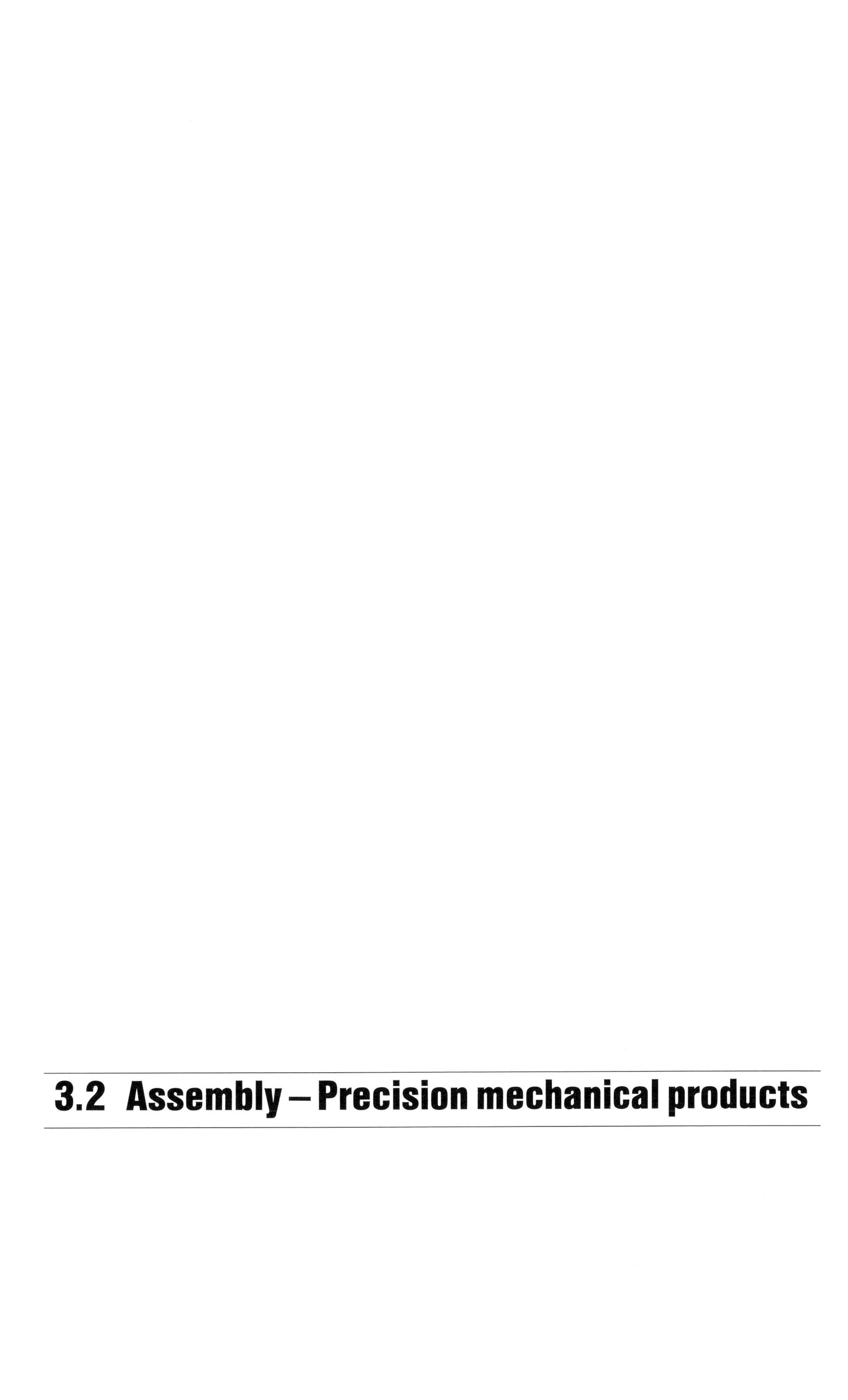

3.2 Assembly – Precision mechanical products

Building watches can be as easy as A, B, C . . .

Seiko watches are produced on automatic assembly lines by three methods: System A for mechanical watches, System B for digital watches and now System C is being developed to give greater flexibility.

ONE OF THE MOST automated operations in Japan is the manufacture of Seiko watches, with production running at around 27 million units a year. Watch production is handled by two companies, Daini Seikosha and Suwa Seikosha, as well as a number of subsidiaries.

As is usual with Japanese companies, the group has set up some small companies to manufacture equipment developed in its mainstream operations. Thus, there is a company producing small business computers, another making automated drafting machines, and another producing atomic absorption spectrometers and x-ray thickness gauges.

Daini Seikosha also makes its own machine tools and assembly equipment. Overall, the company employs 10,500 people in five factories, the largest being Takatsuka with 4,500 employees.

Daini Seikosha's level of automation is impressive. For example, it has integrated pressing and machining operations for base plates with the result that only three men run a complete machine shop turning out 10,000-20,000 components a day. Also, in the shop where the bridge pieces are pressed and machined, automation has led to the reduction of manning from 20 to 2.

In fact, the automation equipment in the press and machine shops is nothing more than a combination of simple standard equipment with special purpose machinery. For example, there are rotary table machines carrying 25-50 pneumatic drill heads to drill, ream and chamfer holes, as well as other similar machines to handle milling. Also there are a number of presses.

Fig. 1. View of System A, showing the main assembly line carrying stacks of pallets in containers and the spur lines off at the sides.

Bowls and chutes

Automated handling takes the form of hopper-fed vibratory bowls and simple chutes. There is a conveyor device from the hopper, taking the parts up to the vibratory bowl, and a chute from the bowl to the machine.

Then, there is a simple arm which picks the plate up by suction to transfer the plate to the press or rotary machine. The arm is rotated through some 45° on a vertical axis, and has a limited amount of vertical movement. After being processed, the parts are transferred by gravity down a conveyor to the next hopper. All this equipment is equally suitable for assembly automation.

Perhaps the most interesting feature of the operation in this division is that although only three men run the large shop, some 80 people are involved in maintenance and inspection.

The production shift operates from 16.30 to 01.00, while the maintenance and inspection staff work during the normal day shift. Whether so many people are really needed on the day shift is not certain, but an important point is that Daini Seikosha has limited the number of people working on the night shift to a minimum.

In addition, it has retrained production workers as inspectors and maintenance engineers. This is clearly a pattern for the future.

Daini Seikosha set about automating assembly of mechanical watch 'movements' – that is, the mechanism – in the 1960s.

The first job was to develop a system suitable for conveying small components through the plant. Requirements were divided up into 11 different projects. Only when one was under way was the next started.

Automatic assembly

Following development of the conveyor system came work on the multi-station machines for assembly, sub-assembly machines and inspection equipment. By that time, general principles had been established so a programme of parts standardisation was started.

Only after this had been progressed did the team set about co-ordinating the projects into one system – System A. In fact some parts, such as the conveyor system, were developed by about 1972 but it was some years after that when the complete system started to operate fully.

Of course, one of the major problems with the automatic assembly of a mechanical watch movement is the small size and flexibility of the parts. These comprise tiny gears, spindles, a ratchet, a spring and hairspring, about five small pressings, the relatively large flyweight and bearing for the automatic rewind, and finally the date ring.

In addition, there are the tiny hands. What is more, to maintain accuracy, many of the spindles run in synthetic jewel bearings which are very hard and which must be inserted with great precision.

On the other hand as with most assemblies there is a convenient starting point. In this case it is the base plate which is a fairly large circular plate that can be located easily in a pallet. The combination of that base plate and the pallet forms the basis of System A.

Daini Seikosha set the following targets when designing the System A equipment:

○ An uptime of 80%;
○ A cycle of 4-6s to produce 100,000 movements/month on an eight-hour shift;
○ A reject rate of less than 1%;
○ A 75% reduction in manning level from 90 to 22.

In fact, the output was gradually increased till it reached 20,000 units/month, but with the trend towards the use of quartz systems, production has since dropped back.

Saving space

To save space and create a simple method to carry buffer stocks, the assembly machines are arranged in spurs off the main track. In the main sections of the track – Loops 1 and 2 – there are 15 spurs in all. There are 9 in Loop 1, and 6 in Loop 2.

The sequence of operations is:

□ assembly of escapement, balance and train wheel;
□ set jewels in the base plate;
□ fit the day and date ring for the calendar mechanism;
□ install the dial, hands and finally insert the movement in the casing.
□ A subsequent 'test' section preceeds final inspection.

There is space for buffering along the main line but, as a bonus, a large buffer store is provided for complete movements – that is after the day and date rings have been inserted.

Although this system is based on hard automation some flexibility has been built in. First, it was decided to produce three different models of the same basic movement on one line. Secondly, full computer control is used. A large Toshiba Tosbac computer controls the complete system, and to cope with the variety, each pallet has a number of magnets in its base.

Each pallet is a plastics moulding. On one side is a row of six holes in which magnets are inserted. Since either the positive or negative pole can be facing outwards, several different combinations can be used.

Sensors read these magnets before each station is reached, identifying the type of movement passing down the line.

There are up to three machines at each station and, according to the signal received from the pallet, two of the machines are locked out of operation. An advantage of this duplication of machines is that minor faults can be corrected without the need to stop the line; but investment in machines is clearly greater.

Different machines

For System A, Daini Seikosha engineers devised a number of different machines,. First, there is an in-line machine which can assemble a number of parts – usually around 10 – in sequence as the pallet proceeds down the line. Then there is the centre column type of rotary machine used to carry out the longer operations, followed by the special in-line jewel setting or jewel installation machine. Finally there is a multi-station rotary machine that can be used for manual or automatic assembly. This can have around 15 stations, but with manual operation, there are only one or two stations at which operations are carried out. With automatic operation, more operations can be carried out, and it is also possible to combine

Fig. 2. General view of spur lines.

automatic and manual operations at one table.

Daini Seikosha has also developed a two-station machine for rivetting. Of course, the actual rivetting station needs to be built to take substantial loads.

To transfer parts into the system, Daini Seikosha uses hopper-fed vibratory bowls with chutes that are diamond-ground to ensure that the tiny parts maintain their correct orientation. Another key feature of the complete system is that machine movement is designed to give smooth acceleration and deceleration, so that vibration is kept to the minimum.

Lot sizes are generally of around 1,000 units. At the station before the line, watch base plates are bonded into the pallets and the pallets are then stacked. A stack of 60 pallets, each pallet measuring 60 × 60 mm, is mounted in a container, which takes the form of an open tower. Many of these containers are held on the conveyor as a buffer store at the beginning of the line, and when the container reaches the line the pallets are pushed out one-by-one on to the assembly line. Throughout the line, the pallets are very close to one another. At the end of the sub-line, the pallets are fed into the container to proceed to the next sub-line.

Burr removal

At each station, the parts are fed from the vibratory bowl through the chute to a pick-up station where a simple arm uses suction to pick up the part and transfer it to the sub-assembly.

Daini Seikosha found that an important prerequisite to the success of these operations was the removal of all burrs; and the tightening up of manufacturing tolerances. In some cases, small details, such as chamfers, had to be changed as well.

Although this might have been expected to increase the cost, the reverse was the case. The reason for this was that tolerances were not just reduced; the machines were revised so that more precise parts were produced, with the aim of reducing the reject level.

Using this approach fewer problems of misfits were encountered. This led to a reduction in rejected sub-assemblies with the knock-on effect that fewer operators were needed to rectify the rejects. Also, fewer operations were wasted in building rejects.

Not every operation in the system is automated. For example, because the hair spring is so delicate and flexible, it is difficult to handle mechanically. Therefore, it is trued manually.

The other major manual operation is rectification. Initially one person working on almost every sub-line is doing rectification. Basically, each sub-line is U-shape, with the assembly machines arranged on the line running away from the main track leaving the line running back to the track as a return line.

However, the last machine on the assembly side is an inspection machine, and parallel to the return line, there is a sub-line. Rejects are transferred to this sub-line automatically. Here the operator repairs the movements as necessary, and then returns them to the return line. One consequence of this arrangement is that the movements may end up in random order, depending on the number of rejects, and the time the operator takes to make repairs. To cope with this random order, of course, the magnetic sensors are used.

Fig. 3. Close-up of the line with its vibratory blow feeders, tools and automatic lubrication.

Test circuits

When the watch is complete, it passes to the test circuit. This is a long chain that runs around a tortuous path for 1.6 km (1 mile). The watch remains on this chain for 48 h, and at each turn, the winding flyweight is moved. In this way, the test track is said to resemble the movement of the wrist of the wearer.

During the test route, the watch is subjected to air pressure to simulate a water test. There is a master clock in the system to monitor the ability of the watch to keep accurate time, and the reserve power of the mainspring is checked.

When the line was planned, only large process computers were available, but nevertheless, Daini Seikosha decided that such a machine should be used to control operations. It is arranged to control the operation of the lines, quality and outputs. Thus, in response to input data, the models to be produced and the appropriate cycle time are determined, together with the machines necessary at each station – machine selection results from sensing the magnets on the pallets, of course.

The results of the automatic inspection are fed back to the computer, and if several successive movements are defective, either a warning is issued to the operator on the sub-line, or the line is stopped – this depends on pre-set circumstances. In addition, the data are fed back to the previous section, to indicate drift from standards.

Daini Seikosha claims that it met its major objectives with this system, and that the manning level was reduced by 66% – from 90 to 30 people. In addition, the total time taken to produce a watch has been reduced by 11%.

However, since the system has been operating, Daini Seikosha has found it possible to reduce the manning level further; now only 10 people work in the shop. There have been two reasons for this: first, as the reliability of the machines has improved so fewer operators are needed to watch for stoppages. Secondly, reduced rectification has been needed. It is important that the company realised that once the system was installed, there were opportunities for continuing to reduce manning levels – this is an essential technique in increasing productivity.

But of course, in the six years, the sales of watches with quartz movements have rocketed. In 1974, 17 million of 18,500,000 watches produced had mechanical movements. By 1979, only 8,300,000 had mechanical movements, whilst another 12 million analogue quartz and 5 million digital quartz units were produced. And the trend is still moving rapidly to quartz watches with digital displays.

To accommodate this Daini Seikosha installed System B to assemble quartz movements at its subsidiary, Morioka Seiko Kogyo. This assembly process is similar in principle to System A, except far fewer opertions are required. The number of sub-assemblies is approximately in the ratio of 3:2:1 for mechanical, quartz analogue and quartz digital watches respectively. In the case of the digital quartz watch there are only around 25 sub-assemblies.

At Morioka Seiki, there are five lines producing quartz movements, and only one operator is needed on each line. The other fundamental difference is that each line is controlled by microcomputers.

Daini Seikosha is now developing its System C, intended to give it flexibility of operation, with some programmability. In this way, it hopes to be able to cope with changes in the market without the need to continually change plant. And since watches are now becoming fashion accessories, this flexibility is becoming more important.

Automatic assembly of LPG valves

Precision and automation indispensible to the watch industry can spin off into other industries. A manufacturer of LPG hand valves is using the technology evolved by the Citizen Watch company.

CROSS-FERTILISATION of technology between different industries can have a significant effect on productivity. In Japan, the watch manufacturers have automated in a big way, and have then made machines for other industries.

Following its automation programme, Citizen Watch set up companies to make NC bar auto lathes and assembly machines. One of the companies that installed a Citizen automatic assembly line was Hamai Industries, a manufacturer of valves for LPG and other gases. Hamai was established fifty years ago, and twenty years ago moved from the middle of Tokyo to a new factory in the western suburbs. Since then it has built a smaller plant on the other side of Tokyo.

At that time, as now, the main products were hand-controlled LPG valves. These are screwed into the top of the gas cylinder, and then turned by hand to allow the gas to flow. There is also a built-in pressure relief valve. Hamai's chance came ten years ago when its main rival became embroiled in a major dispute with the unions and went bankrupt. Hamai decided to automate assembly and machining to take advantage of the increased market. In fact, it tackled the machine shop first, and then installed one automated line for assembly. But recently, a more modern line built by Citizen was installed.

Taxis

Last year Hamai built 7 million valves, which included about one million ball valves and valves for high pressure gas. The remainder were for LPG. The reason that the market for LPG valves is so big in Japan is that about 50% of the homes depend on LPG gas supplied in cylinders. In addition, all the taxis and some small commercial vehicles run on LPG as well. Therefore, the total market is about 10 million valves a year in Japan, of which Hamai has over 50%. It also exports quite a number of valves, some of which are bought on an OEM basis by companies such as Air Products.

The valve itself is based around a cross-shape brass forging, the main stem being threaded for insertion into the gas tank. Opposite that end – on the top, of course – is the handwheel which actuates the valve itself. One of the horizontal arms of the X is the outlet, which is threaded, and opposite that is the relief valve. The pressure relief valve takes the form of a spring-loaded disc retained by a threaded cap. Once the pressure is set, the cap is pinned in position.

The valve itself consists of an upper and lower part, there being a nylon seat in the nose of the lower part. The upper part is screwed into a sleeve which is screwed into the main body. There is an O-ring in a groove in the lower stem which seals the valve to the body.

In the Citizen Watch equipment, there are four rotary assembly machines, and a walking beam pressure setting stage. These are arranged like this:

Stage 1: Rotary machines to assemble valve stem (two in parallel);
Stage 2: Rotary machine to assemble upper and lower stems together, and spring and relief valves;
Stage 3: Linear walking beam machine to adjust relief valve;
Stage 4: Rotary machine to assemble handwheel and nameplate.

The machines are arranged to

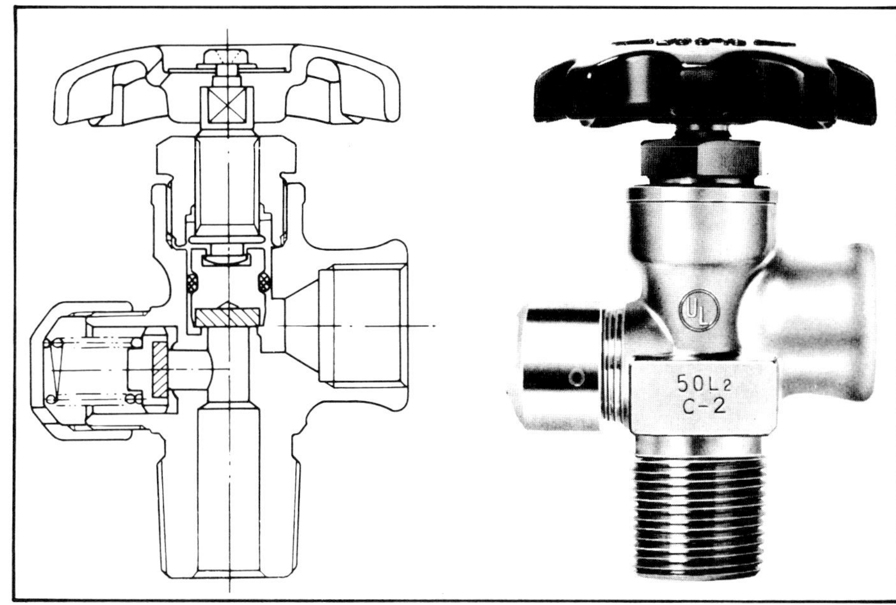

Hamai Industries' main products are LPG valves which are assembled automatically.

Hamai's automatic assembly line for LPG valves consists of three rotary stages, and one linear stage.

operate with a cycle time of about 4s, to produce 10,000 units/day on two shifts.

Transfer arm

At the rotary machines, most components are delivered from vibratory bowls through chutes to the work stations. To start with, a cam-operated arm transfers the valve's lower stem to the first station of the rotary machine. In fact, to suit the feeding system – a gravity conveyor, with the stem supported on two rails – the arm picks up the stem, turns it around, and then lowers it vertically on to the machine.

Just prior to this stage, an O-ring is lowered on to a spigot adjacent to the location for the lower stem. Then, a spot of grease is applied to the stem at the next station. After that, the O-ring is picked up on another spigot, and this spigot then lowers the O-ring into position in the groove of the lower stem body.

At the next station, the stem is placed next to the nylon sealing washer, and subsequently installs the washer. Meanwhile, in another rotary machine, the upper stem has been assembled. The operations are similar, but in this case, the stem is spun into the threaded sleeve.

At the first stage of the second rotary machine (Stage 2) the two parts of the stem are assembled, the lower portion being spun over the end of the upper stem.

Then, the main body joins the system, so that the relief valve can be assembled, and the stem assembly inserted. Before the stem assembly is inserted, a blob of anaerobic adhesive is injected at the threaded body. After the stem has been inserted, the body continues to move around the table – there are 14 stations in all – while the valve, spring and cap of the relief valve are inserted.

When complete, the assemblies are removed from the table, and fed along a gravity conveyor to the next machine, which is linear. A row of 16 valves is needed, because this operation takes a full minute. In practice, there is one row of 16 valves going through the machine, another waiting to go through, and a third row being formed.

Walking beams

The machine has two walking beams connected together, so that two rows can be transferred simultaneously. Thus, as it transfers one row of valves from the setting station to the transfer station, the next row is moved from the waiting station to the setting station.

The machine itself is computer controlled, air being applied to the valve, and an adjuster then turning the cap until the correct pressure is obtained. There is a pressure gauge above each valve so that the operator can see that the pressure has been set correctly.

At the final rotary machine, the body is inserted vertically, at the first station, so that the handwheel can be dropped on at the second station. Then, the nameplate is lowered into position, followed by a screw which is then tightened up. Next, the relief valve receives attention. First, a small hole is drilled through the cap into the body, after which a self-tapping screw is inserted, and its tightness checked. The valve passes a photo-electric cell which monitors that everything is present. After completion, the valves pass to a visual inspection station, where pressure settings are measured.

Hamai still uses a manual assembly line for the lower-volume products, this taking the form of a carousel shaped conveyor, at which a number of operators sit. Hand tools are suspended from rails above the line.

Redundancies

Although Hamai is a small company, it invested heavily in automatic assembly because it saw a big chance in the market – if it could reduce prices – to take over the share of its previous main competitor. For all that, things did not go well; it had already ordered

Manual assembly, previously used for all valves, is still the technique where volumes are small.

the machinery when the energy crisis knocked sales for six. It was forced to make 30% of its staff redundant, which to some extent was fortunate, because once that automation was installed it would not have needed those people anyway. Today the company employs 250 people, with 35 people/shift as direct labour in the main valve plant.

Saburo Hamai, managing director, says that the use of the automatic assembly line has reduced the manning requirement from 100 to 20 people per shift. At Hamai, average wages are around £7,000 p.a., so the annual saving of the line is £1.1 million, and even in Japan, labour costs are increasing at about 7% p.a.

Hamai says that they aim for an uptime of 80% with all their automated equipment, and to make this possible, major maintenance is carried out at the weekends. Obviously, minor jobs are done during the day, but the aim is to get these done before the shift starts, or during the lunch break.

Of course, the investment in this line was substantial, but Hamai was able to raise the money because it was well established and has a policy of keeping its borrowing level low – many Japanese companies operate mainly on borrowed capital. It is also a rather old-fashioned company in the way that it operates.

The company president spends two days a week in each factory, and one day at head office, while the other main director is responsible for sales. The president is usually to be seen checking the quality of the product, or supervising the installation of some new machines. Thus, Hamai prospers because it has a sound product range, an integrated manufacturing system, and management that knows exactly what is going on, and what needs to be done in the plants.

Assembly robots coming into focus – but not for every job

The role of robots in automatic assembly as viewed by industry in Japan, including some of the major robot makers, is reported.

SINCE IT IS now being recognised that robots, even without vision sensing, can be used for a variety of assembly operations, there has been a quickening of the pace in commercial deals here. Fujitsu Fanuc, intent on relying more heavily on robots for future growth, has been busy arranging deals for overseas sales. Of course, the most dramatic of these has been the proposed joint venture with General Motors.

More startling, perhaps, was the announcement that IBM was to sell the Sankyo Skilam robot in the USA as the IBM 7535. The Skilam, of course, is one of the Scara group of robots developed at Yamanashi University by Prof. Hiroshi Makino for assembly. Makino claims that the Scara can be used for all 'down to up' assembly tasks, which he says, account for 80% of all existing work.

In addition to Sankyo Seiki, whose main product incidentally is music boxes, NEC, Nitto Seiko, Yamaha Motor and Pentel provided funds for the development of the Scara. Then, they designed their own units around the basic concept of an assembly robot that has some compliance built in; basically, so long as the object being inserted in a hole strikes the chamfer of the hole, and not the flat surface, it will be inserted successfully.

However, only Sankyo decided to incorporate the motion curves developed by Makino to reduce vibration at the end of the movement. These are polynomial curves, the deceleration period being twice as long as the acceleration curve.

The Sankyo Skilam is a sound design, with a stiff arm pivoting on the pillar. The basis of the Scara is that the arm has a shoulder and elbow joint, both of which pivot on vertical axes, while the normal tool can rotate and move vertically. There are two models, both of which can have three or four axes of movement. The larger one, SR-3 has upper and lower arm lengths of 400 and 250mm, while the SR-4's dimensions are 200 and 180mm.

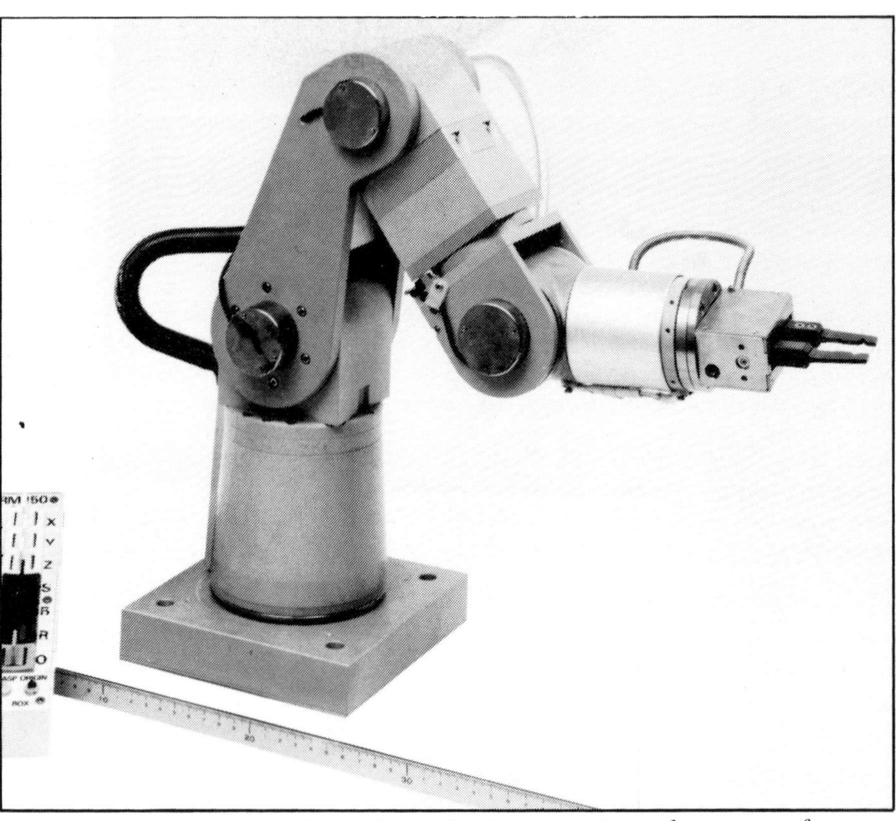

Fujitsu's Micro Arm 150 robot is claimed to have a positional accuracy of 4 microns.

Owing to the use of the polynomial acceleration/deceleration curves, the Skilam can operate quickly, a 3s cycle being normal. Makino claims that the Scara robots are faster than the Fanuc assembly robot, and in practice as fast as the Puma, and in Japan, at any rate, they are about half the price of the Puma – about £12,000. In the past year, almost 400 Scara robots have been built, 200 by Sankyo, over 100 by Nitto – mainly for tightening screws – and about 80 by Yamaha.

So far, one of the biggest applications of the Skilam robot is at one of the factories of Pioneer, the audio company. It is using eight to insert components into printed circuit boards in preference to the NC devices used by most manufacturers, and produced by Matsushita, Fuji and TDK. The claimed efficiency of the Pioneer line is 99.5%.

Pentel has so far built very few, but is using these for a demanding job. The robot picks a conical rubber moulding, approximately 2mm in height from a moulded sheet of 11 by 11in parts and inserts it in the tip of a clutch-type pencil. The cycle is 2s, and the robot is operating 24h a day.

Hirata has also started to build

Scara-type robots, but these are made under licence, and other companies in Japan are evidently following suit.

Demanding

Fujitsu, anxious to use robots in assembly of electronics parts – it has already announced that it plans to use robots on assembly in a new factory for small computers and peripherals – has developed a robot with a claimed accuracy of four microns.

Controlled by a 16-bit microprocessor and four arithmetic data processors, the jointed arm robot is moved in small steps. Fujitsu claims that with this arm, the problem of 'offset' has been overcome. In this phenomenon the servo motor does not overcome the frictional force and the inertia of the arm, but the controller normally assumes that it has.

In the Fujitsu Micro Arm 150, the movements involved as the joints in the arm overcome the friction, inertia and backlash are measured by a rotary encoder. The computer command and the signals from the encoder are compared and integrated. The integrator then increases the movement of the arm to allow for the offset.

The robot arm is actuated in steps of four microns in the Cartesian mode, and it is the size of the step that governs the positional accuracy. Maximum speed of operation of the robot is 1m/s.

The arm itself has six degrees of freedom: it can rotate through 300° on a vertical axis on the base; the shoulder can rotate through 160°, the elbow 170°, while the wrist can bend through 180°. The wrist can rotate through 330° in one axis, and through 300° in another.

Of course, this is a small robot, upper and lower arm each being 150mm long, while the wrist is 111mm long. The arm can turn back right over the pedestal, and the wrist can reach 590mm above, and 140mm below the bottom of the base of the robot. Maximum reach is 411mm.

Music

However, robots are not the solution for all assembly operations. For example, Sankyo Seiki, despite the fact that it makes the Skilam robot, uses no robots in the manufacture of its music boxes. These are produced on assembly lines that are largely automated with cycles of 2-2.5s.

Takashi Kumagai, president of Shinko Engineering Research is one automation specialist in Japan who is not convinced of the merits of robots in automatic assembly. He says that in most applications, the variety involved does not justify the use of robots. 'It is much cheaper to use three of four arms to deal with variety than one robot,' he said.

That is perhaps a heretical view at present, in Japan at any rate, but there are still hundreds of products where variety is small, and lives are long. For example, Shinko is just building a set of eight automatic assembly machines to assembly focal plane camera shutters. The cycle is 5s, and each line deals with only one model, or subassembly, so there is no need for a robot as such.

So far, Shinko has built the first line, in which the spindle and spring are assembled. Of course, these are tiny components, the spindle being about 1mm diameter by some 50mm long. A retaining collar, long coil spring and small washer are asembled to the spindle. Two depressions are formed in the collar to form a location in a hole in the spindle. This collar then supports the spring.

In this line, the sub-assembly is carried on a pallet which is indexed along the line. At each station, a horizontal pin engages in a V-notch in the side of the pallet to ensure location within a tolerance of 0.05mm. The pallet is a simple plate in which two holes are bored, and the spindle is carried vertically standing in one of these holes.

At the first station, the spindles are fed from a bowl feeder along a track. The spindle is moved on to a turntable, which rotates through 90°, so that the spindle is parallel with the track. Then, the spindle is picked up by a gripper on a lever, which turns through 90°, so that the spindle is vertical. It is then picked up by an arm, which transfers it to the pallet. This arm is the standard Shinko cam-operated pick-and-place device.

Transverse

At the second stage, there is an unusual and interesting operation. The transverse hole in the spindle is aligned. To do this, the spindle is rotated, at about 2 rev/s, while a light is directed at the position of the hole. When the hole is aligned correctly, the light passes through it and strikes an optical sensor. Then, the spindle is stopped.

Of course, there is a minute delay between sensing and stopping, but this can be allowed for in the jigging. At the next station, two probes come into the hole from opposite sides of the track, to ensure that the hole is present, and is in the correct position. In fact, the probes are pins with conical ends which are shaped so that as they enter the hole, they straighten up the spindle.

At the next station, the brass ring is picked up pneumatically, and is dropped over the spindle. Then, the spring is dropped on, but since it is gripped at the top end, only the bottom end is fitted over the spindle. In the subsequent operation, a chuck grips the spring, and spins it down fast into position. The spring is of small diameter at the lower end, and then it is stepped out to the full diameter about one-third the way up. So, in this operation, the spring is spun down into the collar.

Only then is the collar staked in position, a pair of arms coming down to form depressions in the collar to lock into the hole. Afterwards, the small washer is dropped on to the spindle, but since there is a hook at the top of the spring, the washer often gets stuck. Thus, at the next station, an arm grips the spring lightly, and orbits it around the spindle for one revolution so that the washer can pass the hook.

Up to now, the spindle has been mounted in the first hole in the pallet, in which it is supported on the end of the spindle. Incidentally, when precise location is needed, the spindle is gripped at the top as well. But at this station, the spindle is picked up and placed in the second hole in the pallet, where the spindle locates the shoulder of the collar. At the following station, a check on the strength of the collar staking is made. This is a simple operation, a head coming down and exerting a light force on the top of the spindle. A digital readout indicates whether the spindle has moved; if it has then the joint is suspect.

In the final operation, the subassembly is transferred to a magazine, simply a strip with a number of holes. The magazine is indexed along parallel with the line in sequence so that there is an empty hole adjacent to the station.

This line has no need for provision for flexibility, although the basic elements of pick-and-place arms, the optical sensing of the hole position and pallet carriers could be adapted to give flexibility. But that would increase cost, and in many applications such as this, it is preferable to stick with conventional automation. However, as the price of robots decreases, so this situation can be expected to change, but still not for every product.

Automation keeps music boxes turning out

Sankyo Seiki relies on many automated lines to produce music boxes at the rate of one million a week.

IT HAS taken Sankyo Seiki of Suwa-shi, central Japan, 30 years to build up music box production from nothing to 50 million units a year – and that represents almost 75% of the world's market. With that sort of volume, automatic assembly is inevitable, but so far, Sankyo Seiki has shied away from the use of robots or even a very flexible system. The main reasons are that the music box is currently a mature product, and the cycle times are short – around 2s.

Apart from its music boxes, Sankyo Seiki manufactures small machine tools and the Skilam Scara-type robot. Its factories are in Nagano Prefecture, roughly in the middle of Japan, near the Japan Alps. At Suwa-shi, where the music boxes are produced, there are several companies involved in precise assemblies, such as Seiko, Olympus and Yashica.

The start of the No. 4 music box line at Sankyo Seiki, with a hopper feed to a vibratory bowl and conveyor tower. At the next station can be seen the stacks of trays containing pedestal assemblies.

Precision

A music box may seem a crude and unimportant product, but despite the fact that most of them are used in toys, their manufacture involves a lot of precision. There are about 24 components in the typical movement of a music box, including a pressure diecast base plate. The components include a spring and clockwork mechanism of plastics gears, an air damper, and the drum and vibration blade.

The drum is typically 75mm long by 20mm diameter, and from its periphery there are a number of small pins projecting. In fact, these are extruded from the surface. Then, there is the vibrating plate, which is slit to produce a number of fingers which are struck by the pins to produce the notes. Sankyo Seiki produces movements with a variety of 1,700 tunes, and in fact, to keep the size to the minimum, each vibration blade contains only enough fingers for a particular tune. However, fittings of the blade to the base are standardised.

According to Hideo Takei, manager of production control, the OEM price of the standard movements is only 50p-75p, yet they include the sort of precision found in watches. For example, the backs of the blades are ground on special fixtures to a 'tolerance of microns' to obtain the desired sound. Takei points out that as a result of the use of automation, the real cost has fallen dramatically. 'Fifteen years ago, the music box cost

The first station, where the base is dropped on to the conveyor, and then the gear/spindle assembly is mounted.

the same amount as a haircut, and 10 years ago it cost the same amount as a bowl of noodles, but now it is much cheaper.' In fact, in Japan, a haircut costs around £3, and a bowl of noodles around £1.30.

Sankyo Seiki started to automate production in the late 1960s, and has recently been refining the assembly automation, which is used to do all the work on the final assembly line, except inspection and palletisation. About 300 people are employed in music box manufacture, including inspection. Manufacturing operations include pressing, machining and assembly.

On the assembly lines, the following parts are assembled to the base:

- a plastics gear/spindle assembly, mounted vertically;
- a spring and cap, mounted above the gear/spindle assembly;
- an assembly of a pressed pedestal carrying a drive gear and air damper;
- the drum and its spindle;
- the vibration plate.

Prior to assembly, some sub-assemblies are built up manually, although the plastic gear and spindle part is assembled in the press shop. The gear and spindle come in from one side of the press, while the strip is fed in from the other. The strip is pressed out, and then the pressing is pushed over the spindle to act as a locking washer, the completed sub-assembly being ejected from the press at the rate of 55/min.

In the assembly shop, which is on the first floor, there are 10 lines working in parallel. Most of these are used for the standard movements, some being reserved for the larger non-standard units. In all cases, though, the concept is similar.

The spring is wound into the cap on a rotary machine.

At the beginning of the line, an operator tips diecast bases into a big hopper, from which a drag conveyor takes them up to a vibratory bowl feeder. In the bowl, the bases are orientated, and fed into a slot running around the periphery which is shaped so that only correctly orientated bases are accepted.

The bases are fed into a tower which raises them to the line. Inside the tower is a chain-driven conveyor, the vertical chain carrying a number of flaps which are inclined downwards from the chain at about 30° to the horizontal. Adjacent to the track there is a window in the tower, and as the chain passes the window, so the base slides off its flap through a chute on to the conveyor. The conveyor is of the continuous belt type, the base being stopped and located at each station.

Vibratory bowl feeders are used to supply simple parts to the line, such as the spring cap, but trays are used to carry the more complex sub-assemblies, such as the pedestal assembly. At the first assembly station on the line, the gear/spindle part is rolled down a chute, picked up and deposited in the base. A simple arm, which rotates on a pillar is used to pick and place the gear. This type of arm is used at most stations. At the next station is a probe which checks that the gear has in fact been installed. There is a sensor of some sort after each assembly station, these usually taking the form of simple probes.

At the second station, the cap is lowered over the spindle/gear, without the spring. Subsequently, there is an eight-station rotary table machine at which the drum and spindle are inserted. After the drum is dropped in position, the spindle is inserted, and finally a small grub screw is fed through a tube by compressed air, this is tightened to hold the spindle in position.

At the next stage, there is another rotary machine, which has four stations, to wind the spring into position. The spring steel is supplied as a strip, it has holes already puched at the points where the strip is to be parted. The hole is formed so that when the strip is cropped off, it provides the end tags.

After the length of strip has been parted off, it is fed through rollers into the slot in the cap. Meanwhile, the base is fed on to the rotary machine and placed in the fixture, so that the vertical spindle projects through to beneath the fixture to be gripped by a driver. The strip is pushed into its location on the spindle, and the driver

At the end of the line, a girl listens to a note of the music box to identify it before palletising.

starts to rotate the spindle. As soon as rotation has started, the table indexes round one station, the driver continuing to coil the spring on to the spindle at the next station. At the fourth station, the sub-assembly is returned to the track.

Buffer store

Here there is a buffer store in the form of a duplicated track. The buffer section takes the form of a bridge, with one track above the other. One section of track is in operation, while the other section is acting as a buffer, either above or below the operational track.

Next comes the all-important section where the vibration plate is mounted. Since the operations take 4s, there are two lines in parallel. At the first station, the vibration plate is pushed into position from the base of a vertical magazine. At the second station, two set screws are passed by pneumatic feeders from a drum to automatic screwdrivers, which insert them.

At this stage, the complete assembly passes from the automated shop into the inspection area but still remains on the same line. Just inside the shop is an operator who checks the height of the vibration plate, and adjusts it as necessary. Then, the movement passes on to a rotary mechanical inspection machine, which winds the spring up, and ensures that the drum rotates – a simple functional check.

Right at the end of the line, there is one more operator who not only places the movements into boxes, but also checks them audibly. She picks up each movement, and places it against her ear to identify the tune; she can to this instantly, which is just as well, as she only has 2s in which to do so!

The simple music box showing the vibration plate, drum and cap containing the spring.

At the end of one line, a Sankyo Skilam robot has been installed to palletise the assemblies. But said Motonori Usui, manager of product engineering: 'Because the woman does more than just palletise the music boxes, she is far better than a robot. It would be too expensive to make a machine to recognise the tunes as well as palletise the movements.'

Takei claims that uptime on the assembly lines is about 94%, and that major rejects – such as with a part missing – occur only once in every 10,000 assemblies. Each line is used to process several different assemblies, and to cope with the variations, there are a number of spare stations.

To change from the assembly to another, it is necessary for the operators to change over a number of switches, and to supply different parts. 'The change is made in sequence down the line, so the conveyor is usually empty for about three minutes,' said Usui. Usually, about 250 different batches are produced monthly, the average batch being 5,000.

The main reason that this line is based on hard automation is that it was designed some time ago, and that the cycle time is so short. But there is another reason; although Sankyo Seiki claims to have 74% of the world's market, demand is changing fast. There are already small electronic devices with ceramic speakers that can produce a similar range of sounds as music boxes. Of course, these rely on differences in software to produce tunes, whereas the music box relies on hardware changes.

Sound quality

It is true that at present, these electronic devices cannot get near traditional music boxes for sound quality, but they are likely to be so cheap that they seem certain to displace the conventional music box. Needless to say, Sankyo Seiki is already producing some of these electronic units, although it is also looking towards the developing nations as new markets for its conventional products.

Robots start to assemble watches

Although it relies on normally automated lines for the bulk of its watch assembly, Seiko is now beginning to use robots.

FOR 20 years, the Seiko watch group in Japan has been working on automation of the watch making process, but it is only just beginning to install robots. With government backing, it spent 10 years developing the 'A' and 'B' systems of automatic assembly of watch movements. Now it is turning to robots to carry out key operations in case assembly – where the movement, face and hands are installed in the case. But hard automation remains the backbone of the newest assembly lines.

The Seiko Group consists of many companies, such as Daini Seikosha and Suwa Seikosha. Within Suwa Seikosha is Shiojiri Kogyo, which is at Shiojiri, on the edge of the Japan Alps. Shiojiri Kogyo makes watches with Alba and Pulsar brand names, at the rate of around 8 million a year. In 1982, it built a new assembly shop, with all assembly operations carried out in one large hall, separated into different shops by partitions. Previously, assembly took place in several shops which were in various parts of the plant.

Improved layout

Yoshiaki Nakamura, general manager, technical department, at Shiojiri Kogyo said that in laying out the new line, the aim was for 'the most efficient line, one of the best in the world.' In fact, he says that the biggest gain in productivity has come from the improved layout, rather than the automation equipment. 'Now that everything is in one shop, handling time is cut to the minimum, and we have been able to reduce work-in-progress.'

In fact, one reason that work-in-progress is low, is that Seiko's version of the Toyota Just-in-Time system, AJ or 'Always Just' (in time) is used. 'There is no surplus stock,' said Nakamura. 'Between each department there are containers for one day's production only. The preceding department must have authority to produce before it can make some more.'

In the new temperature-controlled assembly shop, the preliminary operations take place on groups of machines that are at right angles to the assembly line itself, and separated from it by a partition. Included in this section are coil-winding and preparation of the integrated circuit, which is purchased from companies such as NEC and Toshiba.

Coils are wound from 19.5μm diameter wire on rotary machines which have eight stations. Three vibratory bowl feeders supply the parts to the winding machine, which winds the coil, assembles the cores, and welds the end of the wire in position. Since this process is fairly slow, the coil winding machines work 24h a day.

In the same shop is the ic preparation line. The actual ic is assembled to a substrate which is then bonded to a reel. Complete reels are then transferred to the assembly shop, which is separated from this preliminary shop by a partition.

Assembly of the watches is divided into two main sections. First comes the assembly of the movement, which is almost totally automated. Then, there is 'case assembly', which actually involves fitting the glass, face, hands and movement to the case. Assembly here is almost entirely carried out by girls, but robots are now being developed to do some jobs.

Movement assembly is carried out on a straight track, with some 20 stations in a line 30m long. There are four lines working in parallel, and Nakamura said that in 7.5h, each produces 10,000 units, implying a net cycle of 2.7s. 'The lines operate on only one shift,' he said. Each of the four lines is in two halves, the second half including manual inspection.

The movement is built up on a plastics pallet which travels down

Fig. 1. Vibratory bowl feeders and simple transfer arms are used for automatic assembly of watch movements at Shiojiri Kogyo.

a continuously moving non-synchronous line. When the pallet arrives at a station, it is stopped, and locked precisely in position. If there is a hold-up on the line, other pallets built up behind it.

In the first section of the line, there are a couple of girls loading the line, but otherwise the line is 'unmanned'. Each station follows a common pattern: parts are fed by vibratory bowl feeders to chutes while arms pick the parts up, and insert them in the movement.

The arm resembles a simple robot, and like the other equipment in the line is actuated pneumatically. However, it is not a robot, since it can only make one series of movements. It can move the tool that holds the part vertically and can turn through about 90° on a vertical axis. Thus, the arm picks up the part, transfers it to the line, and lowers it into the movement. At some stations, there are three bowls and three arms. Of course, all the parts, such as plates, the ic and gears are small, typically less than 5mm square.

Where the gears are fed to the line, a special feeder is interposed between the bowl and the chute to ensure that the gear wheels remain vertical – that is with the spindle hole horizontal. As it leaves the bowl, the gear is fed into a slot so that it must be vertical, and then it follows a path of a 'figure of 8' before the straight part of the chute which goes to the pick-up point. The chute is basically C-section, with the opening on top.

By the time the movement has proceeded halfway down the line, it can be operated by a slave system, and there is a manually controlled function check. Subsequently, the ic is pressed out from its reel, and is transferred in two stages by a pair of arms to the movement.

Afterwards, the movement is inverted and transferred to another pallet so that assembly can take place from the other side. One complication at this stage of the line is that some models have date mechanisms, and some do not. The stations used to install the mechanism for the date are idle when plain watches pass through.

When the movement is complete, it is picked off the line and placed in a pallet by a Seiko robot, this being the first application for the newly-developed SCARA-type robot. There are 6 × 9 rows in each pallet, and the robot places the movements in sequence to fill the pallet. Then, the complete pallet is transferred to the line where the batteries are installed, and then the casing is built up.

Fig. 2. The second half of the line involves some manual operations.

Fig. 3. One robot is used at the end of each line to palletise the movements.

Movement

Shiojiri Kogyo employs 670 people, including 350 staff, and 250 on assembly. The movement itself is now assembled 90% automatically, and in the last two years, productivity has been increased by over 20%, Nakamura claims. 'This is mainly because we have less production control people now that the AJ system is in use,' he said. 'In the future, we must use robots, though,' he said. 'For example, with hard automation we cannot palletise, as the arm can only move from one fixed point to another.'

Therefore, Suwa Seikosha set about developing its own robots, under the direction of Kazuo Abe, assistant general manager, production engineering. Abe said that the SCARA type was chosen because 'it is cheap, and can do 70% of ordinary assembly.' He added that the accuracy is better than that of the Puma robot as well. Suwa Seikosha has just started to sell the robots in Japan, at prices from Japanese Yen 3.5-5.0 million (£10,000-14,000).

By the end of the year, three or four robots will be working on each of the four lines at Shiojiri Kogyo, for palletising and for insertion of the hands in the faces. Thus, in all there should be about 20 robots in use. But it is a slow business, since each new application requires a lot of development.

Robot assembly machines

Toyoda Machine Works builds two assembly systems around its new robots.

BRUSHING conventional approaches aside, Toyoda Machine Works has developed its own robots and straight away built some assembly lines around them. One of the lines, which incorporates 15 robots and an assembly machine, is designed to assemble vane pumps for automotive power steering systems. The other is a general-purpose assembly machine for small components, but Toyoda has configured the first one to assemble flow control valves.

Fig. 1. Toyoda's assembly machine for small components incorporates three robots.

Cycle time

Both systems operate to cycle times of under a minute, and are notable for the relatively simple control systems. In addition, the general-purpose machine is compact, and is unusual in that the robots operate the parts feeders themselves.

First, of course, Toyoda had to design some robots, and it was able to draw on its experience as a major contractor for the assembly system for the 'FMS with laser' project sponsored by the Japanese government. In the spring, it announced a range of three basic models aimed specifically at assembly work. They are: an articulated arm type (RA); a modified Scara type (RC); and an overhead Cartesian co-ordinate type (RR). There are several variations, with a RA4 articulated arm unit being built in sizes with 2.5 and 6kg capacities. It can be mounted upright, or upside down on a gantry, as it is in the small assembly system.

The smaller RA4-1 has both upper arm and forearm lengths of 335mm, while both arm sections of the RA4-2 are 500mm long. These machines have four axes of freedom, both arm joints being able to articulate through 90°. There is also the RA6-2, which has two extra axes of freedom at the wrist.

All three RA robots, and the RC Scara type, have quoted repeatabilities of ±0.05mm, which is sufficient for most assembly jobs. However, the RA models are designed to operate faster than the other designs; the RA4-1 at

Fig. 2. The first robot transfers bodies from the container to the pallet, and also operates the escapements for components.

Fig. 3. The second robot selects shims for a battery of six dispensers, pushing on plungers to release the shims.

2.5m/s on point-to-point control (PTP), and the RA4-2 at 3.0m/s. Under continuous path (CP) control, both are limited to a more conventional 1.0m/s.

Scara-type RC models with two and four axes of motion are available. The upper arm and forearm are respectively 500 and 315mm long. The gripper, which has a capacity of 6kg, can be moved at 2.3m/s under PTP, and 1.0m/s under CP control.

Toyoda's RR4-2 is more like an assembly machine, and is clearly a much simplified version of the highly-complex robot Toyoda developed for the government-sponsored project. The robot arm hangs down from a beam on a four-post structure which covers an area of 1,220 x 1,490mm. The structure is just over two metres tall.

Rotation

The arm can move over a table 800 x 630mm, and can move vertically 250mm. In addition, there is a fourth axis of movement – rotation of the tool holder through 360°. This machine is designed for more precise work than the others, and has a claimed repeatability of ±0.03mm. However, it moves at only 1.0m/s, under both PTP and CP control.

Toyoda has developed a general-purpose assembly machine around three articulated RA4-1 robots – two mounted upside down on a gantry, and the third mounted upright at the end of the line. The system is based on a structure which incorporates all the ancillary equipment as well as the robots. There is a table some 3m long carrying a conveyor and various assembly devices, such as CNC nutrunners and parts-feeding chutes. Behind these are a number of parts feeders – in the first installation there are five vibratory bowl feeders. The control console is separate.

Flow control

Toyoda's first machine, which it used in the Okazaki plant in conjunction with the robot assembly line, is arranged to assemble flow control valves. The valve consists of a small housing, approximately 40mm long by 20mm diameter in which a spring, plunger, ball and filter are assembled from one end. All operations are vertical, which simplifies assembly. There are eight components in all. After assembly, the valve is transferred by robot to a test stand which measures the relief pressure and pulsations of flow.

Overall, there are eight stations in the line, including one idle station, and two robots each work at three different stations. An indexing conveyor carries pallets along the line, and then returns beneath the line. This is actuated pneumatically, and each pallet has four recesses, so that it can carry two bodies and two filter assemblies – two are assembled simultaneously.

In addition, there is a conveyor in front of the main conveyor for the supply of components in pallets. There is one pallet in front of each robot normally.

Filters

The first machine in the line is not attended by the robot; it is a punch which presses out circular filters from

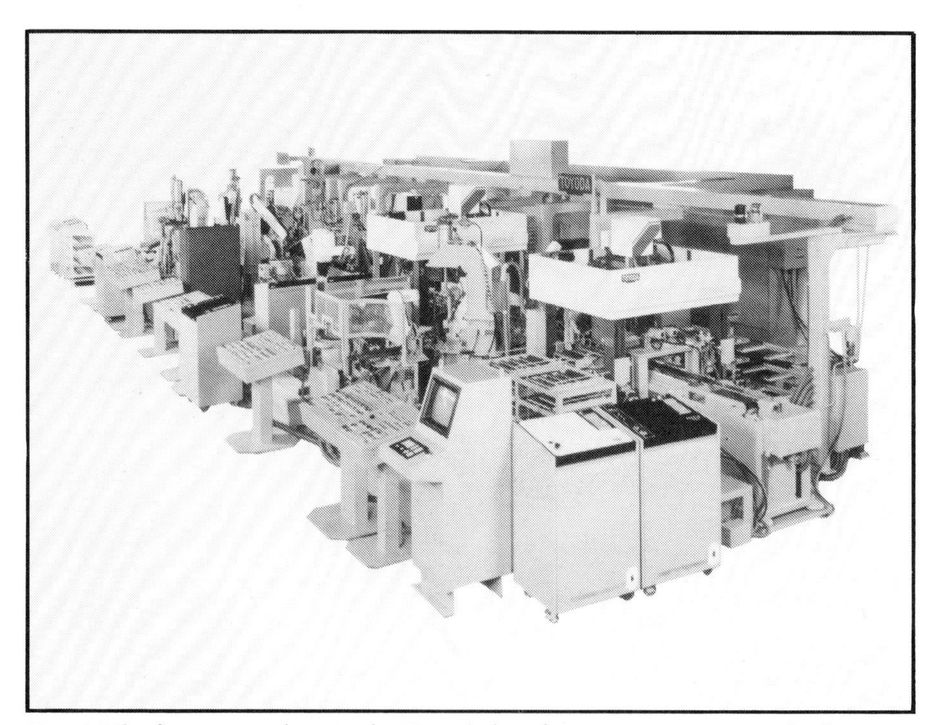

Fig. 4. The four-post robots in the Toyoda line for vane pumps are at the first and third stations, and are separated by an off-line station.

fine steel mesh. The mesh is supplied in three layers, so that the punch can press them out together, and subsequently these are transferred through a closed feeder and a washing station to the first assembly station.

The robots carry double two-finger grippers, which hang down from the ends of a pair of arms. These arms are cranked, so that there is pad on each which can actuate a push-button. At the first station, therefore, the robot arm is first moved to the escapement mechanism for the parts feeder for the filter/screw. The gripper then moves upwards about 10mm, so that its pad actuates the escapement to release a filter/screw, which the gripper picks up. Then the robot moves along a little, repeats the process, and deposits the filters in the pallet.

Fixed arms

At the next station, the filter meshes are pressed into the filters by a pair of plungers. To transfer the mesh from the feeder to the plungers, small fixed arms are used. Then, the pallet moves forwards, and the robot pushes down on a pair of plungers to allow the washers to drop into the filter above the mesh. In the next operation, the rim of the filter body is crimped down over the washer.

As the pallet moves to the next station, the robot picks up two bodies from the container on the conveyor in front of the line, and after passing these close to magnetic proximity switches, it operates rather as a human operator might. First, it takes them to a pair of dispensers, and pushes two buttons simultaneously so that a spring is released to fall into each body. Next, it actuates the push buttons for the plunger feeders, so that the plungers are inserted. After that, a ball is dropped into each body in the same way so that it sits on top of the plunger. Then, the body is placed in the pallet.

At the next station, the height of the plungers relative to the depth in the bodies are measured by differential transformers. In accordance with the result, the second robot takes the pair of bodies and selects shims from a battery of six feeders, by actuating push buttons.

When the robot has returned the bodies to the pallet, it picks up the filter/screws, and places them in the bodies. At the next station, a pair of CNC nutrunners tighten the filter/screws in the bodies. The pallet passes to the next station, where the third robot picks the assemblies up in pairs and places them in the test stand. Those that pass are placed in a container on the front conveyor. Rejects are sent to a chute for rectification.

Fig. 5. The third station, with a container full of shafts, and a vibratory bowl feeder for the pins.

The pump assembly line, with 15 robots and a vane assembly machine, is far more complex. It has 10 stations, and at three of these there are more than one robot, while at two stations solid state cameras are used as sensor systems. The complete line occupies a space 18m by 9m.

There are five types of vane pump assembled on the line, the main differences being in the shape of the cylindrical housing, which has a number of lugs on the periphery. The control valve is inserted in a bore at right angles to the axis of the pump, which consists of a shaft, rotor, cam ring and vanes. There are 35 components in each pump, and the line has been designed to operate to a 45s cycle, for an output of 10,000 units/month on one shift – a planned operating efficiency of 80%.

Pallets, each driven by linear motors pass along a straight track. Each pallet has four recesses – one for the housing, and three for small components. There are two of the overhead RR-type robots in use, and these straddle the line, the nine RC (articulated) and four RA (Scara-type) robots being ranged on the sides of the line. The controllers are arranged on one side of the line, well clear of the safety guards.

At the first station, there is one RR robot, equipped with two grippers. Containers, each carrying eight housings are delivered to the robot, which mounts them on the pallet. Then, it changes the gripper to pick-and-place three circular plates on the pallet.

This second station is the most complex, with two RA-type robots working off the line on one side, and another working on the other side. Parts are supplied by vibratory bowl feeders. First, the RA6-2 robot takes the housing from the line and reverses the body before placing it on a fixture beneath a press. Then, the small, RA4-1 robot picks up a small O-ring on its scissor like gripper, and places it in a recess in the housing. Next, the flow-control seat and circlip are inserted, and then the other robot picks up the 30mm diameter oil seal and places it in the housing. A press pushes the oil seal home against its shoulder.

When these operations are complete the large robot transfers the housing back to the pallet, right side up, so that the RA4-1 robot on the other side of the line can insert the O-ring in a recess in the circular plate in the pallet. To stretch the O-ring over the periphery of the plate, the robot has a gripper with five fingers, which stretch the O-ring out as it is inserted.

At the third station is another overhead robot as well as a Scara type. Shafts are supplied 72 to a pallet, while pins are supplied by vibratory bowl feeders. First, the Scara robot inserts a

Fig. 6. At the sixth station, a special cylindrical holder is held by the gripper to insert a wavy washer.

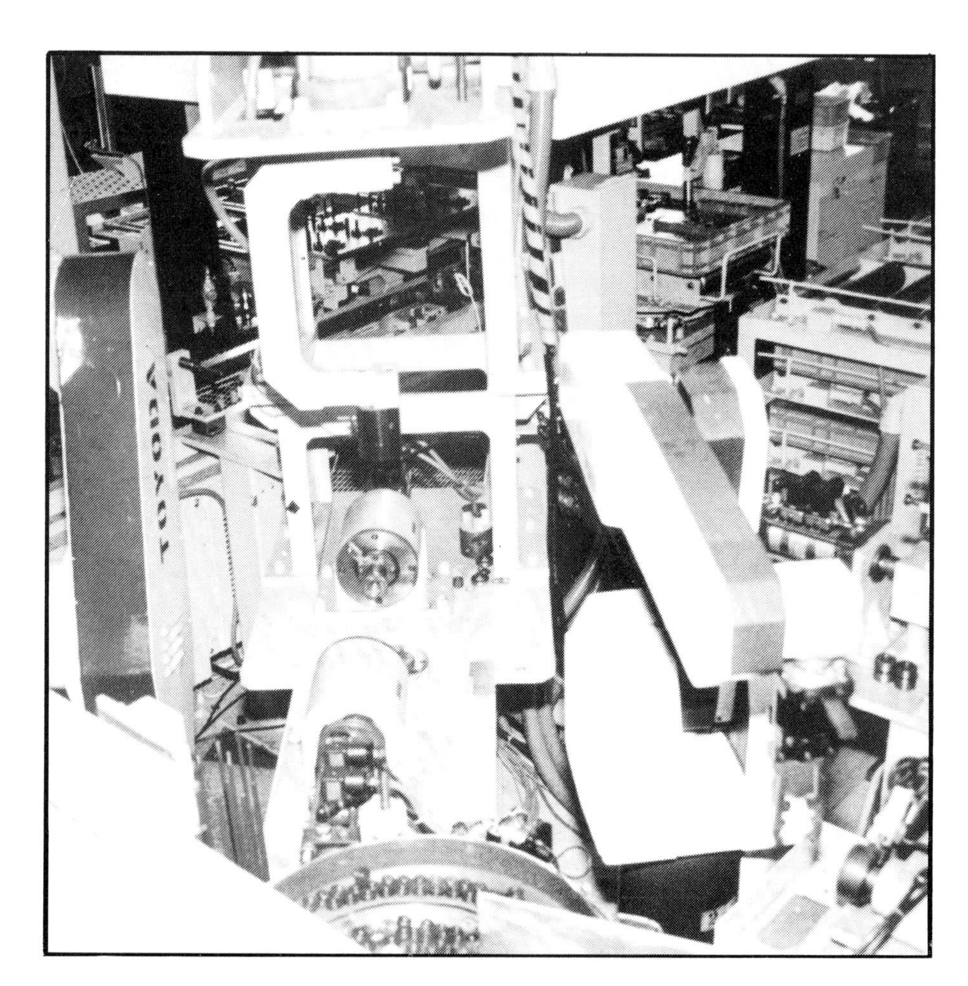

Fig. 7. At the seventh station, a press and two robots work close together. There is another robot on the other side of the line.

pin vertically in one of the plates on the pallet, while the overhead robot, which can operate with greater precision than the other Toyoda robots, picks the shaft up, and inserts it in the housing. Next, it fits a second pin, and then picks up the sub-assembly of the plate and pin, and assembles that.

At the fourth station, there is the vane assembly machine alongside the line. Vanes are installed in pairs in slots in the rotor, and then the rotor is indexed around for two more to be inserted. When these operations are complete, the sub-assembly is installed in the housing by a Scara type robot.

To indicate orientation, a dot is punched on the cam ring and rotor. Since the cam ring is symmetrical, the rotor can be installed with the dots either aligned or at 180° to one another. Therefore, at this station, an Automatix vision system, based on a ccd – charge coupled device – camera is used to examine the alignment of the dots. Correct alignment is needed for the insertion of a pin, so if the rotor is 180° out of position, the rotary table turns it around to the correct position.

At the fifth station, one Scara-type robot inserts a snap-ring and plate, while an articulated type robot is used at the sixth station to insert a wavy washer and the end cover over it – one of the most difficult jobs in the line, according to Fumihiko Ookoshi, manager, robotics, at Toyoda. To insert the washer, which is approximately 70mm in diameter, the robot picks up an extra gripper. This gripper is basically a cylinder, with a groove machined in the periphery in which the two fingers of the robot can register. There is a shoulder in the bore, to locate the washer, and a couple of slots so that the gripper can locate over the parts feeder.

Articulated

At the seventh station, there are two articulated type robots, to insert a plug and O-ring, bolt and washer in an inclined bore in the housing. The end of the plug is at an angle of about 45° to the shank. First, the robot nearest the line takes the housing from the pallet, and places it on a fixture off-line. Then, the robot places a plug in a fixture in a two-stage operation, orientation being checked and corrected first.

Then the other robot, taking parts from vibratory bowl fed chutes, fits the O-ring to the plug. The washer and bolt are fitted next, and the housing is taken back to the line. On the opposite side of the line is a CNC nutrunner, and the robot holds the bolt while the nutrunner starts tightening.

Before the assembly proceeds to the next station another Automatix vision system is used to read the grade letter on the housing, the result being fed to the robot at the ninth station. Meanwhile, at the eight station, a snap-ring is inserted by an articulated robot.

At the ninth station, there is one articulated and one Scara-type robot. The Scara type robot selects and installs the flow control valve – previously assembled on the small robot line – while the other robot installs the spring and plug. The plug is subsequently tightened by a CNC nutrunner. At the last station, the complete assemblies are passed to an inspection machine before being palletised by an articulated robot.

The lines have just been installed at the Okazaki plant, following trials at the machine tool factory at Kariya in the summer. Ookoshi said that the small machine should be suitable for many different assemblies, so long as the parts were fairly small. As for the large machine, this was a special-purpose line, and owing to the use of the vision equipment, was expensive. 'Each Automatix system costs Yen 50 million (£140,000),' he said, 'but our tests show that they work well.'

The small machine is unusual in that the robot actually operates the parts feeders. Why? 'We did that to eliminate the need for any other devices,' replied Ookoshi, who added that the sequential controller which starts the robots on their cycles, was the only piece of ancillary equipment needed to control the system.

On the pump line, photo cells and proximity switches are used to check that each part is actually assembled, and again the robots are controlled by sequence controllers – no supervisory computer is needed. Each robot can operate with three different programs, and up to 300 steps. In each program, the robot moves between 50 points on average.

Whereas the valve line represents an investment of Yen 80 million (£220,000), the pump line costs Yen 350 million (£1,000,000). Although the valve line is designed to operate at a 30s cycle, and the pump line at 45s, times of 40 and 50s respectively had been achieved in trials. Of course, at this stage downtime is a big question mark.

In any event, Toyoda intends to operate both machines on two shifts, with one man/shift at each. Previously, four men/shift were needed for the valves, and 11 men/shift for the pumps. In this case, the investment will be recouped by the reduction in labour costs in four to five years, which is rather a long time. However, these are the first lines of this type, and as the output of the robots and machinery increases, so the cost will reduce. In addition, of course, vision equipment is expected to become much cheaper in the future. But because Toyoda is in the business of selling assembly systems, the new lines are also good advertisements, and so can justify that investment. In any case, it is gaining experience with a system of the future now.

3.3 Assembly – Electrical products

The flexible parts feeder which helps a robot assemble automatically

T. Suzuki and M. Kohno, Hitachi, Japan

A flexible part feeding system has been developed in conjunction with a six-axis robot and vision to assemble 13 components into a portable tape recorder.

THE AUTOMATION of assembly has been applied mostly to mass-produced goods such as home appliances, watches and car components. It has almost reached its culmination by means of so-called hard automation, which consists of many single-task working stations.

However, the recent trends of increasingly diversified tastes of customers have been forcing manufacturers of such products to adopt more flexible assembly facilities, which are adaptable to frequent changes of products with short change-over times. The flexible assembly machine is also needed to improve the productivity of small batch production, which has hitherto been less automated by conventional manners.

Under such circumstances, industry has found industrial robots as the most effective tool for handling parts and assemblies, and some new concepts of flexible assembly systems and elementary technolgies have so far been reported. These include flexible part feeding devices[1], universal grippers[2-3], and new concepts of assembly line configurations[4-5]. However, they are still at laboratory stages and have not yet been successfully applied to practical production at factories.

A development project for a flexible assembly system for small mechanical products such as portable tape-recorders has been carried out at the Production Engineering Research Laboratory of Hitachi. A pilot system has been developed, and early experiments performed.

The system comprises a jointed-arm robot and a programmable part feeding machine equipped with vision.

Hardware

The system concept is that several parts are supplied and assembled at one station where the robot picks them up in a programmed sequence. This is particularly necessary for a flexible assembly-machine which must assemble various products consisting of different numbers of parts.

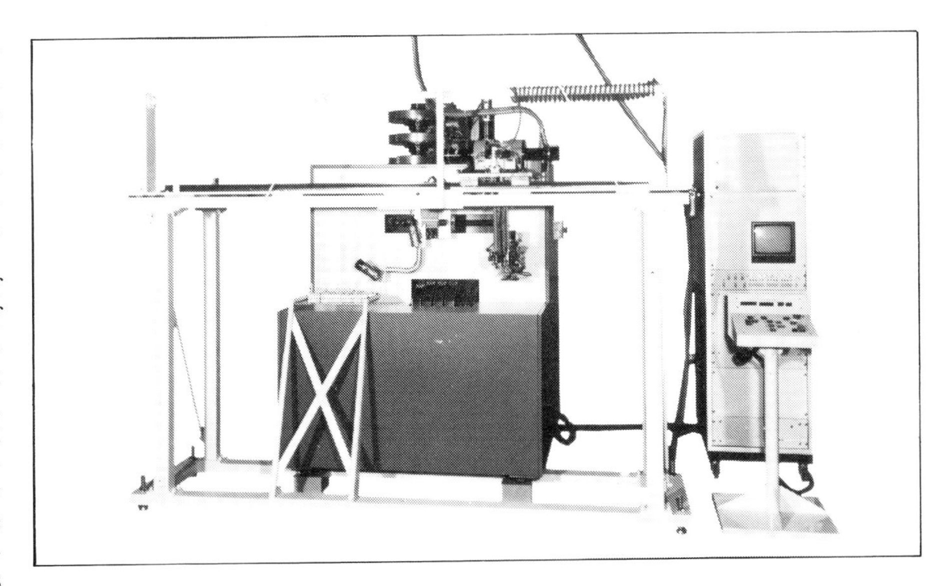

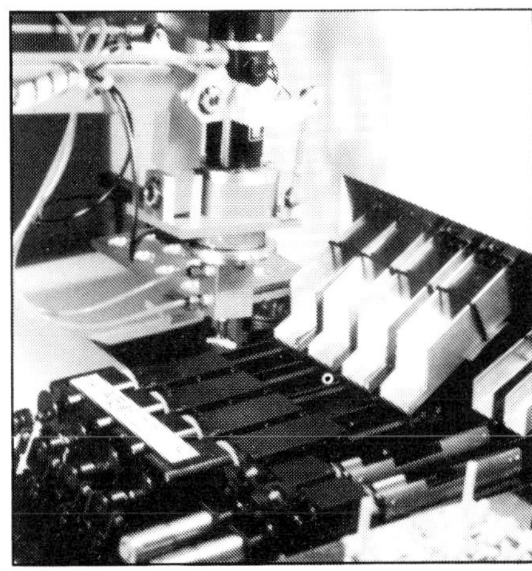

Fig. 1. Views of the flexible assembly machine showing the multi-level bowl feeder, visual system and robot arm.

In conventional assembly lines, which are composed of single-task stations, the change in the number of parts assembled would mean additional or redundant stations. In the Hitachi flexible assembly-system, however, changes in the number of parts mean nothing more than the alteration of programmes, which, in some cases, can be done only by programme selection among multiple programmes.

The flexible assembly-system, shown in Fig. 1 consists of two main components: a programmable part feeding machine and a robot. During early experiments, 13 different parts, some of which are shown in Fig. 2, were fed and assembled. They are

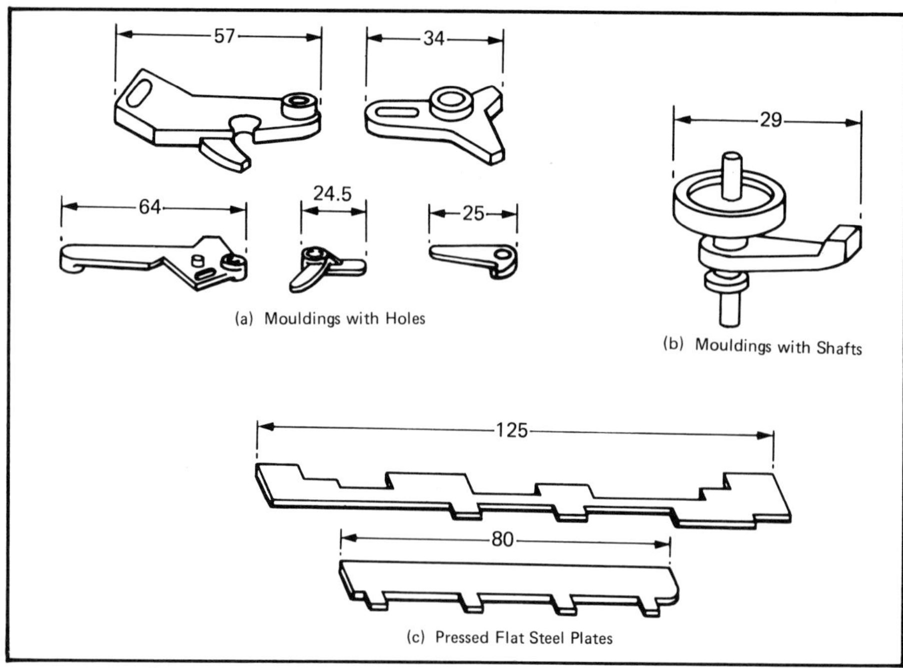

Fig. 2. Parts handled by the programmable assembly complex fall into three basic groupings.

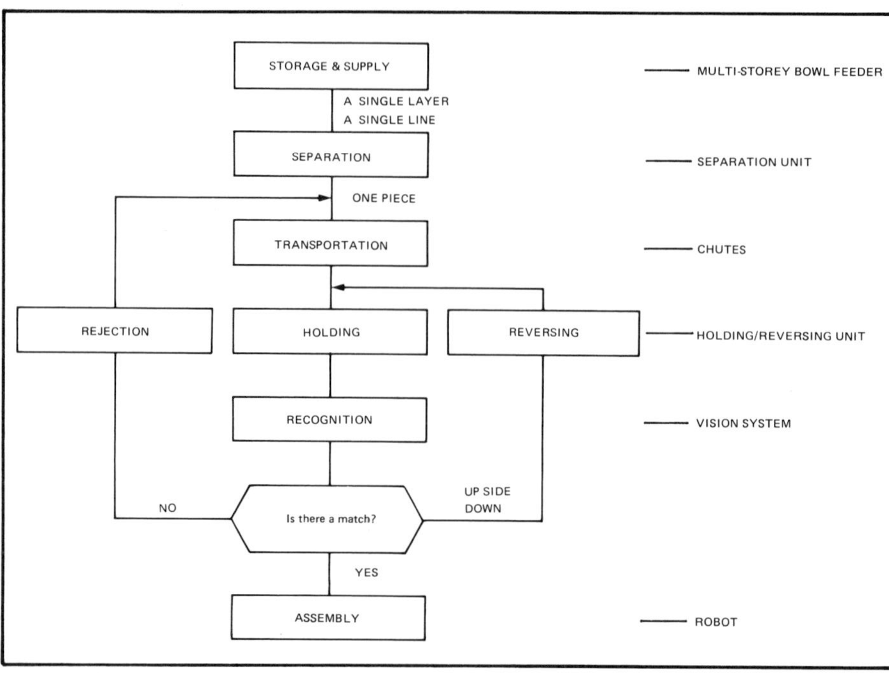

Fig. 3. Flow-diagram of part-feeding and assembly operation shows the six stages of the process.

divided into three categories: plastic mouldings with holes, plastic mouldings with shafts, and flat steel plates. A flow diagram of the machine is shown in Fig. 3.

The part feeding machine has a feeding section and a recognition section, the former being composed of a multi-storey bowl feeder, separation units and chutes, while the latter has a 256-bit line scan camera, an image processor and a part holding/reversing unit.

The parts are stored in the bowl feeder which, unlike conventional bowl feeders, has no orientation mechanisms on its tracks. Thus, the parts are fed disorientated, and then the separation unit drops one piece at a time into the chutes.

At the bottom of the chute, the separated part reaches the part holding/reversing unit. In this unit, two dimensional pusher press the component against two-dimensional datum planes, and the part settles into a predictable stable posture.

The vision system identifies the posture of the part and position information is transferred to the robot controller. According to this information the robot picks up the component and places it onto the chassis mounted on a conveyor.

A 8-bit micro-computer controls the part feeding machine, including the image processing, while another 8-bit micro-computer controls the robot. The operational sequence of the whole system is controlled by the second computer.

Since the system has no tooling hardware appropriate to specific features of particular parts, parts changes require no adjustments to the mechanism, only an alteration to the software. Consequently, the part-feeding machine is practically universal to a range of parts.

The robot is equipped with a newly developed gripper mechanism able to accommodate a variety of small mechanical components.

Another feature of this system is that it is suitable for applications where many parts are supplied to and assembled at one station, and where the robot can pick them up in a programmed sequence. This is particularly necessary for a flexible assembly machine, which must assemble various products comprising different numbers of parts.

Robot

The robot has an electrically-driven jointed-arm structure. Robot arms are generally classified into four categories: cartesian, cylindrical, spherical and jointed-arm. From the viewpoint of operational volume/robot size ratio reliability of parts and number of parts required, the jointed-arm is superior to the others. In terms of control on the other hand, jointed-arms require the most complicated co-ordinate calculations.

However, the recent development of micro-computers and the reduction in their cost has enabled the jointed-arm robots with fairly high functions to be made at moderate cost.

Robot drives fall into three categories: pneumatic, hydraulic and electric. The pneumatic still lags behind the others in the capability of its servo control and is used only for simple, fixed-operation robots employed as a low cost actuator. The electric drive has an advantage that power supply is widely available. The advantage of the hydraulic is that it can generate high power from a small actuator, but it requires intensive maintenance. For this flexible assembly system, the robot handles only small components and conventional dc servo motors and powerful enough to drive the robot.

To broaden the robot's operational volume, the arm is suspended downwards from a traverse base which can move on a pair of rails. The arm has three degrees of freedom: shoulder rotation (rotation of the upper arm), elbow rotation (rotation of the fore-

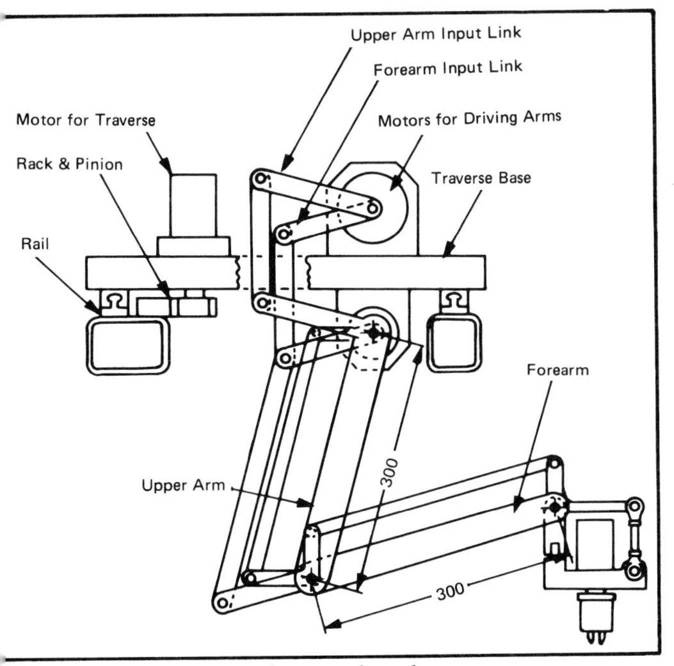

Fig. 4. Arm-driving mechanism has the motor above the traverse base.

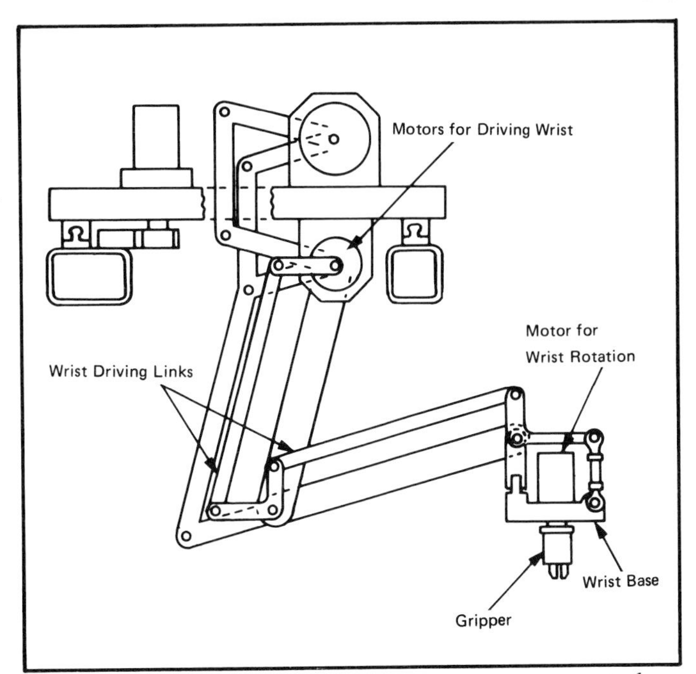

Fig. 5. Wrist-driving mechanism has its motor mounted below the traverse base.

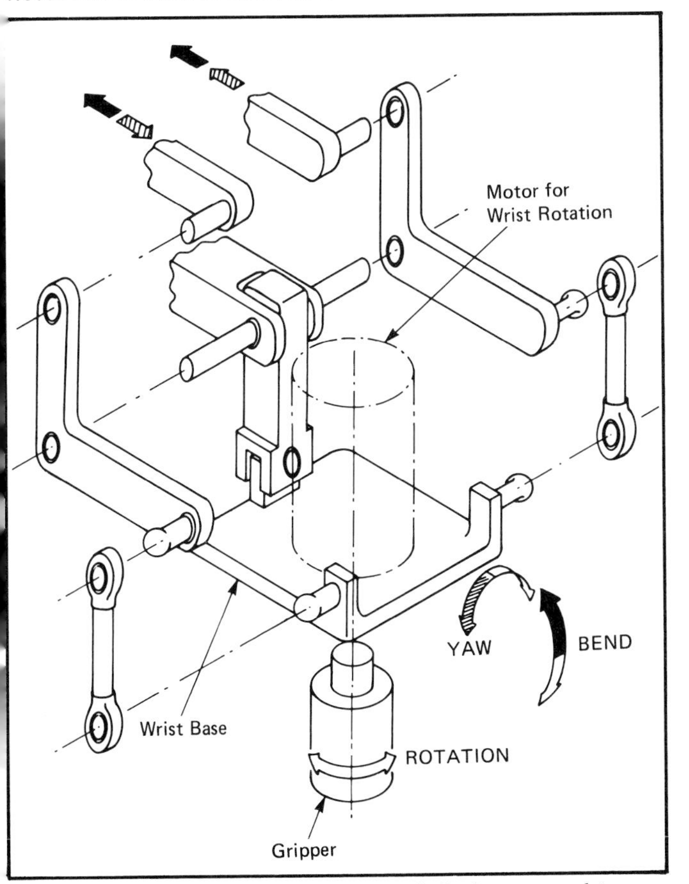

Fig. 6. Wrist mechanism uses a simple linkage to achieve motions of yaw and bend.

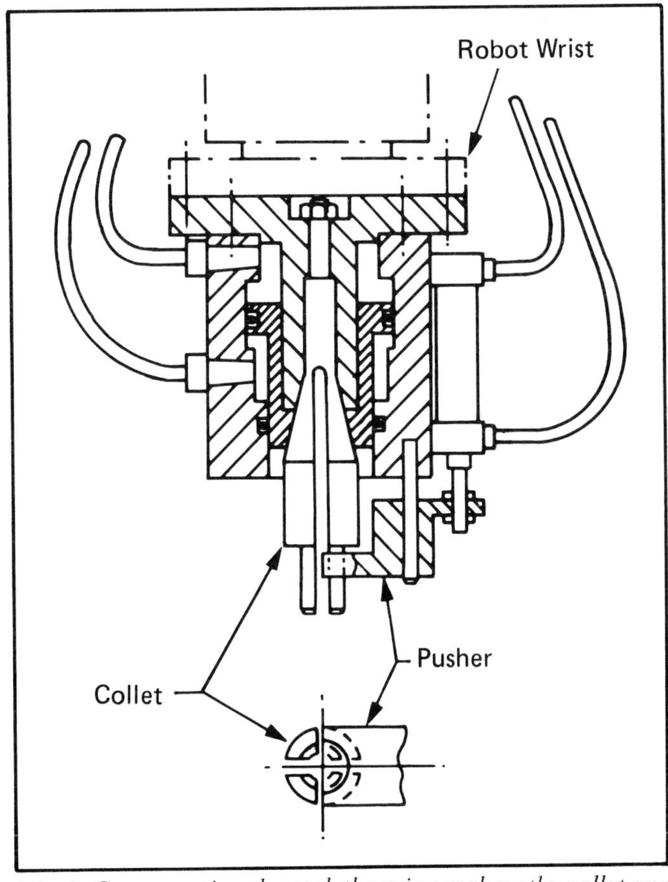

Fig. 7. Cross-section through the gripper show the collet and pusher used for collecting components.

arm), and traverse.

A wrist mechanism with three degrees of freedom of motion has been attached to this arm to give six degrees of freedom as a whole, excluding finger movement.

The mechanical configuration of the robot arm is based on parallelogram linkage mechanisms, as shown in Fig. 4. Two dc motors, driving the upper arm and the forearm, are attached to the upper side of the traverse base to reduce the weight of the arms. Driving forces are transferred to the arms through parallelogram linkage mechanisms.

The base, on which the robot arm is mounted, travels on a pair of rails and is driven by a rack and pinion mechanism from a dc servo motor.

Wrist

The mechanism for the wrist motions, is based also on parallelogram linkage, as shown in Fig. 5. The motors for these motions are attached to the underside of the traverse base, and two series of parallelograms link the movement of the motor to the end of forearm.

The two-degrees-of-freedom spatial linkage mechanism, shown schematically in Fig. 6, is built at the end of the arm. This mechanism generates bend and/or yaw motions of the wrist according to the movement of the two linkages.

When the input links move in the same direction, the wrist bends, as indicated by black arrows in the figure. When the input links move in opposite directions to each other, or when the two input links move with a phase difference, the wrist yaws, as shown by the hatched arrows. The third motion of the wrist, rotation, is generated by a dc servo motor attached to the wrist base.

Since the motors for driving the wrist are fixed to the traverse base and the motions are transferred through the parallelograms, the wrist maintains its posture against the floor regardless of the arm position, unless the wrist motors are activated. Consequently, the posture and motion of the wrist can be controlled independently of the arm motions.

Gripper

The key factor of a flexible assembly robot is design of suitable grippers which accommodate a variety of mechanical parts. One route to coping with part differences is to prepare several grippers corresponding to the parts and to change them in the course of assembly operations. This technique has been applied to some experimental assembly systems[8-9], but it requires time to change the grippers.

Since there are no mechanical hands which are as adaptable as human hands in handling varied parts and assemblies it is better to limit the variety of parts and design a gripper mechanism adaptable enough to handle the limited range of parts.

A gripper capable of handling all these parts without re-tooling has been designed and built; cross-section is shown in Fig. 7. It comprises a collet chuck and a double-acting pneumatic cylinder.

The parts are gripped as follows (Fig. 8). Mouldings with holes are gripped in the bore by the external diameter of the collet (Fig. 8(a)); mouldings with shafts are gripped at the shaft by the internal circle of the collet (Fig. 8(b)); flat plates are gripped at slits between the collet segments (Fig. 8(c)).

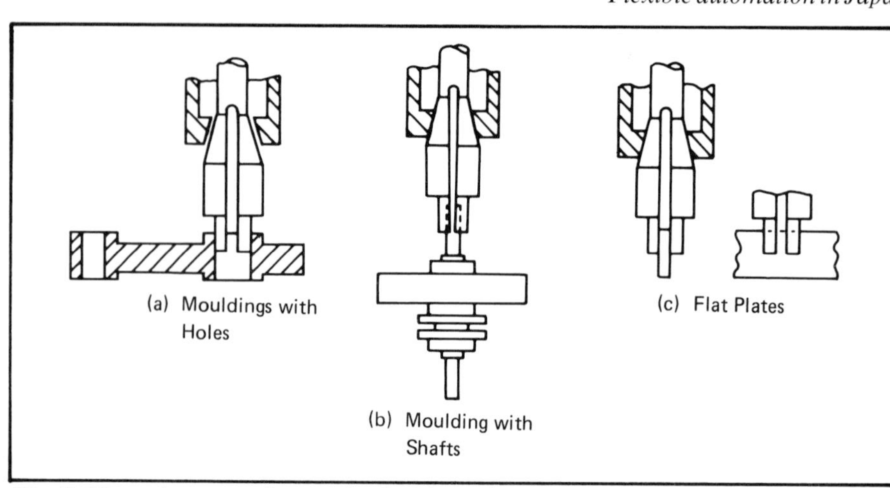

Fig. 8. The gripper can be used to hold mouldings with holes and shafts, as well as flat plates.

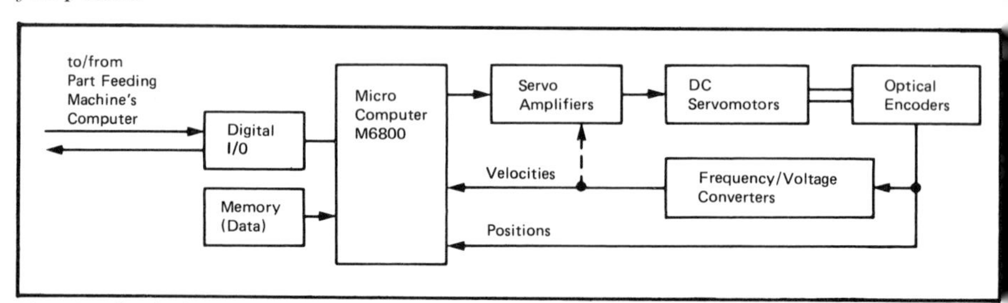

Fig. 9. Block diagram shows how software servoing is linked into the part feeding machine's computer.

Table 1. Specifications of the Robot

Configuration	Jointed-Arm
Degrees of Freedom	6
Maximum Tip Velocity	1000 mm/S
Arm Length upper Arm Forearm	 300 mm 300 mm
Operation Range Base Traverse Upper Arm Forearm Wrist Bend Wrist Yaw Wrist Rotation	 1800 mm +40°, −25 +30°, −40° ±45° ±30 ±180°
Load Capacity	2 kg
Positioning Repeatability	±0.3 mm
Drive	DC Servomotors
Controller	Micro Computer HMCS 6800

Control

An 8-bit micro-computer controls the robot. The programmable part feeding machine has its own micro-computer to process visual information while sequential operations of the whole system are controlled by the robot's micro-computer. Both computers are linked to each other through their digital input/output interfaces, and all possible locations and postures of the parts at the part holding/reversing unit are stored in the robot micro-computer as subroutines of the motion programmes. The appropriate subroutine is selected according to the coded information sent from the part feeding machine.

The robot uses point-to-point control with linear interpolation. Circular arc paths are approximated by small segments of straight lines, and the circular interpolation can be added to the current software if necessary.

Precise operations, such as the assembly of small mechanisms, generally require very accurate motions. Accurate linear motions are needed particularly for peg-hole insertion, the basic operation in assembly, and they save labour when long straight motions of the arm need to be taught. If the linearity is accurate only a point at each end of a straight path is necessary.

To improve accuracy of linear interpolation, software servoing, in which the computer is located within the feedback loop of the servo system, has been applied to the robot control. Fig.

'ig. 10. The six-axis robot with its ;ripper is suspended from overhead ails.

9 shows the block diagram of the system. At end sampling, the position and velocity signals are fed back to the micro-computer, which calculates the next outputs (i.e. the velocity signals to the servo amplifiers) from the current values, thus eliminating the deviations which are caused by the change in the dynamic characteristics of the robot mechanism and, therefore, are inevitable in conventional servoing. The sample cycle is 20 milliseconds.

A special feature of this control system is that the velocity feedback signals are taken from the incremental optical encoders, also used to take the position feedback signals, through the frequency/voltage converters. Tachogenerators are not therefore required.

Some specifications of the robot are listed in Table 1, and Figure 10 shows a view of it.

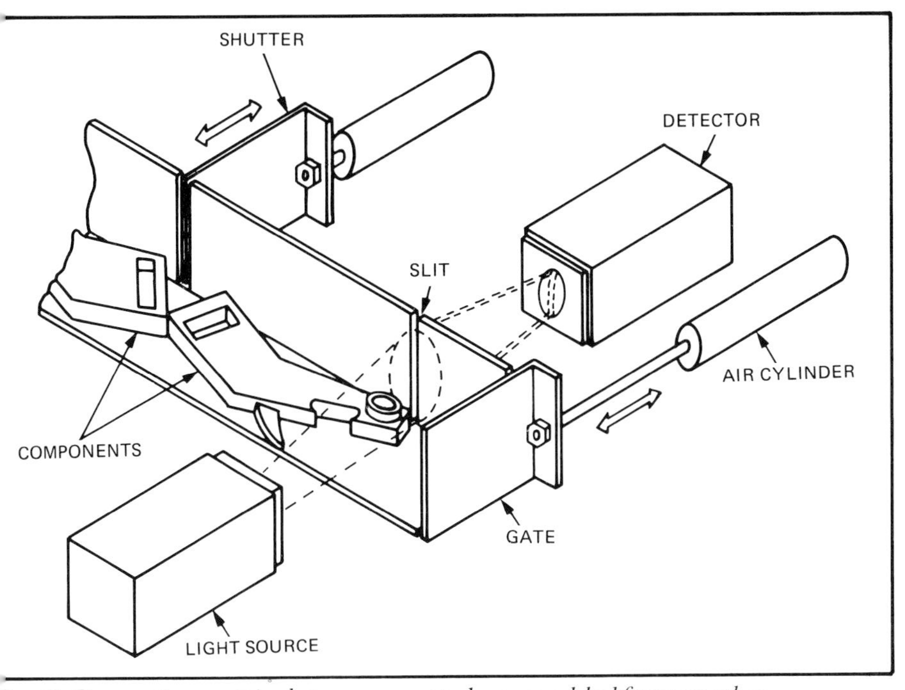

'ig. 11. Separation unit isolates one part to be assembled from another.

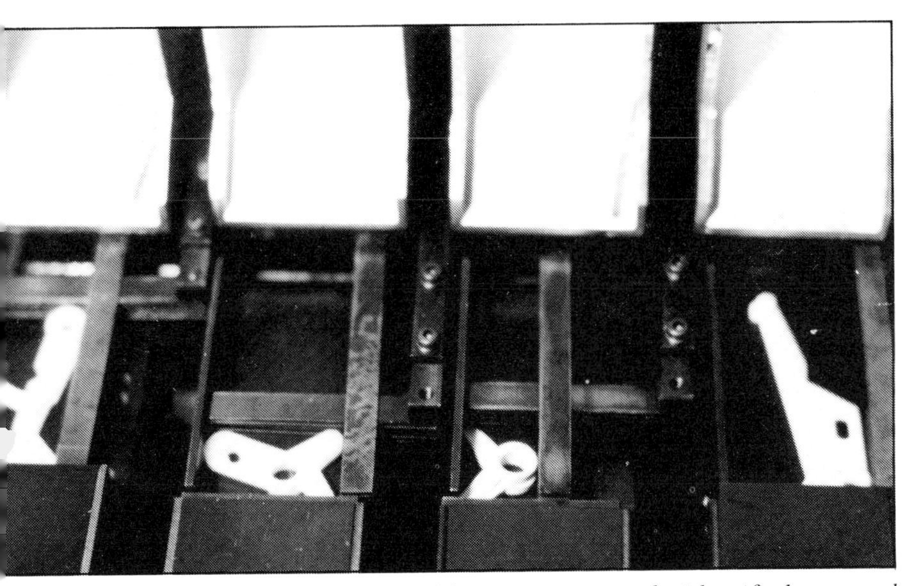
Fig. 12. The holding/reversing unit enables components to be identified or turned over if they are the wrong way up. Components can also be rejected if they are of the wrong type.

Part-feeding

The part-feeding machine uses a multi-storey bowl feeder for compactness. The feeder has five bowls and each bowl contains a different part. The function of the multi-storey bowl feeder is confined to feeding parts one at a time and in single file to the separation unit. Thus each bowl is equipped with a wiper and a dish-out. The wiper and the dish-out can accommodate some change in the shape of the parts; they can be adjusted for larger changes in shape, if necessary.

The separation unit is at the exit of each bowl (Fig. 11). The parts enter into the separation unit, where the leading edge of the first part is detected by the photoelectronic sensor. The leading part of the line is then separated by the shutter from the others. Next, the gate is opened by a signal from the micro-computer and the separated part is dropped into the chute.

In this unit, the following conditions are required to ensure separation:

○ to limit the area of detection by passing the beam through a slit;
○ to detect by adjusting the strength of the beam.

The separation unit has almost the same facility as the bowl for accepting parts of different shape.

The part then drops down chutes from the separation unit into the holding/reversing unit. The chutes consist of vinyl tubing and aluminium square tubing. They are cheap and adaptable for size within a limited range.

The holding/reversing unit (Fig. 12) has three functions:

□ From the chute the part is pressed against the two-dimentional datum planes, firstly in the X direction and then in the Y direction, by a pair of comb-shaped pushers. These put the part into a known position to be easily identified by the image processor. Being held by the pushers, the part is checked by the robot without fail.

□ When a part is positioned the wrong side up, it is pushed into a turning case, which vertically rotates the part 180° and pushes it back to the datum planes. Then the first routine is repeated and the part is identified again. By this means all the correct parts entering the unit can be used without rejection regardless of their postures.

□ If the wrong kind of part is found in the unit the bottom plate opens and the part is rejected.

The unit can handle any part providing it is smaller than the rectangular area formed by the pushers

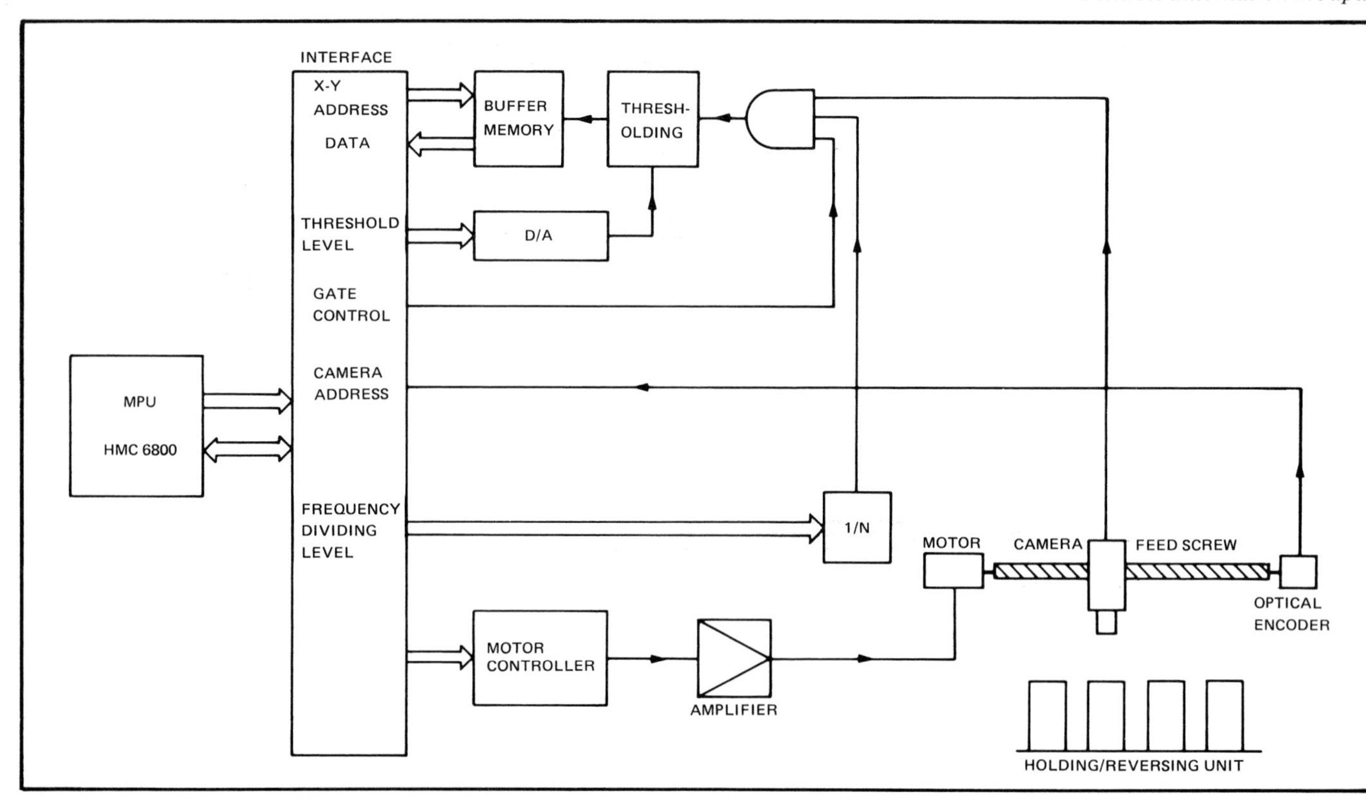

Fig. 13. Block diagram of the vision system.

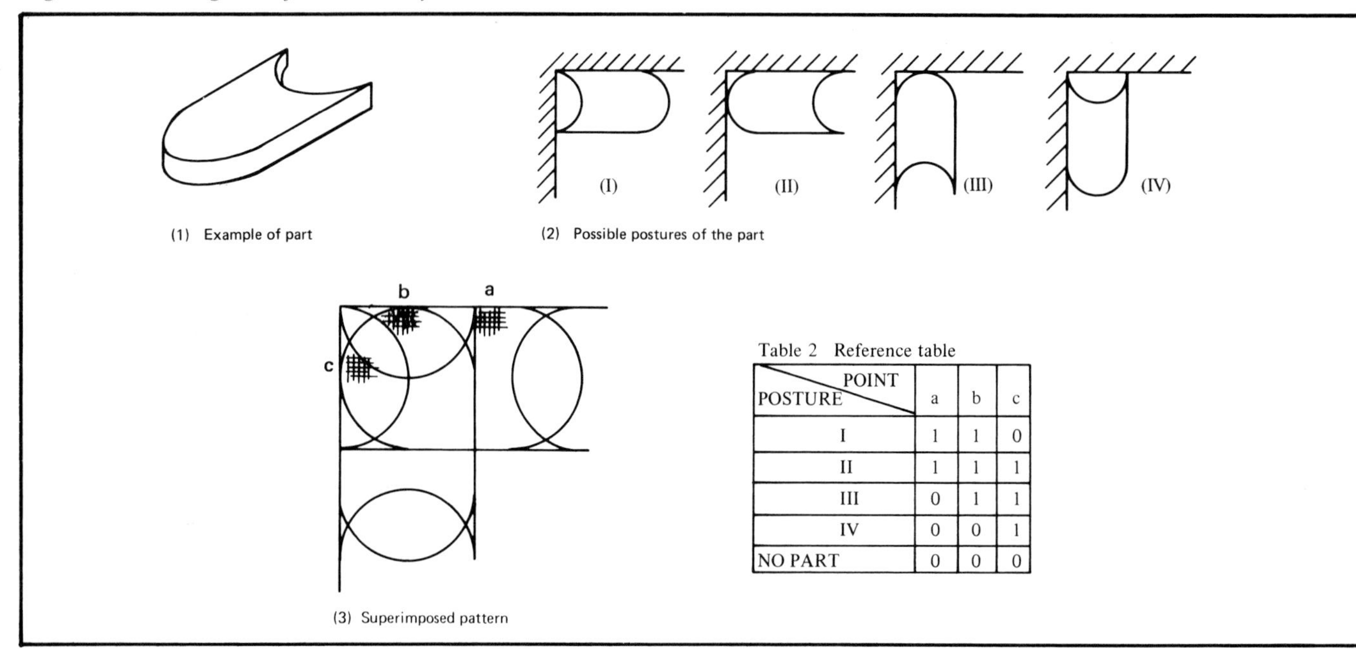

Table 2 Reference table

POSTURE \ POINT	a	b	c
I	1	1	0
II	1	1	1
III	0	1	1
IV	0	0	1
NO PART	0	0	0

Fig. 14. Simplified parts matching system.

and datum planes. It is possible to deal with five different types of part sequentially with one unit.

Vision

The vision system comprises a Reticon LC600 solid-state line scan camera and an image processor controlled by an 8 bit micro computer (Fig. 13).

The camera travels over the holding/reversing units by means of a feed screw mechanism. An optical encoder is attached to one end of the screw and provides the image processor with the camera's location signal.

The traverse movement allows the line scan camera to compose two-dimentional images while the image processor only inputs images of the area where parts are expected. The camera's optical array consists of 256 elements spread over 70 mm, providing a resolution of 0.3 mm in the Y direction. The screw lead is 93.3 mm and the optical-encoder generates 2,000 pulses/rev. The resolving power for X direction is therefore about 0.05 mm.

These resolutions are sufficient to characterise the different parts to be assembled. The analog video output of camera is compared to a threshold level – which is programmable in 256 grades – to produce a train of binary pulses. These pulses are temporarily stored in a buffer memory and then only useful data are transferred to the micro-computer.

The method used to identify component posture is simple and practical. It is shown in Fig. 14. Since the sequence of assembling components has already been programmed, the part in question is already known. The component is in any case restricted to several postures in the holding/reversing unit.

In this example the possible postures of the component are four (Fig. 14-(2)), which being supermposed, compose Fig. 14-(3). If the hree points of a, b and c are distin- uished into black and white, the osture is recognised as shown by Table 2.2.

At the conclusion (using n points or reference) 2^n postures at maximum re recognised, including the case of o component being placed. The eference points should be set away rom the edges of the components in order to avoid errors. Also to avoid errors caused by dust or noise, adjacent pixels around a reference point are examined. If seven or more pixels appear in the same colour, black or white, the reference point is egarded as this colour. The time equired for this process of the 8-bit micro-computer is less than 1 ms, which does not influence the cycle ime of the assembly operation at all.

The whole system has been tested hrough experiments of an assembly operation of a home appliance mechanism. In this experimental operation, five kinds of parts are supplied and identified completely. The feasibility of this sort of flexible part-feeding system has therefore been proved. The binary image of a part, which is held by holding/ reversing unit, is shown in Fig. 15. The three intensified dots in the figure are the reference points.

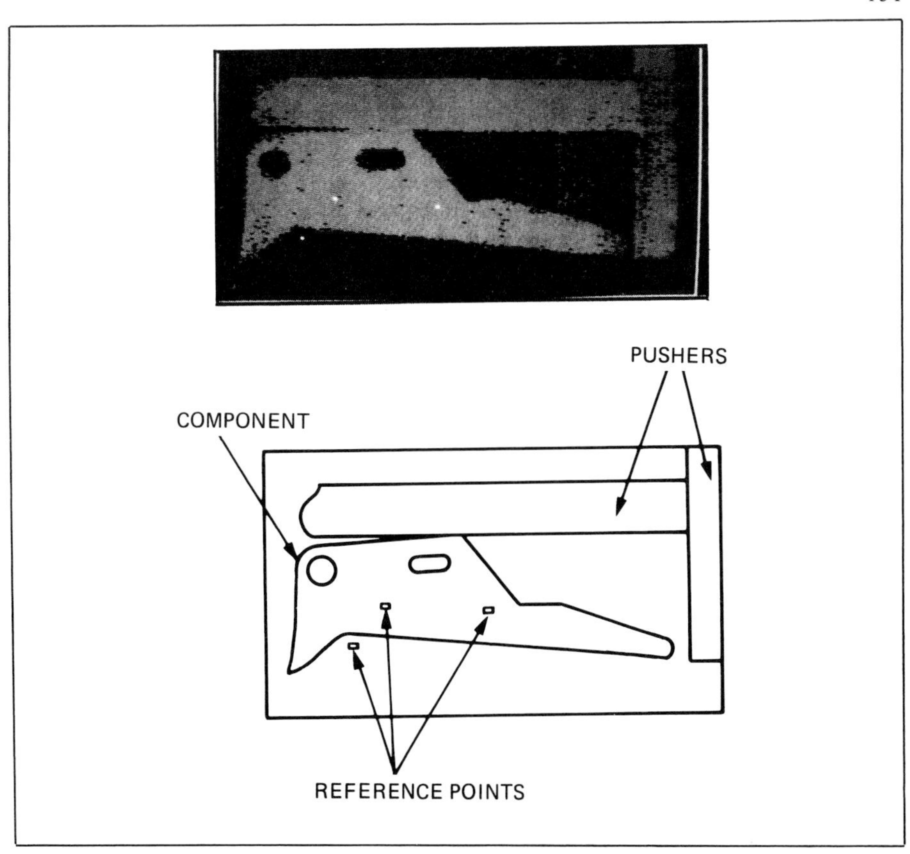

Fig. 15. Binary image of a typical part positioned in the holder/reverser.

Conclusion

A flexible assembly system comprising a flexible part-feeding machine and a robot has been developed and the following results have been obtained:

(1) It has been proved that the part-feeding machine developed is adaptable for the changing shape of the parts.

(2) A simplified matching method, which identifies the posture of the parts in less than 1 ms by a micro-computer, has been found practical.

Introductory experiments have been carried out, and further experiments are being performed to improve the system towards more practical applications.

References

1. *Warnecke, H. J. et al.: "Pilot Work Site and Industrial Robots", Proc. 9th International Symposium on Industrial Robots, Washington D.C., March 1979, pp. 71-86.*
2. *Okada, T., and Tsuchiya, S.: "On a Versatile Finger System", Proc. 7th International Symposium on Industrial Robots, Tokyo, October 1977, pp. 345-352.*
3. *Hirose, S., and Umentani, Y.: "The Development of Soft Gripper for the Versatile Robot Hand", ibid, pp. 353-360.*
4. *Abraham, R. G.: "Programmable Automation of Batch Assembly Operations", The Industrial Robot, 4(3), September 1977, pp. 119-131.*
5. *Abraham, R. G.: "State-of-the-Art in Adaptable-Programmable Assembly System", May 1977.*
6. *Suzuki, T. et al.: An Approach to Flexible Parts Feeding System", Proc. 1st International Conference on Assembly Automation, March 1980, pp. 275-286.*
7. *Sugimoto, K.: "An Approach to Structural Synthesis of Robots", Proc. 9th International Syposium on Industrial Robots, Washington D.C., March 1979, pp. 641-655.*
8. *Nevins, J. L. et al.: "Exploratory Research in Industrial Modular Assembly", CSDL Report, No. R-996, August 1976.*
9. *Kondoleon, A. S.: "Cycle Time Analysis of Robot Assembly Systems", Proc. 9th International Symposium on Industrial Robots, Washington D.C., March 1979, pp. 575-587.*
10. *Pugh, A. et al.: "A Microprocessor-Controlled Photo-Diode Sensor for the Detection of Gross Defects", Proc. 3rd International Conference on Automated Inspection and Product Control, Nottingham, April 1978, pp. 299-312.*
11. *Kohno, M. et al.: "A Robot for Assembling Variety of Mechanical Parts", Proc. 10th International Symposium on Industrial Robots, Milan, March 1980, pp. 501-510.*

An experiment in robot assembly – building electric motors

Fujitsu Fanuc are gaining experience in robot assembly using their own robots. This is a report on the developments taking place at the Hino factory.

FUJITSU Fanuc has started to use its new assembly robot model 0 to assemble dc servo motors. It plans to use the new system on a large scale starting in the late summer in a factory now being built.

In the experimental layout, one Fanuc M model 1 robot works in conjunction with three small assembly robots. Thus, the arrangement is of a cell, of semi-circular layout, with the large robot in the middle to handle the workpieces between stations – where precision is not required. There are also three carousels carrying workpieces, while much of the work is done on a table which incorporates a simple press.

The operations involve fitting of the oil seal and bearings to the rotor; insertion of the rotor assembly in the flange or end housing; lowering the stator over the rotor; fitting the cover. At subsequent stations in the cell, studs are fitted to the assembly and nuts are fitted and tightened.

The first of the assembly robots can reach the main press/assembly table, as well as another table with a press. Thus, it can place the oil seal and bearing on the rotor at this subsidiary table. While this pressing operation is going on, a rotor/seal/bearing assembly is taken from a carousel and deposited on the main table for further assembly. In fact, the M model 1 robot then lowers the stator over the rotor on to the end flange.

Stud insertion

When operations at this station are complete, the sub-assembly is moved to the next station, in front of another assembly robot. Here, four long studs are inserted through the housings and are tightened in place. At the final station, an A model 0 robot picks up a nut and washer from the chute of a vibratory bowl feeder, drops the washer over the stud, and then winds down the nut. This process is repeated for the other three studs.

Fanuc claims that with this cell, which can operate for 24 hours continuously, output is 300 motors/24 h. With normal manual operation, each man can produce 30 motors/shift. The concept is to be adopted for a complete factory, which will reduce assembly costs by 30%, according to the company.

Fanuc's assembly robots – robot A model 0 and 1 – are designed to operate with a repeatability of ±0.05 mm. They are of the cylindrical co-ordinate type, generally with five axes of freedom.

The model 0 is mounted on a column, which can rotate through 300 deg., and can raise the horizontal arm 300 mm. The arm has 300 mm of telescopic motion, and there are two directions of rotation at the wrist – about the vertical and horizontal axes. Capacity is 5 kg. The model 1 has a capacity of 15 kg, and its linear travels are 500 mm instead of 300 mm. Both machines have magnetic bubble memories.

So far, Fujitsu Fanuc has made only a handful of these new robots, which are a little more expensive than the other new assembly robots in Japan at about £15,000. But, they are larger than those machines, and so can be used for assembly of quite large components.

Layout of the experimental assembly cell for dc servo motors at Fujitsu Fanuc's Hino factory.

Keeping television assembly in its place

Automation and manual assembly are kept well separated in Matsushita Electric's television plants. Production at the company's big Ibaraki plant is reported.

ALL THE PROPAGANDA from Japan about its electronics industry talks of high levels of automation, with robots everywhere. The brochures generally show plants where people seem in short supply, assembly being left to automation.

Matsushita Electric Industrial, manufacturer of products with the National, Panasonic and Technics brandnames is a case in point. It employs 85,000 people, making everything from small and cheap radios through audio and TV sets to small computers, facsimile machines and solar energy cells.

By way of introduction, the company spokesmen say that Matsushita has 3,000 robots, but in fact these are nearly all of the sequence control type, with limit stop controls. Then, the official photographs of the TV assembly line show sets coming down a line that seems almost unmanned. However, final assembly of TV sets is left entirely to people, not machines.

In fact, Matsushita has some highly automated lines, such as that for dry cell batteries. There, the line starts with a roll of metal strip, and follows through forming and assembly processes to packing at the rate of 600/min. Then, in the vacuum cleaner plant – these are not upright machines, but small units with separate elephant's trunk tubes – the level of automation is high.

In the television assembly plants, though, there is a conventional approach to automation – but with some interesting variations. Matsushita has five TV plants in Japan, as well as 12 smaller ones in other countries, including one in the UK. Total output is 400,000 sets a month, of which 250,000 sets are colour TVs, and 150,000 are monochrome. Exports account for 150,000 colour and 135,000 monochrome units. Nowadays most of these go to the Middle East, but it is clear that the large number of exports is the key to Matsushita's low costs.

The biggest of the TV plants is at Ibaraki, near Osaka, where Matsushita's manufacturing operations are clustered. It has a nominal capacity of 95,000 sets/month, and as always in Japan, the maximum capacity is that when some overtime is being worked. In other words, it is recognised that plants are often under utilised, so overtime working is used in peak periods.

At the Ibaraki plant, there are 2,800 employees, half of them being involved in production, the remainder being in design, engineering, sales and service departments. All models in the range can be made at Ibaraki, these running from the tiny 1.5 in. monochrome set through the normal 12-20 in. models to projection units with screens of 45 in. and upwards.

Assembly is divided into three stages. In the first stage, the small parts, such as resistors, are inserted into the printed circuit board (pcb). This stage is highly automated. In the

Matsushita has five lines of NC insertion machines in its high-volume insertion shop.

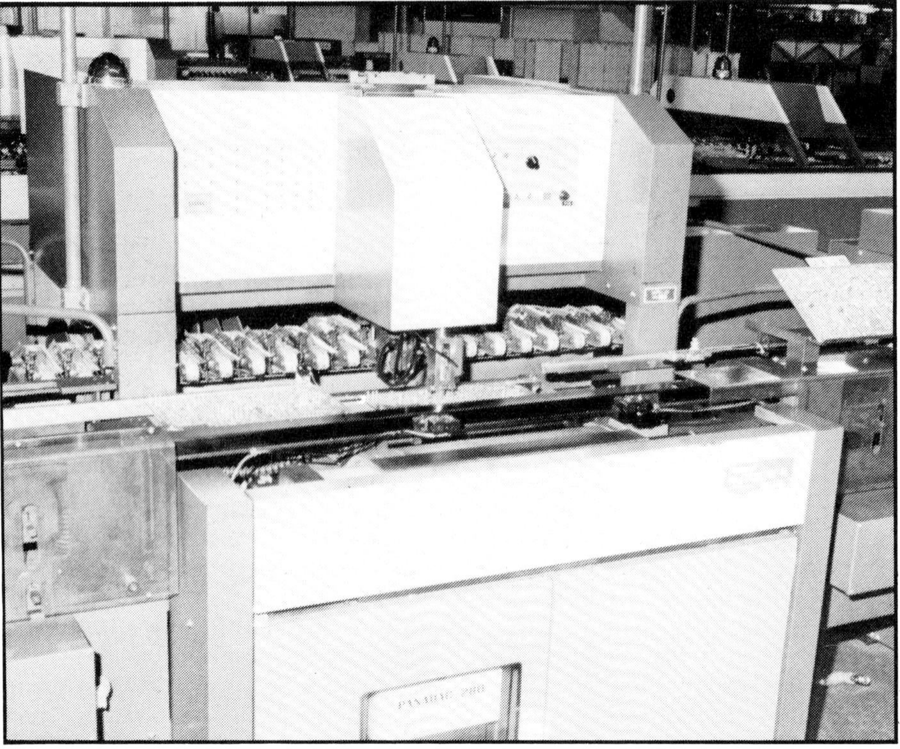

Most of the insertion machines have X-Y tables to move the board and an insertion tool that draws components from magazines parallel with the track.

second stage, larger components are assembled to the pcbs. Most of the operations in this stage are carried out manually, but some robotic devices have recently been introduced. In the third stage, the crts and cases are assembled to the chassis. On these final assembly lines, all operations are carried out manually.

Insertion

Matsushita has developed its own machines for inserting parts in pcbs, and several of these are available to other manufacturers. These fill the two pcb assembly shops. In one shop there are five lines consisting of four or five machines, all these being of linear type. These handle the pcbs produced in high volume.

In the second shop there are a number of machines arranged independently to handle the pcbs produced in smaller volume. In many cases, lot sizes are under 1,000. Both these shops operate on two shifts, unlike the shops for subsequent operation, which operate on only one shift, mainly because most of the employees are women, as is normal in this industry.

The insertion machines in the high volume shop operate on a cycle of just under one min/board, there being 50-100 components assembled by each machine. Most of the machines consist of an x-y table which moves the board around, and a row of magazines on a carrier parallel to the line. In each magazine is a row of taped components, which are installed parallel with the line as well. The magazine moves linearly along the track to position the appropriate magazine adjacent to the head, which then picks up the components, turns it through 90 deg, and inserts it in the pcb which has meanwhile been manoeuvred into position.

Naturally, Matsushita has standardised on pcb size, there being one size for the main boards, and another for sub-system boards. At the beginning of the line is an AH-11 machine, which normally inserts jumper wires – U-shape wires whose two ends are pushed through a pair of holes in the board – diodes and resistors. Both resistors and diodes are of the type which have wires projecting longitudinally from each end.

The wire for the jumper wires is in the form of a reel on the machine, which crops off a length of wire, forms it into a U, and then presents it to the insertion tool. The machine at the typical first station inserts 60-odd jumper wires before following on with the other parts to complete the 95 operations in under one minute. Each insertion takes 0.6s.

At the second station, there are four arms operating on three pcbs, this being an M-12 machine. The three boards are grouped together, and are moved simultaneously in the x-y axes. Generally, one arm does a number of operations before the next one takes over, but in some cases, they change over after only a couple of movements, to cut out the time needed to move the boards. At this station, 58 items are inserted in each board.

At the third station another M-12 machine performs similar functions, but the fourth machine is designed to handle resistors which have their wire terminals preformed into two legs. The machine at the fourth station handles taped resistors. It inserts 60 parts per cycle.

However, the last machine differs from the others in that the board remains stationary, and the arm moves across the board as necessary. In fact, there are two arms working at two pcbs at this station. These resistors have no external wires, and are attached to the back of the pcb, so there is a turnover station just before the machine. The resistors are coated with adhesive, and the arm merely pushes the resistor on to the board to locate it. Immediately after the machine there is an ultraviolet curing chamber.

Above each arm are a pair of hoppers containing resistors in different sections in rows – 50 different types for each arm. The hoppers are moved by a cam and link system so that they rock gently up and down. Projecting upwards into each section of the hopper is a plastics tube. As the hopper moves downwards so the tube pushes up through the collection of resistors. This action lifts up some of the resistors, and when the hopper moves upwards, one of these usually aligns itself with the tube to fall into it.

These tubes direct the resistors to a manifold joining together to feed one chute to the tube linked to the arm. A controller opens the appropriate gate to allow the desired resistor to fall down the chute, where to help the sticky component on its way, a tiny belt conveyor is installed.

The insertion head itself is mounted on two parallel bars forming a carriage running across the track. The head can move laterally across the track along the carriage. The carriage itself is carried on blocks which are mounted on a pair of parallel rods – one each side of the track. Thus, the head can move in the x-y axes. In addition, the head can move the insertion device vertically, to place the resistor, supplied through a plastics tube, on the pcb.

The head is actuated by dc servo motors driving through small toothed belts. This machine is being used by Matsushita only where the parts are of similar shape. According to plant manager, Itsuo Sukemune, it is not suitable for other applications, but in this case, its simplicity and low inertia make it ideal.

Throughout the shop, and the next stage of assembly, the pcbs are stored in carriers which can each hold about 20 pcbs. Between insertion machines there are generally about 8 pcbs as a buffer, but at the beginning of the chassis assembly line there are about

Most of the chassis assembly work is done by girls.

200 boards – at the cycle of 56s that is enough for about 3h, which is a lot by Japanese standards, but which reflects the fact that the assembly lines operate on one shift, and the insertion lines two shifts.

Although the insertion machines are all controlled by NC devices, each line in the high-volume shop is dedicated to just one pcb which it produces continuously. Says Sukemune: 'This is simpler than switching from pcb to pcb. We don't see any prospects for completely flexible assembly with pcbs in random order in the future.'

In the low-volume shop there are 45 insertion machines which are used to produce over 100 different pcbs. These machines are loaded manually, with the operators transferring boards from machine to machine as needed. It is interesting that seven people are needed to look after the five automated lines in the high-volume shop – that involves some 25-30 machines – whereas 13 people are needed on each shift to look after the 45 machines in the other shop. Overall, therefore, 40 people are needed to control insertion of something like 90,000 parts/h. Sukemune says that this accounts for 80% of the parts of a TV set.

In the chassis assembly shop there are 8 lines, each U-shape, and each doing an identical job. In practice, three of the lines have less automated transfer systems because sets with more than an average numbers of parts are assembled on them. Generally, there are 14-20 girls working on each line, to a cycle of 54s.

On the more modern lines, non-synchronous conveyors are used. The pcb is mounted on a pallet which is transferred by a continuously moving conveyor. The girl applies a brake to stop the pallet to work on that chassis.

At the beginning of each line is the main innovation in assembly – five robotic arms. Otherwise, the only jobs carried out by machine are automatic soldering, and insertion and tightening of screws – by conventional machines.

Each robot is a simple three-axis machine, of the Cartesian co-ordinate type in which the arm is moved vertically on a column, and extends horizontally. In addition, the gripper can move vertically.

At the first station, the robot, fed from a bowl feeder and tube, inserts screws to secure the pcb to aluminium brackets. The second robot tightens the screws. At the third station, a robot inserts 5 connectors, at the fourth, 6 square transformers, and at the fifth some resistors. In each case, simple chutes, loaded manually with 40-50 parts, are used to supply parts to the robot arm. These chutes are inclined down to the track, the last portion being horizontal of course, and terminating in a stop. The robot arm picks and places the components in turn from the ends of the chutes to the pcb. These are very simple machines, produced in-house by Matsushita, which are giving a worthwhile increase in productivity – especially since the connectors are difficult to install by hand, as a lot of pressure is needed.

In the final assembly shop, there are also eight lines, each with 20-25 operators. Thus, overall, there are about 350 people involved in TV assembly at the Ibaraki plant. The average age in the assembly shops is 24, compared with 28 in the plant as a whole. For Matsushita as a whole, the average age of employees is 30.5, and average wages are £8,000 p.a., which is probably 50% above that in assembly, wages in Japan being related to age.

For the future, Sukemune is looking for simplified design to improve quality and productivity – through the inevitable use of larger integrated circuits. Over the next decade, Matsushita intends to install 100,000 'robots' in its plants, but Sukemune confirmed that most of these would be very simple devices for pick and place operations. Echoing a view often heard in Japan, Sukemune said: 'We will increase the amount of automation in the future, but it is to improve quality as well as productivity. We don't do it to reduce the number of workers.'

To some extent, this is a propaganda ploy, but it is true that the Japanese are anxious to improve quality with automation. The real point is that the decision to install automation is not made on a simple comparison of the cost of the man versus the machine, but also of the effects on quality and flexibility – and that, of course, is sound management.

Fanuc switches to robot assembly

Large number of robots used to assemble motors in new factory, while new systems are on show at Osaka.

IN A dramatic move, Fanuc has adopted large-scale robot assembly at its new servo motor factory. Situated next door to the plant where robots are built near Mount Fuji, the new two-storey plant is used to machine and assemble motors. Although the statistics are as impressive, as is expected from Fanuc, there are some doubts whether the plant is as productive as it might be. But there can be no doubt that in the way in which robots are used to assemble components produced in relatively low volume, makes this a trend-setting installation. tion.

One reason for doubt, of course, is that the factory is still in the early stages of the learning curve, since production started in September 1982. The other reason is that, at present, demand for servo motors is low, partly because the machine tool business has been hit hard by the recession, and partly because robot demand is not increasing as quickly as had been expected. But in any case, the sections of the lines are not perfectly balanced, as is inevitable with a robot line.

The plant has been set up with 101 robots, 52 in the machine shop and 49 on assembly, to produce spindle motors and both ac and dc servo motors. There are 60 workers, and capacity is put at 10,000 motors of 40 kinds per month. This involves 900 different parts in lot sizes from 20 to 1,000. Fanuc built the plant, which is 200 × 60m in size, at a cost of £17 million which is a little less than was invested in the first Fuji factory.

There is a large automatic warehouse which extends from the ground floor up through to the first floor, and when parts have been machined they are stored in the warehouse before being withdrawn for assembly. They are also stored in the warehouse between stages in the machining and assembly processes.

On the first floor, there are four rows of assembly robots with a total of 25

Figure 1. Control console in Fanuc's new motor factory.

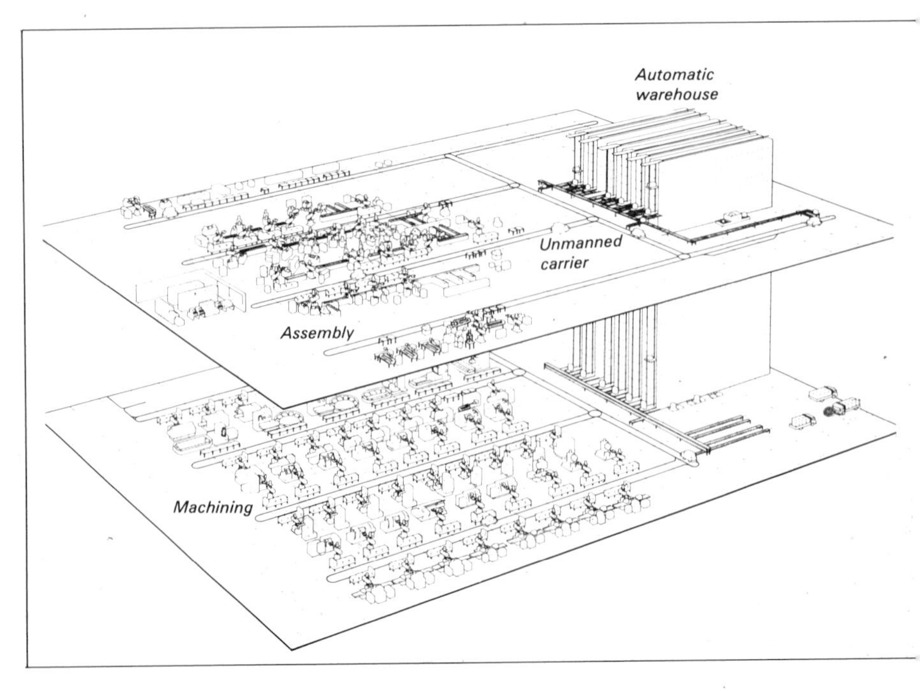

Figure 2. Layout of machining cells on the ground floor, assembly cells on the first floor, with an automatic warehouse serving both.

cells incorporating 49 robots – a mixture of M and A Models. The robots are arranged linearly, and workpieces are transferred between cells by conveyor. The whole plant is controlled by computer which also dictates the production schedule in response to orders from the head office.

Although the machine shop operates for 24h a day, with the operators on one shift only, the assembly shop is operated only for 8h/day. The reason for this, says Fanuc president, Seueimon Inaba, is that some of the assembly is done manually. The level of automated assembly is put at 65%. 'We would like to remove the men altogether,' he said, 'but that will take two or three years.' Until this is possible, he is reluctant to operate on two shifts.

Magnetising

In fact he says that although the machine shop is geared up to produce 10,000 sets/month, the assembly shop can currently produce 6,000/month, Altogether, there are six lines running down the shop. In the first row are some machines for magnetising the shells, then there are four robot assembly lines, and finally, there is the manual assembly line. The first robot assembly line builds small dc servo motors, the second large dc servo motors, the third line has not yet started to operate, but is for ac servo motors, and the fourth row is for the large spindle motors. In the future, production of dc servo motors will decline in favour of ac servos; the lines have been designed to accommodate both.

Figure 3. General view of the assembly shop, prior to operation, with pallet carrier and control unit adjacent to each cell.

Figure 4. Part of the Fanuc shop, with rows of rotors between cells.

Because the spindle motors are produced in low volumes, different grippers and minor components are needed for different models, and the tooling is changed over accordingly. But in the other lines, the robots and machinery are designed to handle a range of units without changes in tooling.

The actual operations performed by the assembly robots differ according to the size. There are three 'cells' in the line for the spindle motors, and in each of the other lines there are an average of eight.

At the beginning of the line, the rotor/shaft assembly is balanced, then the ball bearings are pressed on to the shaft at each end of the rotor. An oil seal is pressed into a flange, and sealant is applied. Then, the rotor is mounted in the flange, and the shell is lowered over the rotor on to the flange. Subsequently, the end bell is assembled to the shell and pressed in position. Four tie-bolts are then inserted, washers and nuts are lowered into position, and the nuts are tightened. Then, the motor is passed to the manual line for adjustment.

Currently, there are two major jobs that are done manually. First, the rotary encoders must be adjusted to the correct datum by hand. Then, the flange is adjusted so that equal current is fed to both coils, and some wires are fitted by hand.

In all cases, the concept is similar. Five unmanned trolleys carry parts on standard pallets along four aisles between the automatic warehouse and the various cells. The capacity of each pallet is 20-50 parts, according to size of the part. At each assembly cell, there is a table which can carry three standard pallets, and the control unit for the cell is adjacent to the pallet table. The handling robot draws parts from one pallet station, and there is another station for loading and unloading from the trolley, and one spare station.

Various

These pallets are used for the parts that are added at the various cells, but of course, the sub-assembly itself moves between the cells on small conveyors. These are like the oval carousels used as workpiece feeders in Fanuc machine shops, the robot at one cell placing sub-assemblies at one end, and the robot at the next cell removing it from the other end.

As a preliminary operation, the shells are magnetised on the separate line, and are returned to the central warehouse. Meanwhile the rotors are withdrawn from the warehouse and are placed at the first cell. At the first station, the rotor is picked up by a Fanuc M-Model 1, and is inserted in a special purpose gripper for a balancing machine. At the same cell is a machine that cuts an undercut; at this machine, the robot inserts the rotor directly in the chuck.

Then, the rotors are transferred to the second cell where the bearings are fitted. In this case, the Model-M robot transfers the rotor to the press, and the assembly robot is used to transfer the bearing to the spindle so that the press can push it down. Then, the M robot picks the rotor up, turns it through 180° so that the second bearing can be pressed home.

At the next cell, the oil seal is pressed into the flange, and sealant is applied to the face. In all these cells, the principle is the same; the M-Model robot is used to handle the sub-assembly between the conveyor and table, while the A-Model robot is used to handle the small part for assembly. However, since the shells are heavy, they are generally lowered on to the flange by the handling robot.

Fanuc has developed a special feeder for tie-bolts, nuts and washers; the tie-bolt feeder consisting of chutes through which the bolts are fed horizontally, and a simple device to pick up the bolt, and turn it through 90° so that it is orientated correctly for the robot.

The assembly robot used at this cell has a special double gripper which can carry both washer and nut. The nut is picked up by a central chuck, and then moves to the stack of washers, where a small pair of cranked fingers pick up a washer. Then, the robot moves to the assembly table, and drops the washer over the tie-bolt. The small fingers extend so that they are out of the way, and the chuck tightens the nut.

Adjustment

Once assembled, the motor is taken back to the warehouse before being passed to the manual assembly line. After adjustment, it is checked functionally on an automatic machine, which records the results. Then, the completed assembly is taken back to the warehouse prior to packing.

In the complete plant, there are 60 employees, including 21 in the machine shop – on one shift – and 19 direct workers in the assembly shop. All handling is done mechanically in the assembly shop. Nevertheless, the robots seemed to be working slowly, and the machines seemed to be idle for quite a lot of the time that I was watching the plant. Certainly, the line for spindle motors started to operate only after I had been watching for about 10min. Inaba explained that because the plant has been operating for only a short time, the robots are moving slowly. 'The priority is not to stop the robots,' he said, adding that as they gain experience, so the robots will be operated faster.

In the central control room, there is a display which indicates how well the cells are operating; red indicates a breakdown, orange that there is a shortage of parts or some other delay, and green means that the cell is functioning correctly. When I visited the plant, there were nine orange lamps showing, one of which was in the warehouse, and another at one of the unmanned trolleys. Inaba conceded that this was 'not very good', and countered that when he visits the plant, he can usually see green lamps only!

All these factors tend to suggest that the use of robots may not be the most efficient way of doing the job. To this suggestion Inaba says: 'We have high variety and low volumes, so we need robots. This is different from companies with big volumes and low variety. All the Japanese electrical companies are now trying to increase productivity, and in theory, our new

Figure 5. At the spindle motor line, the robot inserts the rotor into an oven before placing it on a stand and placing the spindle in position so that the press can push it in.

Figure 6. Citizen Watch Synectrons arranged to assemble gears to a housing.

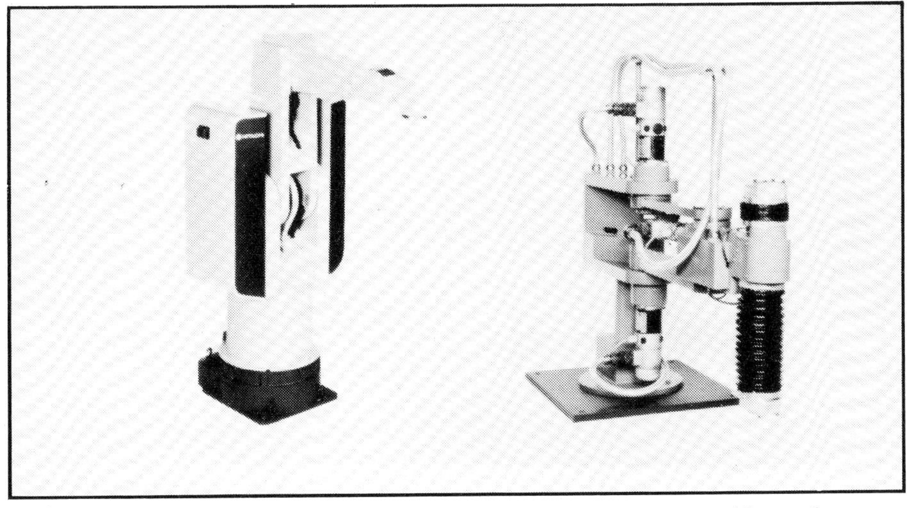

Figure 7. Hitachi's A6030 and A3020 robots are aimed at assembly work.

Figure 8. Toshiba assembly robots, one with a small camera, being used to position wires and solder them to terminals.

plant increases productivity by three times. We used to have 32 robots and 108 men to produce 6,000 motors; we shall have 60 men and 101 robots producting 10,000 motors.'

He asserts that the manufacturing cost of the servo motor will be cut by 30% including depreciation. The value of the output when the plant is in full swing will be about £40 million a year, or more than twice the capital investment.

In addition, of course, Fanuc will be able to gradually switch from dc to ac servo motors as demand changes; it will not need to make a sudden switch, as would have been needed with hard automation. Since the robots are operating for only one shift, any faults can be eliminated, and programs can be optimised outside normal production hours.

Again, this looks like the classic long-term approach of Japanese companies. If the plant is producing only 3,300 motors/month, its level of productivity per man will be the same as in the old plant, yet it will have the potential to operate with broadly similar manning levels with much higher output. Clearly, in that sense, this is a flexible plant.

It will be about three years before the assembly shop is able to operate completely unmanned, although if demand warrants it, the shop may operate on more than one shift before that. In the end, therefore, the plant will pay handsome dividends, and in the meantime, Fanuc is extending its knowledge of assembly by robot, which will help it to increase robot sales.

Citizen

Citizen Watch, which has been producing its own assembly equipment for some time, is now offering the Synectron range of simple robot assembly devices. Actuated by dc servo motors, these machines are designed to carry objects up to 2kg, but are really intended for assembly of small parts, generally weighing a few grams.

A cartesian co-ordinate machine, the Synectron is available in two sizes, with either 300mm or 500mm movement in both x and y planes. There is 50mm movement in the z axis, while the gripper can rotate through 360°. Accuracy is put at ±0.2mm, and the arm can move at up to 250mm/s.

At the Osaka Machine Tool Fair, Citizen was showing four of these robots in a row. They were being used to assemble a number of gear wheels to a small plastic housing, the biggest gear being about 25mm in diameter. The first robot was used to transfer the base to the pallet on the conveyor. The second robot picked and placed four gear wheels in position, while the third robot inserted a couple of circlips. The last robot transferred the assembly to a tray. The cycle time was around 30s.

These were simple operations, except for the installation of the gears. This was a compound gear train, and it was necessary to ensure that the teeth of mating gears were in mesh before the second gear was pushed down. To overcome this problem, the gear was turned slightly as it was lowered into position. Although the line seemed to be performing well, it would clearly be a costly installation. The Synectron costs about £12,000, so the robots alone would cost £48,000, apart from the conveyor line and ancillary equipment, which is one reason why such automation can be justified only where the volume is high.

The same comments apply to the flexible device intended for drilling and bolt tightening produced by Japan Automation Machinery. The ML450 is a rotary table with a diameter of 450mm, and an x-ray pick-and-place arm. There is also a device with column and pusher for parts feeding. The table is indexed around, taking 0.6s to index one position – 90° – the workpieces being placed on a 340mm pitch diameter. The arm can traverse over a square 100mm × 100mm.

Thus, it can be programmed to either drill a number of holes in different parts without any special tooling, or it can be used for simple assembly jobs. It can operate to cycle times of 1.6 to 10.6s, and so long as it can be integrated easily into a line, is a useful tool, even though the price is around £9,000 in Japan.

Price barrier

However, Hitachi has made these devices look expensive with its new cheap assembly robots, which only cost around £2,500. This simple four-axis assembly robot is therefore about one-third of the price of most competitive machines. Intended for use in assembly of electronic consumer products and automotive assemblies, the new A4010 robot is certain to increase the number of robots used for assembly work dramatically.

It is similar in concept to the Scara machine, but the arm is mounted on top of a console which contains the controller. The arm can articulate on a vertical pivot on the console, and there is also a vertical elbow joint. At the end of the wrist is a vertical telescopic column giving vertical travel. It can handle parts weighing up to 1kg, and the complete assembly weighs only 30kg. Repeatability is put at ±0.2mm, while maximum speed of operation is 730mm/s.

The A4010 is the cheapest of a range of three assembly robots, the others being the A3020 and A6030, which cost around £11,000 and £18,000 respectively. The A3020 is similar in principle to the A4010, but it is twice as fast, with a maximum speed of 1,500mm/s. In addition, its level of repeatability – ±0.05mm – is clearly aimed at assembly rather than simple pick-and-place operations. The A6030 operates at the same speed, but is a jointed arm machine with a capacity of 3kg.

The two larger machines have magnetic bubble memories with a capacity of one megabit to store 500 steps and 2,000 instructions. They are due to become available in April, and Hitachi plans to build up production to a staggering 400/month.

Meanwhile, Toshiba has developed robots with six axes of freedom and a ccd camera acting as a visual sensor. Two of these robots are to be used to solder wiring terminals to components in the base of an electric fan – a product made in huge volumes in Japan. The robot is similar in concept to the Puma.

Solders

The tiny camera lens is mounted in the middle of the gripper on one robot, and it first identifies the position of the wire and terminal. Then, the robot arm is positioned to pick up the wire, which has been identified by colour, and then places it at the correct terminal. The other robot moves across and solders the wire in position. Toshiba engineers say these robots are about a year away from production.

These machines are developed from Toshiba's assembly robots, which are in use experimentally in one of its factories. These earlier designs have five basic axes of movement, and are similar to the Puma except that the wrist incorporates an extra vertical joint. Thus, the wrist and insertion tool is always kept vertical.

All these developments suggest that robot assembly is moving nearer, both for large and small volumes, even though companies such as Matsushita Electric still see little scope for robots in their assembly shops. Instead, their engineers prefer simple devices with sequence controllers, because these machines are more reliable, yet give some flexibility. It still is a case of horses for courses.

Integrated systems appeal to the Japanese

Brian Rooks, Editor of *Assembly Automation*

An ability to integrate complete production systems is claimed by one Japanese company to be the reason for an expanding order book. Some of the activities of Hirata are reported.

HIRATA Industrial Machineries company, to give its full name, was founded by Tsuneichi Hirata at Kumamoto, Japan, in 1946 and incorporated as a limited company in 1951.

It now employs 430 people at factories in Osaka, Kanto and Kumamoto. Its products range across a spectrum of material handling equipment needed for automated systems, such as conveyors, elevators, guided vehicles, palletisers and storage systems and cranes. It also builds automatic machines for pin insertion, screw driving, nut running and electronic assembly and other assembly machines. But, perhaps its forte is the ability to supply integrated production systems. Domestic appliance, automotive components, cement and building material, agricultural, marine, food and chemical goods are the industries where Hirata systems can be found.

Customers

Many world famous names appear on the company's list of customers. Sanyo, Sony, Panasonic, Honda, Toyo Kogyo and Nippondenso rank among its Japanese customers. Zenith Radio Corporation in the USA, Blaupunkt in Germany and Thorn-EMI in the UK appear on its overseas list.

The assembly systems are based on three main product lines – the Machine Base series of x-y linear tables, the Hi-CNC controllers and the more recently introduced range of assembly robots and controls.

The robot, so-called Arm-Base is of the SCARA type. It is a 4-axis device controlled by a Z-80 based controller and programmed by Hirata's HARL robot language. It is fast, 2360mm per second, and accurate, ± 0.05mm. The claim is that these characteristics make for easy installation and application. Programming is aided by the need to select only the desired commands or program modes displayed on a LCD with just six software designated keys and standard numeric keys for data input.

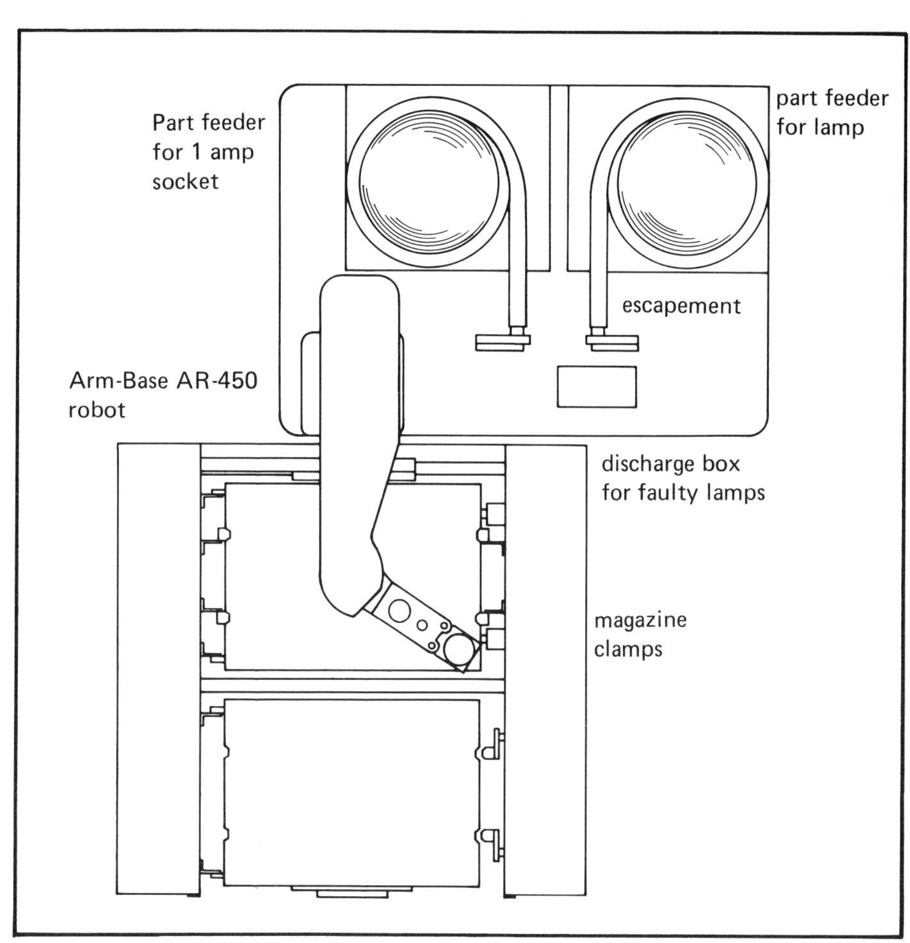

General layout of the robot cell for assembling the lamp assembly for a photocopying machine.

The language HARL has a grammatical structure similar to BASIC. There are just 10 commands which enable the appertaining conditions to be judged and appropriate actions implemented. Both vertical and horizontal speeds are programmable either on a continuous path, linearly interpolated or point-to-point basis.

Mechanically the arm has the fundamental characteristic of the SCARA robots of vertical load rigidity. Thus it is suitable for press insertions, screw driving and the like. Payload can vary from 7kg for a two-axis machine to 2kg for the four-axis version.

Hirata has studied, designed, built and installed many production integrated systems incorporating its Arm-Base robots. Large capacitors, transformers, conenctors are inserted into PCB for several customers. These applications involve Hirata's other robotic product, the Machine Base x-y tables. Video cassette and audio micro-cassette recorders are some of the other products that benefit from the characteristics of the robot. Handling nuclear fuel elements, automatic screw driving and nut running, and palletising are now also finding a use for Hirata's robots and automatic machines.

Last autumn the Kumamoto company delivered a system to a Japanese customer for building the exposure lamp assembly of a photocopying machine. This uses one of the AR-450 type Arm-Base robots working in conjunction with a specially developed pallet magazining system.

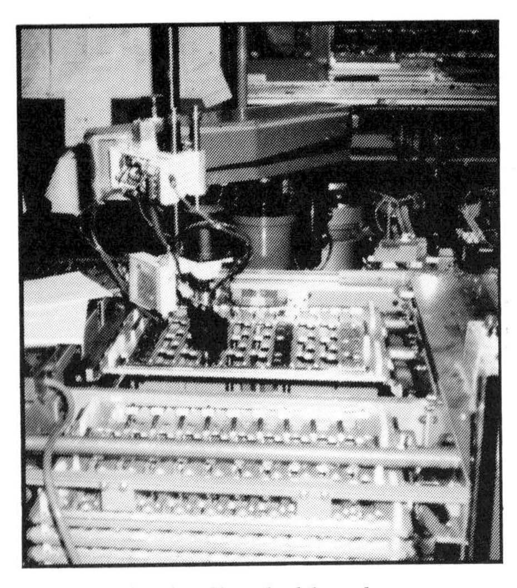

The stack of pallets held in the magazine.

The basic function of this unit is to insert a number of lamp and lamp holders into a PCB. The lamp and the lamp holder are fed by two vibratory bowl feeders which orientate them ready for pick-up by the robot. The holder is inserted first, followed by the lamp itself. However, before the latter takes place a continuity test is performed on the lamp filament. If a defective one is found the robot will go into a jump routine and deposit the faulty lamp into a discharge box.

The PCBs are held on pallets. There are two types, one with six insertion points and one with 11. The robot switches from one to the other under program control. The cycle time taken by the arm to insert each lamp is 8s.

Manually

Each pallet houses four boards and the pallet magazine holds 10 pallets. The boards are set onto the pallet and loaded into the magazine manually. However, once that has been accomplished the machine operates automatically.

There are two stacks in the magazine. That nearest the robot holds the blank boards, and the one farthest away receives the fully assembled pallets. Once all the lamps have been inserted on one pallet it is indexed over to the finished stack of pallets. A pneumatic jack and rack, and pinion arrangement are used to index the two stacks up and down, respectively.

Hirata already has a presence in Europe and aims to further its image at AUTOMAN. A pair of the Arm-Base robots will be demonstrated in a three station assembly system at the NEC this month.

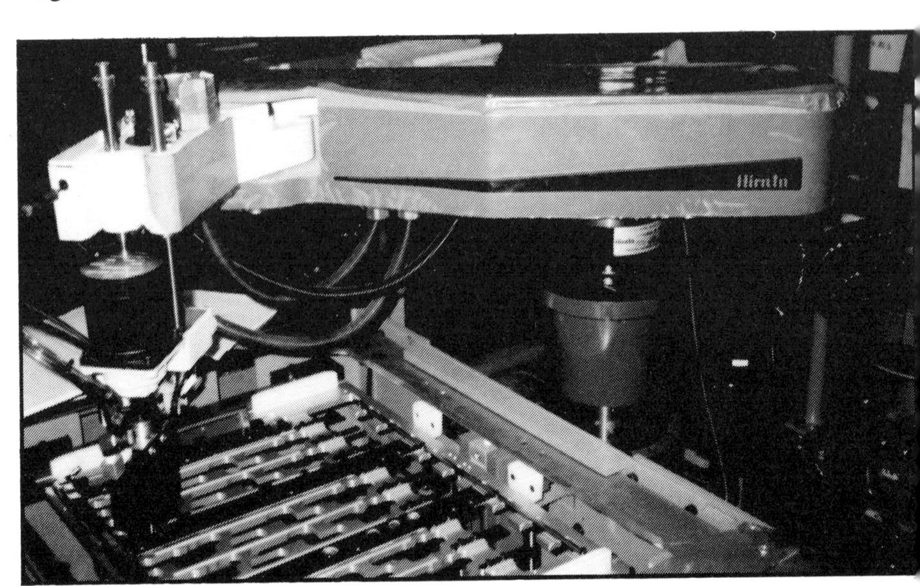

The Arm-Base robot inserts the lamp holder and the lamp.

Hitachi follows the robot road

Robots are intermingled with hard automation in Hitachi's new vcr assembly line.

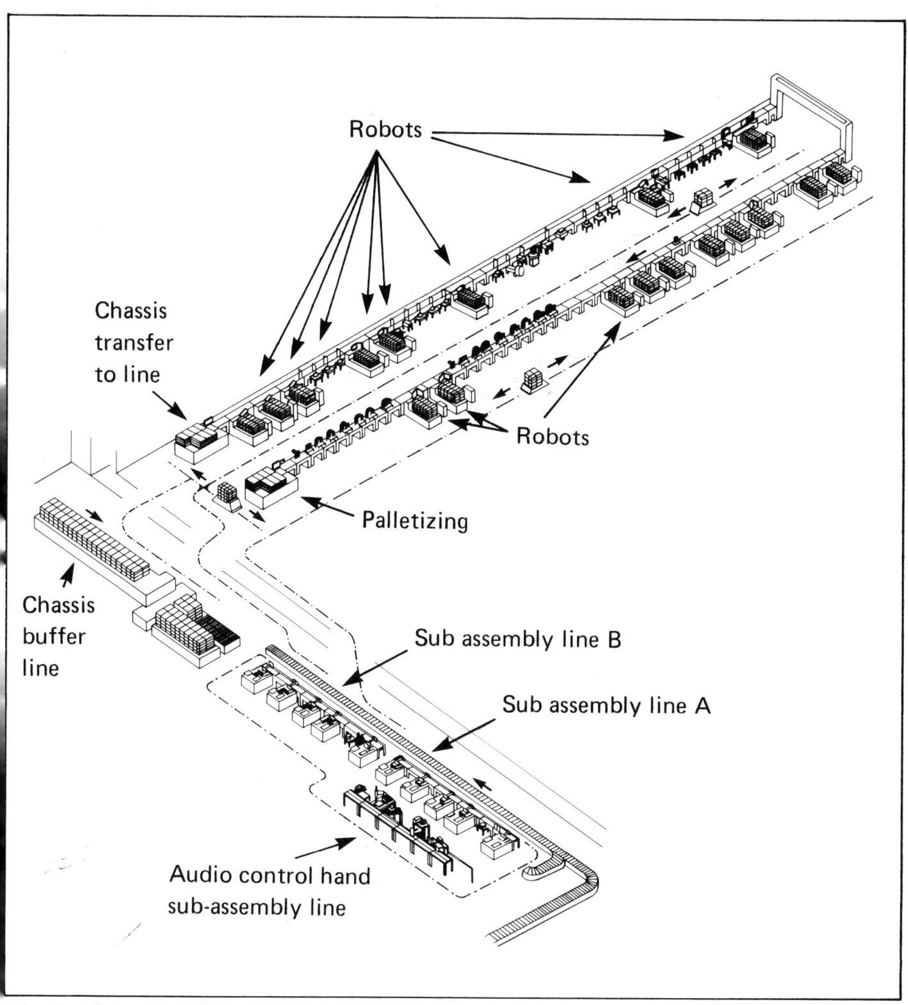

Fig 1. The general arrangement of the Hitachi vcr assembly line.

HITACHI, which installed an almost completely automated line for the assembly of audio mechanisms, has switched to the use of robots for some of the assembly in its vcr chassis assembly line. The line consists of 11 robots and 52 simple pick-and-place arms. However, before installing any robots, Hitachi engineers conducted an assembly study, and rated components according to their suitability for robot assembly. As a result, a number of minor alterations were made to the design of components.

In order to assess how suitable components were for automatic assembly, Hitachi devised an evaluation system, in which features of the component were included, such as the ease with which it can be picked up, whether it has a natural location on the chassis, whether sub-assembly is needed, and whether installation is vertical or not. Marks were given for the various parameters, with a maximum of 100.

When the vcr chassis was first evaluated, its rating was 63, and it consisted of 460 components. The first job in adpating it for automated assembly was to reduce the number of components, so that awkward sub-assemblies were eliminated. For example, the tape guide initially consisted of a sleeve and two collars; this was redesigned into one piece. A location guide was added to a motor, plate and pulley assembly, while the flange of the capstan motor was revised. In this case, parallel sides were needed to simplify gripping. Then, big chamfers were added to the noses of the screws, and washers were enlarged.

Lead wire

Wires are always a problem with assembly, so a lead wire and connector from the loading motor were replaced by a rigid plate and connector. Then, the original ac head was mounted on a pressed plate. To improve location, the end was squared off. A similar modification was made to the B slider. With these and other similar modifications, the number of components was reduced to 350, and the score was increased to 73. Then, it was decided that with a combination of pick-and-place devices and robots, automatic assembly was possible – in practice, a few jobs still need to be done manually.

For robotic assembly the A3020, Hitachi's cheapest robot, which is similar to the Scara, was chosen. The first section of the arm is 400mm long, the forearm is 300mm long, and their ranges of articulation are 200° and 155° respectively. The arm can also rotate the gripper through 360° and move it 200mm vertically. Operating speed is put at 1.5m/s, and repeatability is ±0.05mm.

Folded

The vcr mechanism is complex, with a large number of components mounted on two sides of a folded aluminium chassis. On one side, the main components are the capstan motor, cylinder head, head assembly, two guide roller assemblies, an idler arm, and the spindles that engage with the cassette reels.

On the back of the chassis is the loading motor gear and belt drive, and the drive belts and pulleys. The capstan motor spindle carries two

Fig. 2. Pallets of components stacked at the beginning of the line.

pulleys, the one nearest the chassis driving with a considerable reduction ratio the large flywheel in the main cylinder head drive. The second pulley drives a belt that stretches across the flywheel to the clutch. Above the flywheel and belt is a strip on a pair of legs. Then, there is the eject slider and spring, the mode slider and spring, and the brake slider. Everything is packed close together, but the sub-assemblies are laid flat on the chassis or positioning pins, or plates.

Once the line had been designed, Hitachi uses the GPSS simulation program to determine whether it would actually perform as expected. Stoppages were categorised as follows:

○ Minor stoppages, such as mis-assembly, parts jamming and so on. These normally result in a stoppage of only 30s.

○ A stoppage that requires some rectification to the machine, involving a stoppage of more than 5min. In that case, once the 5min have elapsed, an operator is supposed to take over from the machine.

The occurrence of a serious breakdown was calculated using the formula:

$$T = T_o \times (MTBF + 5\text{ min})/MTBF$$

where T = cycle time including major breakdowns; T_0 is the cycle time with minor breakdowns only; and MTBF is the mean time between failures for the whole line.

After the simulation had been done, buffer sizes and distribution was decided. The simulation showed that the actual cycle would be 5.3s, but with 0.3% average machine breakdown time, the net time would be 7.5s. However, with practical experience on the audio line, it was decided that the actual time would be 7.9s. In practice, the line operates with a cycle time of 5.3s, machine breakdownn averages 0.3%, and the net cycle is 8s.

The line is U-shape, and consists of 11 Hitachi A3020 robots, nine manual stations, 52 automated assembly stations, and 18 buffer stations. It is installed in a shop 150 × 245m, the actual line being 75 × 11m. Parts are supplied by unmanned carts to the buffer area at the beginning of the line, and also to the parts feeding stations behind the robots – inside the U. The line is 88% automated, most of the actual assembly operations being performed by pick-and-place arms. The line is designed to produce 100,000 units/month.

Motor assembly

The robots are used to assemble such parts as the motor assembly, gears, a slider clutch, flywheel, reels, the a/c head, and an arm. Manual assembly is used for a drive-belt, and some other awkwardly-shaped components, and the pick-and-place devices do the rest of the work.

Chassis are supplied to the beginning of the line in pallets, four to a pallet, and are placed on the line in turn, by an x, y, z pick-and-place device. The pallets are supplied in stacks five high, to a station parallel to the beginning of the line. There are normally two stacks of pallets in this buffer, and one pallet in the next station. From there, the pallet is moved transversely across to the loading station. When the pallet is empty it is moved straight back beneath the stack of empty pallets. In all, there are likely to be about 125 chassis waiting – enough for about 10 minutes' production.

At some stations along the line vibratory bowl feeders are used, but at many stations parts are supplied in pallets. There is usually space for about 24 pallets at each station – they are generally supplied in stacks eight high. In each pallet, the components are arranged carefully in moulded mats – a prerequisite for robot assembly.

Equally important is precise location of the chassis at each station, which is designed to be ±0.1mm. As mentioned previously, the robot can place components to ±0.05mm.

There are a group of six robots in the first dozen stations, two more towards the end of the first leg, and three towards the end of the line. Thus, the line is a mixture of different assembly methods, chosen according to the workpiece.

The robots are equipped with

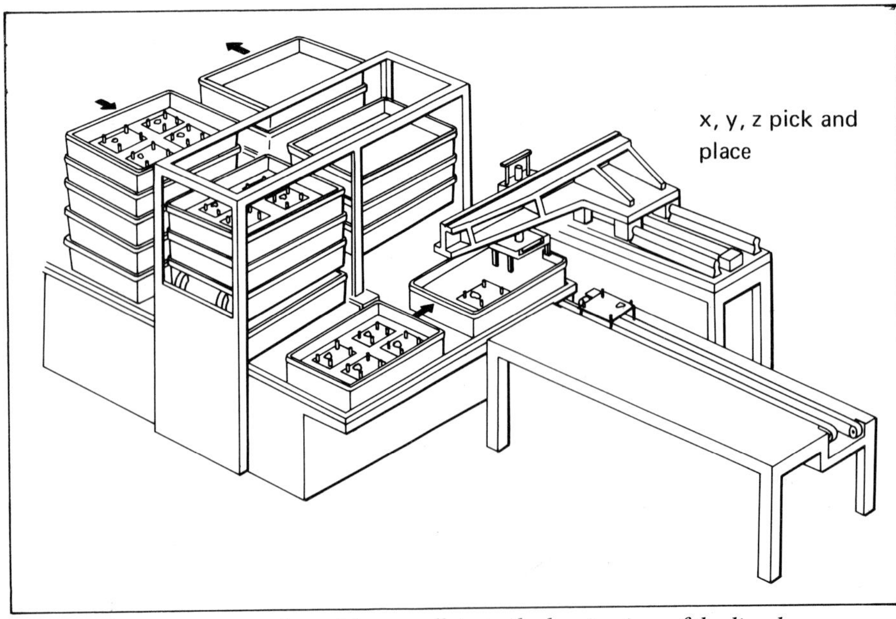

Fig. 3. Chassis are transferred from pallets to the beginning of the line by a pick-and-place device.

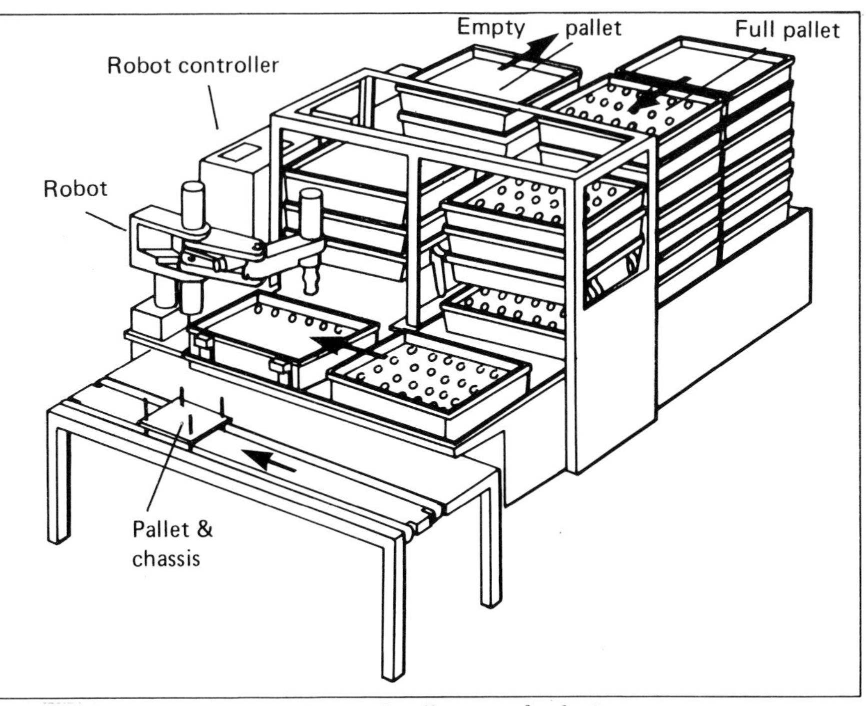

Fig. 4. Layout of the Hitachi robot and pallet transfer device.

special grippers to handle the components, and Makoto Matsunaga, manager production engineer, says that in assembly, FMS means that the model can be changed without renewing the line. Therefore, he does not consider the need for a special gripper for each station a limitation.

The robots are generally used to pick components from pallets behind the line, and install them in the chassis. Typical components weigh a few grams, and are around 30cm². The first robot deposits a small plastics component into the chassis, and at the next station, an arm drops some oil on the assembly.

The second robot picks a motor assembly from a pallet, and places it in the gripper of a special-purpose arm on the opposite side of the line. This device turns down through 90° and then lowers the motor into position. It operates quickly, and since the robot can immediately start to pick the next part, this combination of robot and arm is both quick and accurate.

Steel bracket

The third A3020 robot is used to insert and fasten some screws, and the fourth lowers a pressed steel bracket on to three plastic legs projecting from the chassis sub-assembly. The next robot installs a fairly large gear/pulley assembly, but first it pushes it against a proximity switch to ensure that the part is correctly located.

Perhaps the sixth robot is doing the most interesting job. It fits the flywheel, which is approximately 60mm in diameter, with a hole of 3.4mm diameter, over the spindle. The clearance is only 10μm, and surprisingly, the robot drops the flywheel from a height of about 20mm on to the spindle. 'We followed the procedure adopted by the operators,' explained Matsunaga. Clearly, that is an interesting lesson, since many managers would baulk at expecting a robot to assemble with such precision.

A little further down the line, a girl is sitting fitting the belts over the capstan motor pulley and the flywheel. Although she works very fast, she can only just keep up with the line.

However, the second belt is fitted by machine. This is practical, evidently, because the pulleys are nearly the same size. This is a complex device, developed from a similar one used on the audio deck line. There is a hollow drum which rotates on a horizontal axis, and this is loaded manually with belts. Projecting into the drum is a coarse screw of about 60mm diameter. It extends over a conveyor that leads to a pick-and-place arm.

Thrown upwards

The belts are thrown upwards by the motion of the drum, and some fall on to the screw, and are wound out from the drum. They are guided along, and are allowed to fall on to the conveyor one by one. The belts are held between the sides of the housing, so that they are kept in a neat oval shape, and are indexed forwards. Transfer to the chassis is by the pick-and-place device.

Two other arms are used to install the mode slider spring – a hairpin spring, whose ends must be located under a pair of moulded lugs. One arm drops the spring in position – in fact a small angle out of position – and then two probes come down at an angle, and push the ends of the spring under the lugs. In fact, the positioning of the spring, taking into account its tendency to bounce, is a real problem.

To install coil springs, an arm with a pair of legs is used. The spring is fed to the arm, and the spring is transferred down the slightly splayed legs, so that it is tensioned. At the bottom, the hooks of the springs are lowered over the pins – this seems to work well.

The other robots are used to install wheels and sub-assemblies in a conventional manner, although the x, y, z, arms do most of the precise jobs. Of course, there is also a turnover fixture about halfway down the line, so that assembly can take place on both sides.

To ensure that parts are assembled,

Fig. 5. A girl works between the enclosed robots and pick-and-place devices.

Fig. 6. One of the Hitachi A3020 robots inserting a component in the chassis.

Fig. 7. One of the pick-and-place arms fitting components supplied from vibratory bowls.

proximity switches are used at check stations. There is a programmable controller at each station to control the robot, pallet stacker and other equipment.

Stock

There seems to be quite a lot of work-in-progress along the line, with the vibratory bowl feeders usually containing three-five days stock. Pallets contain 12-80 components each, and in some cases, 25 pallets were waiting at a station, so there might be 500-1,500 parts waiting. However, that is not much for a daily production of 5,800 assemblies.

Matsunaga claims that the line is 99% reliable, adding that a machine with a mean time between failures of 100-200min was considered good. But he conceded that some stations had a figure of around 5min.

Nevertheless, the results are impressive. Previously, 170 people were needed to man a line that took 23.7min to process a chassis, working to a 90s cycle time. Including 15 maintenance men, the new line requires 33 people, and can process a chassis in 3min, on a cyle of around 5s. Incidentally, the cycle time has been reduced from 8s since the line started.

It is not a completely robotised line, and Hitachi is not likely to use such a line for complex assemblies in the foreseeable future. But to increase flexibility, Matsunaga is anxious for the robots to be equipped with some form of pattern recognition equipment. 'The bowl feeders can supply only one kind of part,' he said, 'but with pattern recognition, the robot could pick up a variety of parts.' Thus, either one robot could insert several parts in an assembly, or it could be used for a variety of different chassis, so that several could go down the same line, cutting overall investment costs. In the meantime, though, there is plenty of scope for exploiting the use of existing robots and handling equipment in assembly – but truly flexible assembly is still a long way off.

Assembly becomes more prominent

Recent applications show that robots are being used increasingly for assembly.

WITH exports of robots worth Yen 40 billion (£120 million) in 1983 – twice the level of 1982 – Japan's robot industry is expanding rapidly. The number of makers still seems to increase, too, without any sign of the shakeout predicted by some experts. And, it is the increasing use of robots for assembly that is fuelling this comparison.

Recently, Mitsubishi Electric installed several of its small RC and RV series assembly robots in ME plants. The RC is a cylindrical co-ordinate type, with three or four axes of movement, while the RV is an articulated arm with five axes of movement. Both are small, the RC136P machine having horizontal and vertical movement of 320 and 250mm, while the RV133P has an arm with a total length of 600mm. The RCP136P has a capacity, including the gripper of 3 or 5kg, while the RV133P is limited to 1.5kg.

In three installations, the robots are used: at rivetting cells; to handle workpieces through a cell for grinding and automatic sorting; and to assemble terminal blocks. In each case, the cell is arranged in a semi-circle. For rivetting, a tube, small pressing and a spindle assembly are picked from parts feeder by the RV type robot to a fixture at a small press with an indexing fixture. The cell operates to a 15s cycle and can produce 40,000 assemblies a month, eliminating the need for 0.5 persons.

In the second cell, the RC type robot takes a rotor from a magazine, and mounts it in the grinder, after which the rotor is measured automatically, and the robot is then directed to palletise it according to the size. In this case, the robot is doing both handling and sorting jobs, so it saves 0.7 persons, working to a cycle of 20s to produce 24,000 units/month.

In the third cell, one RC136P and one RV133P are used to assemble terminal blocks. In this case, the RV type robot does most of the jobs, the RC type being reserved for screw tightening. First, the RV robot takes the terminal block from the magazine, and then picks a terminal from a parts feeder, and installs it horizontally in the Bakelite block. This operation is repeated four times. Next, the robot transfers the terminal block to a fixture at the next station, where the RC robot inserts and tightens eight M5 screws.

To save space, the robots are placed close together, while the operations are kept simple because the robot orientates the terminal block as necessary between stations. The operations take 40s, so that 15,000 can be produced monthly.

At its Manufacturing Development Laboratory Mitsubishi Electric are developing, in conjunction with Professor Takano of Tokyo University, an unusual robot of a five-bar design. The robot has two arms, extending horizontally from the shoulders of the body, and these are jointed at the elbows. However, the 'hands' are both attached to the gripper to form a four-bar system. The fifth bar is a fixed linkage. As a result of this configuration, the arms are light, because all the servo motors are stationary, and are rigid laterally. It has a capacity of 3kg, and can operate over a 1,370 × 780mm table at up to 2.5m/s.

Optical sensing

The robot is used in conjunction with a parts feeder with optical sensing to install plastics levers in vcr chassis. To give flexibility, the vibrationary bowl feeders despatch parts to a station one by one, without any control of orientation. Then a line scan camera controlled by an 8-bit microcomputer is used to identify the part and check orientation; if the part is upside down it is reversed, and if there is more than one part at the station, they are rejected.

A special gripper with a pair of plungers is used to pick up the lever by registering in a pair of holes. Resolution of the vision system is put at ±0.35mm, and owing to the chamfers on the holes, the robot can pick up levers so long as positional accuracy is within 0.91mm of the correct position. As a result of the introduction of this system, it has proved practical to use the robot to assemble three levers of different shapes in an 18s cycle.

Meanwhile, Fanuc has been improving its FMS for motor assembly with the addition of some vision sensing. The systems, based on solid state cameras, are used at two stations: one finds the position of the lead wires on ac motors, while the other is used to check the position of tie-bolts.

Fanuc's robots can now be equipped with sensor interfaces, through which data is transmitted from the sensor system to the robot controller. Templates of shapes are stored in the memory of the controller, and to ensure that the system can recognise the part in different positions and orientations, the system is taught. To do this, the part is placed in different positions, and images are produced and stored. The data are in x, y, z co-ordinates, so the sensor interface then has to convert the data so that it is suitable for the cylindrical co-ordinate or articulated robot in use.

In the first application, the sensor system is used to find the position of the wires for the stator winding which must be pulled through a bracket immediately after the bracket is assembled by robot. The ac spindle motor is mounted vertically at the station, with the lead wire projecting horizontally from the top face. The bracket is actually a cap which is lowered on to the motor by a robot with magnetic gripper, so that it covers the lead.

One camera is mounted above the line, and another is mounted horizontally, so that the two images can fix the position of the end of the stator winding wires three-dimensionally. Thus, the robot can be controlled to lower the bracket in a zig-zag path so that it first covers the lead wires, but is offset from the motor, and when at the correct height moves inwards so that the wires come out through the hole. Then, the bracket is placed centrally on the motor.

Since the introduction of the sensor, the job has been done by robot instead of manually, with an 'almost 100% accuracy rate,' it is claimed.

On the dc motors' line, tie-bolts are fed through the rotor shell and the end bell, washers and nuts are assembled,

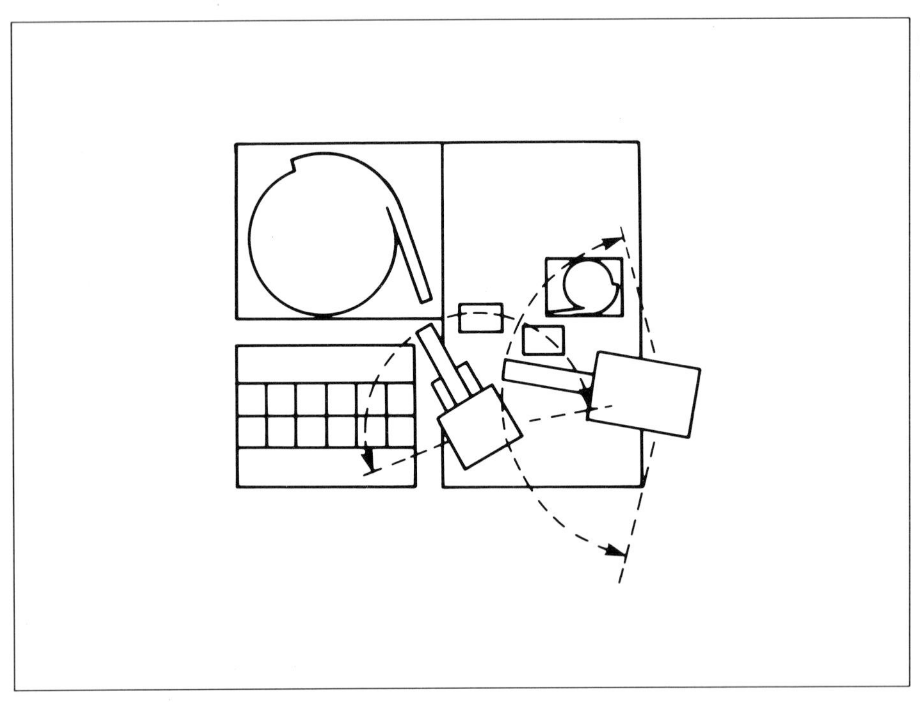

Concept of Mitsubishi Electric's parts feeder and robot assembly system for levers in a vcr chassis.

and the nuts are tightened. However, because the stud is long, and the holes in the end bell are elongated, the actual position of the end of the bolt can vary a little. The use of one camera to find the position of the bolt in the slot ensures that the nut is lowered directly on to the bolt, thus preventing errors.

Murata Machinery, which sells Prab robots in Japan, has recently installed an E1 cylindrical co-ordinate robot to handle workpieces at two CNC lathes. The robot has three main axes, plus three at the wrist. However, the wrist movements are limited to 180° of movement – either on or off. In addition, so that the robot can handle at two machines mounted on a line, it is carried on a traverser giving several metres travel at 0.9m/s. The robot is equipped with a double gripper so that it can carry workpieces of either 30-80mm or 70-200mm in diameter. Capacity is 45kg, including the gripper, so that in practice it can carry two 10kg workpieces.

Components

All components are steel, and batches are generally of more than 500. Rectangular pallets can be stacked in up to seven layers to give a buffer of 140 parts. This simple method of parts storage is practical because the robot can lift the pallet. Since cycles are 2-3 min, this gives a stock for at least 4.6h.

In theory, one robot can service up to three machines, but there would be little time for error, or to allow for mismatch in cycle times of different workpieces. In practice, therefore, actual utilisation of the robot is 50-60%, which is a far more acceptable figure than the 25-30% commonly found where one robot is used to service one machine tool.

Hitachi has built an experimental robot in which shape memory alloy is used to give servo-assistance so that smaller motors than normal can be used to actuate the arm.

Shape memory alloys are special alloys –some are of titanium and nickel – which are at their normal shape at higher temperatures, the actual temperature depending on the use. At lower temperatures, the alloy is pliable, and can be deformed to a curved shape for example. Then, when the alloy is heated – to 80°C in one case – the material returns to its original straight shape, exerting considerable force in doing so. When the material cools, it can be pulled back to its cold shape by a spring, or by a shape memory alloy working in the opposite phase.

Shape memory

In the Hitachi arm, the shape memory alloy works as a linear motor, actuating the joints through pulleys and cables. In the normal state, with the arm retracted, heat is applied to one shape memory alloy member so that it remains warm. Another shape memory alloy member is cooled by air so that it is extended to keep the arm retracted. To extend the arm, cool air is applied to the actuating member, and heat is applied to the retracting member. In this way, fast response is obtained. However, at present, it is difficult to control the shape memory alloys precisely, but since very small and light servo motors can be used, this is a promising avenue for development.

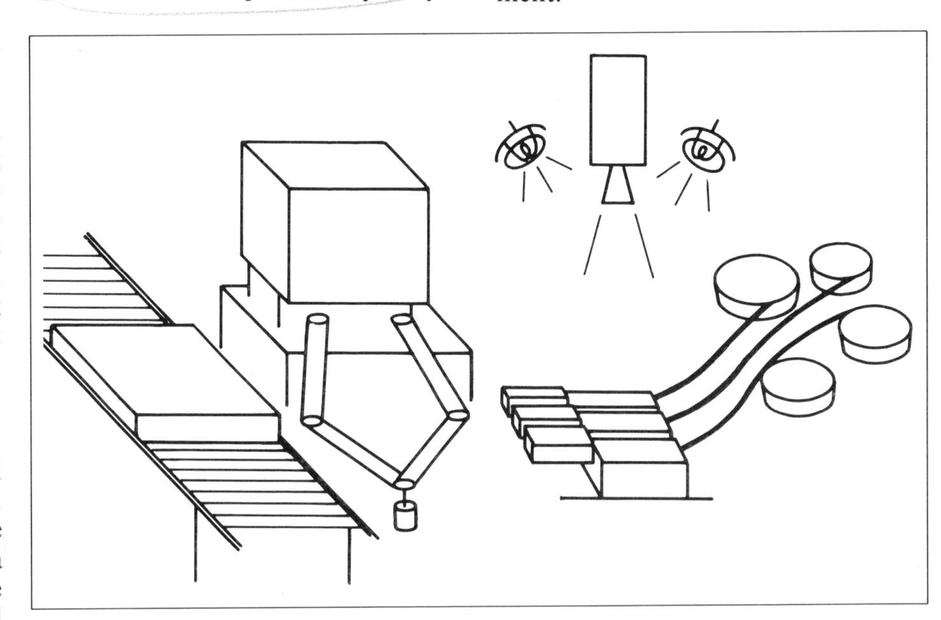

Assembly cell for terminal blocks by Mitsubishi Electric.

3.4 Electronics assembly

Automated assembly of integrated circuits gives flexibility, productivity

H. Ozaki and N. Sakai, Mitsubishi Electric, Japan

Increasing demand for integrated circuits has put emphasis on the need for flexibility and productivity in any system developed to assemble them into systems. Added problems are created by the IC's fragility, size variation and minuteness.

TODAY, the application of electronic techniques is expanding over wide areas of consumer and industrial fields, such as automobiles, cameras, watches, and so on. So there is an increasing demand for electronic components containing various types of transistors, ICs, and hybrid ICs.

To meet this demand, automated assembly systems for these electronic components are required, using high productivity and flexibility. But there are several problems associated with these systems, such as the careful handling of very small and fragile parts, the precise positioning of parts, and the frequent switch in the type of product being handled.

One approach to automating the assembly process is by utilising pattern recognition technique, computer control and servo positioning mechanism.

There are, however, many types of assembly operations for electronic components as shown in Table 1. One of the main tasks is to place the chip elements onto a substrate or metal sheet called a lead frame. In IC or transistor assembly usually one chip is required for each device, while for hybrid IC assembly many chips are required.

Many ICs are formed on a silicon wafer simultaneously with the wafer process. They are then tested and any defective chips are marked on the surface. The wafer is cut into separate ICs by dicing or the scribing and breaking method.

During assembly (Fig. 1) only good chips are selected, picked up and placed onto the lead frame and fixed with solder in a process called die bonding. Then the electrodes of the chip and leads are connected with fine wires. After wire bonding, each lead frame device is packaged.

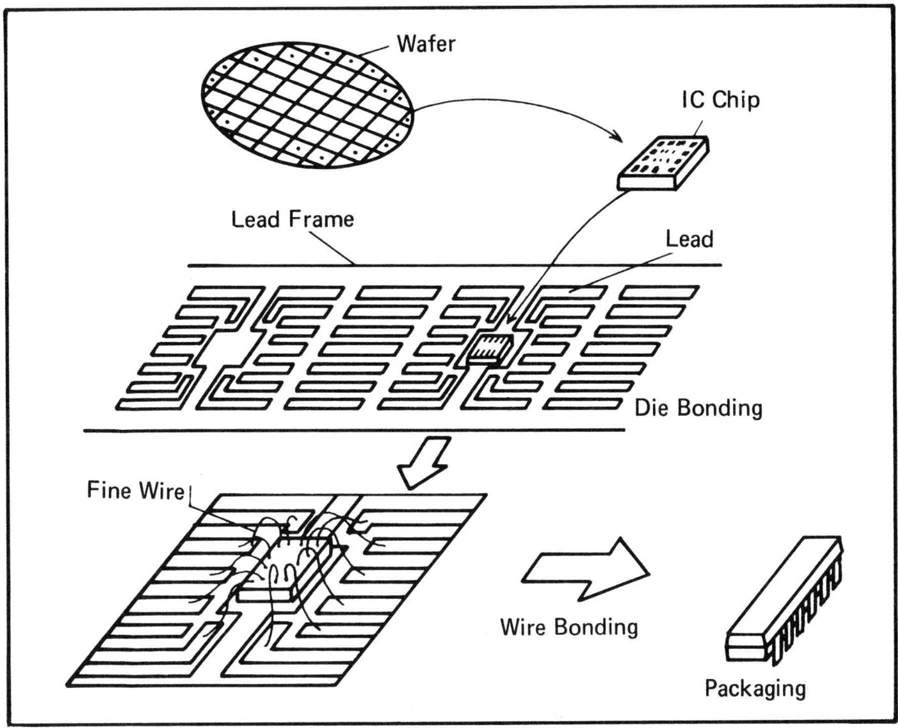

Fig. 1. The IC assembly process in which the wafer is split into chips and then die-bonded to the lead frame before wire bonding and packaging.

Hybrid assembly

In the assembly of hybrid IC (Fig. 2), various kinds of chip elements are deposited on a substrate. The substrate is made of ceramic, on which conductive, resistive and insulation compositions are formed through printing and baking. Solder paste is printed on the limited areas of the substrate where chips should be set. IC chips are then placed onto the substrate upside down.

The chip capacitor has metal terminations on each side and these are connected to the conductor by solder. IC and capacitor chips are deposited on the substrate and held slightly fixed with solder paste. Then the substrate is heated to melt the solder paste so that all the chips on the substrate are fixed completely with the solder.

Electronic component assembly has several special features which make it different from ordinary assembly. First of all there is the chip element itself: A chip is extremely small measuring only 0.4 mm x 0.4 mm, it is fragile, and there are many chip sizes and variations of specification.

Secondly there is the feature of chip placement during assembly. This must be done at a high production rate (in Japan, ICs are produced in a rate of 200 million parts a month); there are many kinds of products; chips need to be handled carefully; there must be precise positioning of IC chips on the

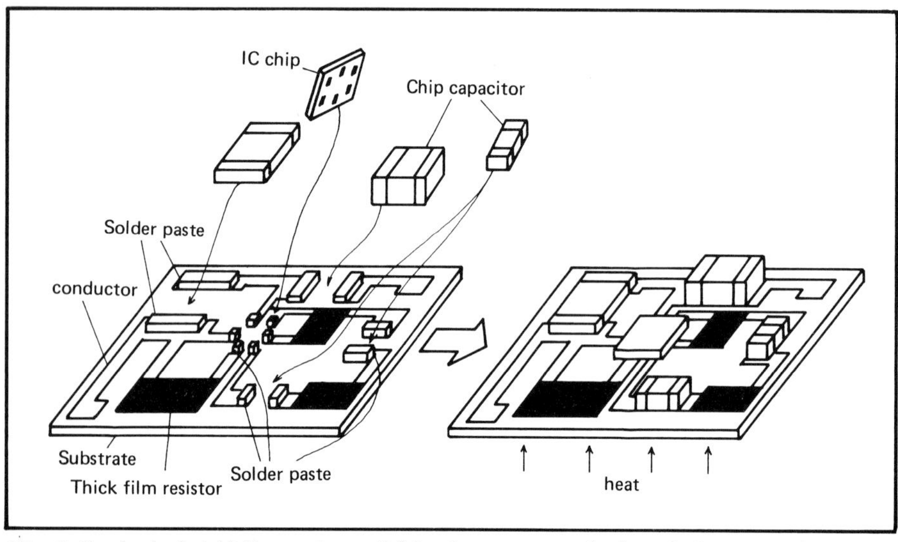

Fig. 2. In the hybrid IC a variety of chip elements can be bonded onto the substrate.

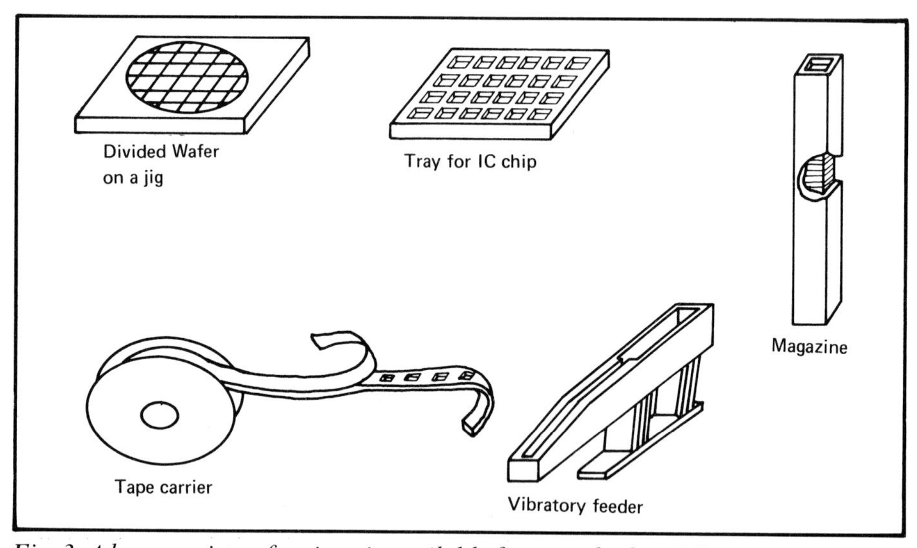

Fig. 3. A large variety of options is available for parts feeding ICs.

Object	Component	Assembly operation
Transistor	Leads Chip (Die Fine Wire	Die bonding → Wire bonding
IC	Lead Frame Chip Fine Wire	Die bonding → Wire bonding
Hybrid IC	Substrate IC and Tr. chips Chip capacitors other chip type components	Die bonding, Wire bonding Chip placing, Soldering
One Board-Electronic Circuit	Printed Circuit Board IC, TR, other discrete components with leads Chip capacitors, Resistors other chip type components	Inserting, Soldering Chip placing, Soldering

Table 1. A large variety of assembly methods are used in integrated circuit manufacture.

lead frame or substrate; and finally it is important to be able to select good chips from a divided wafer in which defective chips are mixed.

The chip placement process is the main aspect of the automatic assembly of fine electronic components. The process is divided into parts feeding, handling and positioning.

To automate these functions, a large variety of methods is available (Fig. 3, Fig. 4). For example, in parts feeding the semi-conductor chips can be fed as a divided wafer, arranged on a tray, or packed in a tape carrier. At the same time chip capacitors and other chip-type elements can be fed loose from a vibratory feeder, packed in a magazine, or packed in a tape carrier.

Vacuum chuck

During handling, the chips can be picked up and placed in the appropriate area by a vacuum chuck, gripper, pusher or a combination of all three. In the vacuum chuck the tip of the nozzle attracts a chip component when the nozzle is evacuated; it releases the chip when it is vented. The vacuum chuck uses a simple construction and has a wide application.

The gripper holds a chip by pushing against the side of the chip with its fingers, while in the pusher a rod pushes a chip and holds it with a stopper.

For positioning, two methods are available: mechanical positioning and numerical control positioning. In the mechanical system the positions of pick and place are pre-set in one of two ways.

One of these is the multi-station type, in which each chip is deposited by a simple pick and place unit at each station. This method is suitable for high volume production. However, it is less suitable when frequent product changeover is required; then setting up is necessary for the many pick and place units. This method is less suitable also when there are a great number of chips on a substrate, for example 100 pieces.

The alternative route is a single station type in which many chips can be deposited at a time, using a jig. This method is suitable for mass production of electronic circuit boards which have a great number of chips along with other discrete parts. This method has high productivity, but there are limits in chip sizes and shapes, and change-over time is not so short.

In the numerically controlled positioning system a pick-and-place device is used with a numerically controlled XY table, and the chips are deposited one by one on a substrate according to programmed information. This method has flexibility because of the short changeover time and easy operation using computer control and the XY table. This method is best suited for short-run application. Moreover, if pattern recognition techniques are adopted with this method, it should be possible to position chips with even greater accuracy.

The system needs to be planned according to the production scale, lot size, the number of chips on a substrate and state of chip supply.

But usually in electronic components assembly, both high productivity and flexibility are required since product design may change in the near future. And, to increase efficiency of the system, changeover time should be short and without any manual operation.

Flexibility is the ability to assemble various kinds of products and it is required in order to adapt for following items: the size, shape and specification of chips; the size, shape of a substrate or lead frame; the number of chips placed on a substrate; and the required position of chips on a substrate.

Various techniques are available to achieve flexibility, such as pattern recognition, computer control and servo-mechanisms. Moreover, these techniques can be applied for some other special functions such as selecting good chips, detecting chip position or detecting conductor pattern on a substrate, according to the needs of assembly of electronic components. A pick-and-place head adaptable for various chip sizes is useful, too.

How these requirements were met, especially those of high productivity and flexibility is demonstrated in the following two examples. One relates to automatic IC die bonding system and the other to automatic chip placement system for hybrid IC.

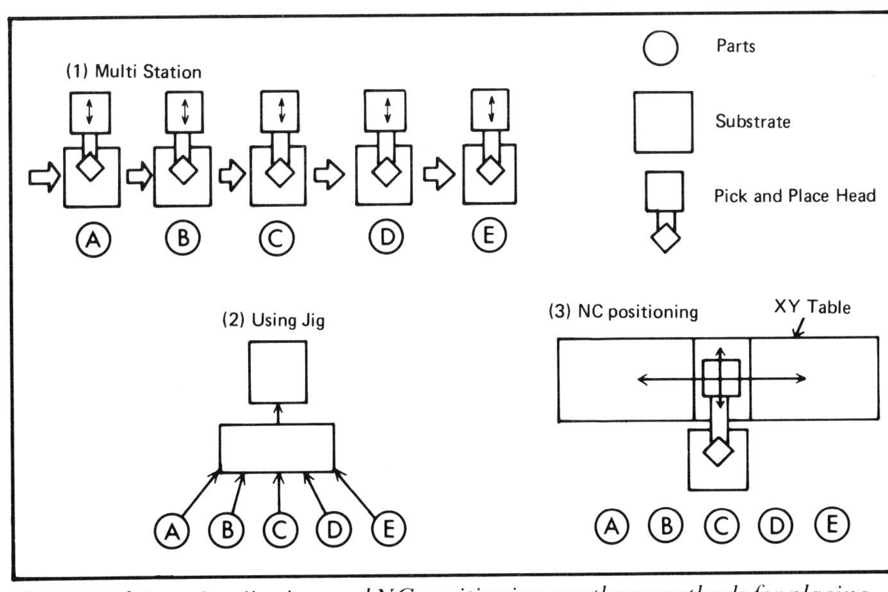

Fig. 4. Multi-station jigging and NC positioning are three methods for placing chips.

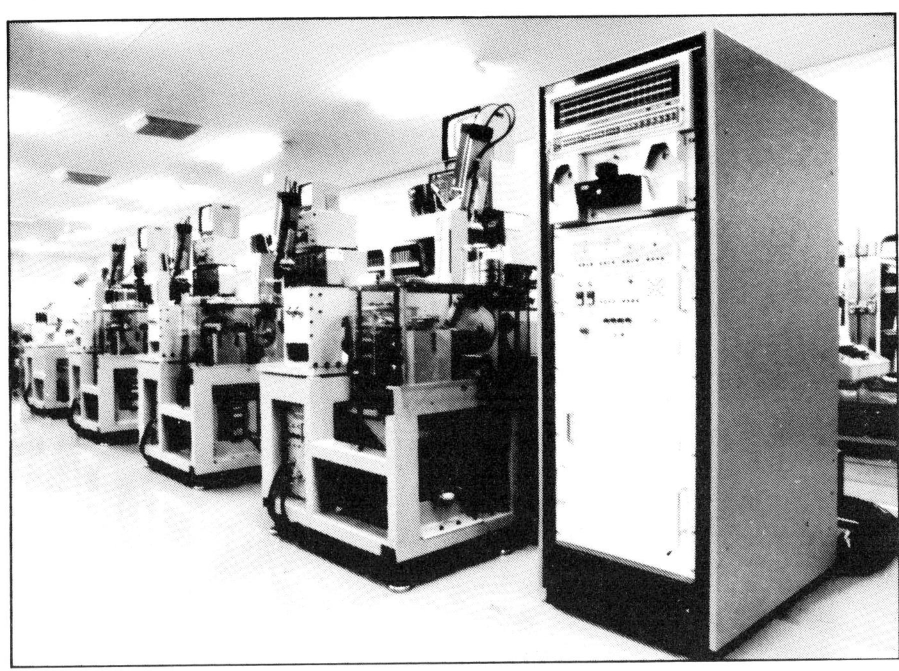

Fig. 5. Automatic IC die-bonding machine.

Die bonding

Fig. 5 is a general view of an automatic IC die bonding system. In developing the system, the following requirements were targets to be met:

○ Only good chips must be selected and bonded. In order to select good chips, defect marks on chips must be recognised.

○ Chips must be bonded precisely onto the lead frame. The precise position of a chip must be detected and aligned at the pick-up point.

○ Availability of the system must be high. It is important to minimise the time for changeover and for selecting good chips.

Fig. 6 shows the construction of this system. Outlines of the functions are as follows:

□ An IC wafer, divided into chips, is supplied on an XY table. A mini computer accepts and processes the binary image information from the first TV camera which has a wide visual field to observe as many chips as possible. And, using pattern recognition techniques, all defect marks of defective chips are detected and memorised.

□ The mini computer measures chip size and chip arrangement using binary image information from the second TV camera and the movement of the XY table. The second camera has a narrower and more highly magnified field of view. The chip size and chip arrangement are memorised in the computer.

□ The computer identifies only good chips one by one according to the data of chip arrangement, chip size and defect marks. The mini computer detects the position error of the chip from the information of the second camera. According to the position error, the computer moves the XY table so the chip may be positioned precisely at the point of pick up. The pick-and-place head handles the chip and deposits it onto the lead frame.

Features

The main features of this system are:

Flexibility. The IC chips are supplied to this system in special jigs according to a production schedule. The jig can be adaptable for any type of IC. As this system can recognise the configuration of chips and their arrangement, the system can operate automatically according to the required data without any additional information. In addition, a special vacuum head has been developed to pick up various

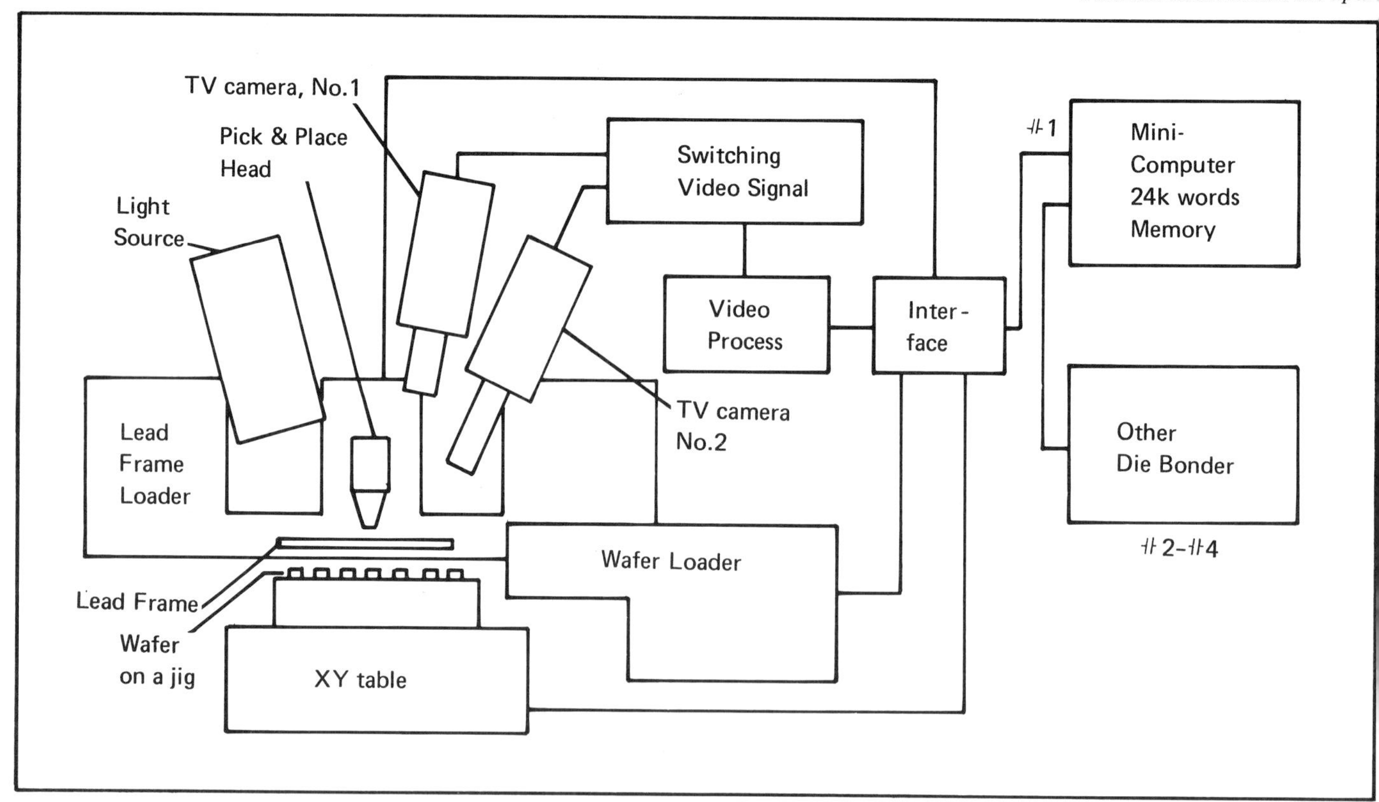

Fig. 6. Schematic of the die bonding machine which uses an XY plotter table.

Fig. 7. Chip placement system for hybrid ICs.

sizes of IC chips and eliminate the changeover time for the vacuum head.

Productivity. When the wafer on the jog contains defective chips, they must be identified and rejected. In the case of a system which examines the chips one by one, it will waste valuable time identifying defective chips. In this system, the defective chips can be quickly recognised and the bonding operation can be performed only using good chips. Thus, the production rate of the system can be kept high. Another technique to increase productivity is to adopt soft soldering instead of eutectic method. This can reduce bonding time by eliminating scrubbing operation.

Specification of this system is as follows:

Chip size: from 0.8 mm x 0.8 mm to 6 mm x 6 mm
Machine cycle time: 1.5 second
Positioning accuracy: ± 50 µm (at pick up point)
Method of bonding: soft solder

Fig. 7 is a general view of an automatic chip placement system for

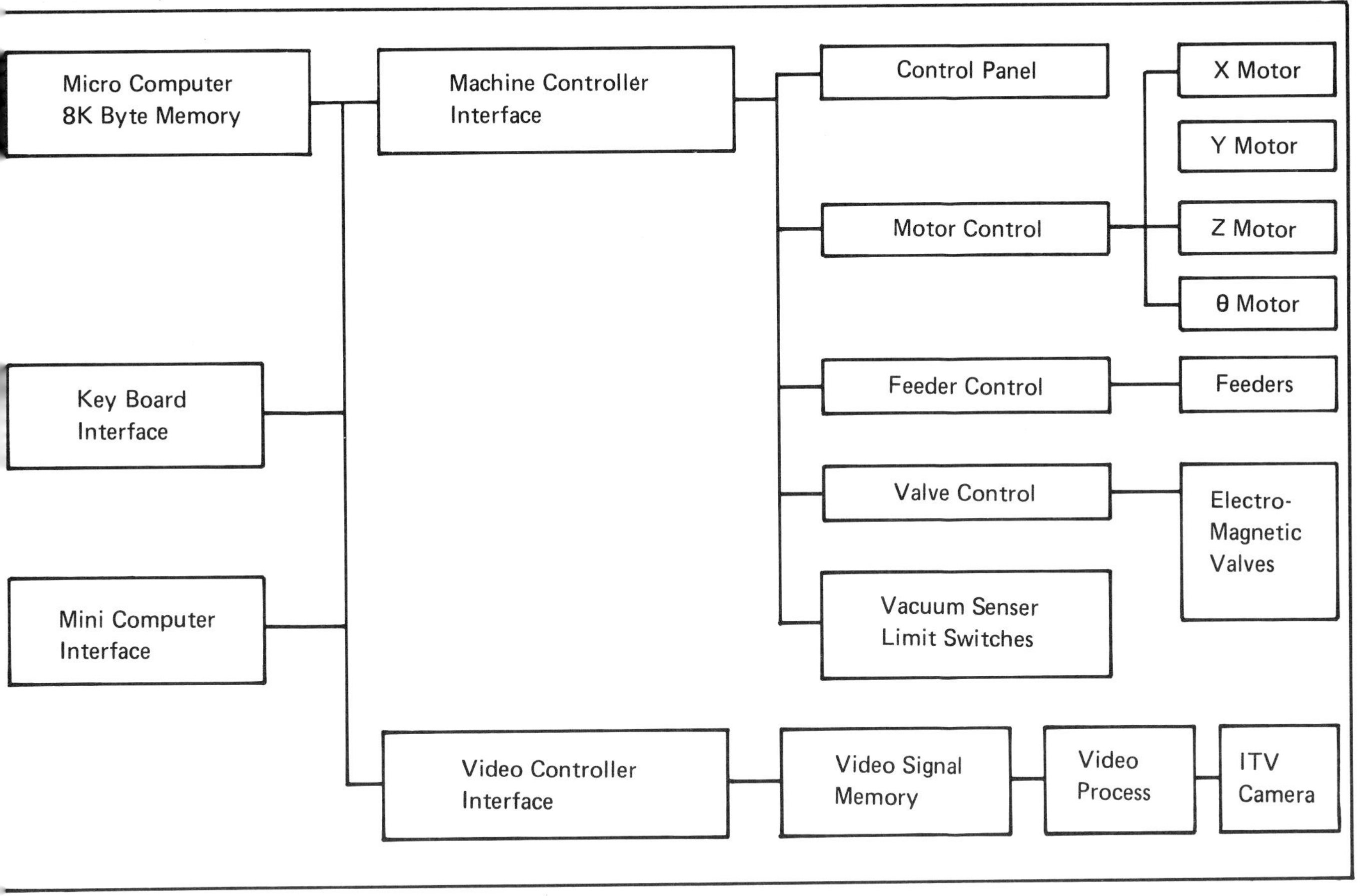

'ig. 8. Block diagram of the hybrid IC assembly machine.

ıybrid IC. The various systems avail-ıble for the chip placement of hybrid Cs depend upon production scale, lot ;ize, or other manufacturing con-litions.

However, the system was developed o meet the following conditions: 'requent product change; the chips on .he substrate though fewer in overall ıumber contain a variety of IC chips, :hip capacitors and other chip type :lements; and finally, some IC chips are provided on a tray while others are fed by a number of vibratory feeders.

The following requirements were targets to be met:

○ Changeover must be easy and completed in a short time.
○ Chips must be precisely deposited to the required pattern on the substrate. There needs to be some correction in the positioning of the pattern to allow for the contour of a substrate.
○ Position of preprinted solder paste on conductor needs to be checked. If the paste has unallowable error, the substrate must be rejected.
○ The mechanism must move quickly to increase productivity.

Construction

Fig. 8 shows the construction of the system. In the system the pick-and-place head with its TV camera is fixed on an XY table. Chip feeders, trays and a substrate are arranged under the head.

The system operates as follows. The TV camera on the XY table is positioned above the programmed point of the pattern on a substrate. The microcomputer accepts and processes the binary image information from the camera, and checks whether the solder paste is printed rightly or not.

Then, for a normal substrate, the computer measures the position of the pattern by detecting points from a characteristic shape. According to the programmed data and the acquired pattern position, the chip placement system operates sequentially. The pick-and-place head on the XY table is moved to a selected tray or feeder, picks up a chip and places it at the correct position on the substrate.

The features of this system are ones of flexibility and productivity.

Flexibility. The following factors are programmable: position and orientation of chips on a substrate; the height of pick-up point and positioning point above every chip; the sequence of chip placement; the pressure applied to the chips while the chip is being placed; and the chip arrangement on a tray.

However, it is not at this stage easy to change parts feeders, so the system is not yet flexible enough. A more efficient parts feeding system suitable for frequent changeover needs to be developed.

Productivity. In this system, an increase in the speed of the mechanism was investigated. A maximum speed of 1 m/sec on the XY table was achieved. The velocity of the Z axis (vertical motion of the pick-and-place unit) is limited in order not to damage the chip when the unit strikes against the chip or the chip hits against the substrate. In this system, the height at the pick-up point and the positioning point above every chip are fed to the micro computer and a suitable velocity pattern is generated to eliminate lost time.

Specification of this system is as follows:

Chip size: from 1 mm x 1 mm to 6 mm x 6 mm
Cycle time: about 1.5 sec./chip
Positioning accuracy: $\pm 60\,\mu m$

References

1. Dallas, E.: "Chip placement systems for hybrids", Electronic Packaging and Production, pp. 50-60, February 1980.
2. Inoue, T.: "Automatic Die-Bonding System for Semi-conductors with TV cameras", SME Technical Paper, AD. 77-719, 1977.

Reprinted by permission of the Society of Manufacturing Engineers, Dearborn, Michigan, U.S.A.

TV camera system inspects and sorts power transistor chips

Kunihiko Edamatsu, Tetsuo Kiuchi, Yoshiaki Isono, Shiro Naruse
and Akira Momose, Fuji Electric Co. Ltd., Japan*

An automated power transistor chip sorting apparatus and position recognition algorithm have been developed at Fuji Electric. The apparatus inspects each chip on a wafer cassette, and picks out the good chips and arranges them on a chip tray cassette, at a recognition rate of 0.3 seconds per chip.

PROGRAMMABLE automation systems such as NC machine tools and industrial robots are flexible and easy to train for jobs performed under constrained conditions, but they are not adaptable to different objects because they have no sensors.

Several intelligent systems using minicomputers with sensors have been reported[1,2], which are of high flexibility and adaptability. We have developed an intelligent robot – a transistor chip sorting apparatus that consists of a TV camera with a microscope, an image processor and mechanisms.

Power transistors are formed on a silicon wafer by repeated oxidisation, diffusion and photo-etching. Their electrical quality is checked and the unacceptable transistor chips are marked with a black point. The wafer is cut into individual clips and fixed on a nylon sheet. The sheet is stretched and each chip is separated. This is a wafer cassette. Human workers select good chips from the wafer cassettes and arrange them on chip tray cassettes.

The apparatus we have developed can replace these human workers. The first step is to search for and inspect a transistor chip on a wafer cassette. The second step is to compute its position and orientation. These methods are based on the multi-pattern recognition technique[3,4].

Wafer cassette

Fig. 1 shows a wafer cassette. The TV camera does not take a whole picture of the wafer cassette but takes the part as shown in Fig. 2, because of

Fig. 1. Wafer cassette picture distinguishes chips.

Fig. 2. TV monitor image includes only part of the cassette for fine resolution.

resolution. Therefore the image processor must estimate the position of the part in the whole wafer. Because the centre chip in Fig. 2 has a bad mark, the image processor judges the chip to be no good.

The recognition algorithm consists of three techniques: searching, inspection, and positioning. These techniques are easy to use because a high positioning accuracy is not necessary.

Fig. 3 shows a whole flowchart of the algorithm.

Searching. The first step is to search for the first chip on the wafer cassette. The image processor searches for the first chip, moving the X-Y table for the wafer cassette as shown in Fig. 4a.

When the area of the chip projection, A, satisfies:

$$A_{min} \leqslant A \leqslant A_{max} \quad (1)$$

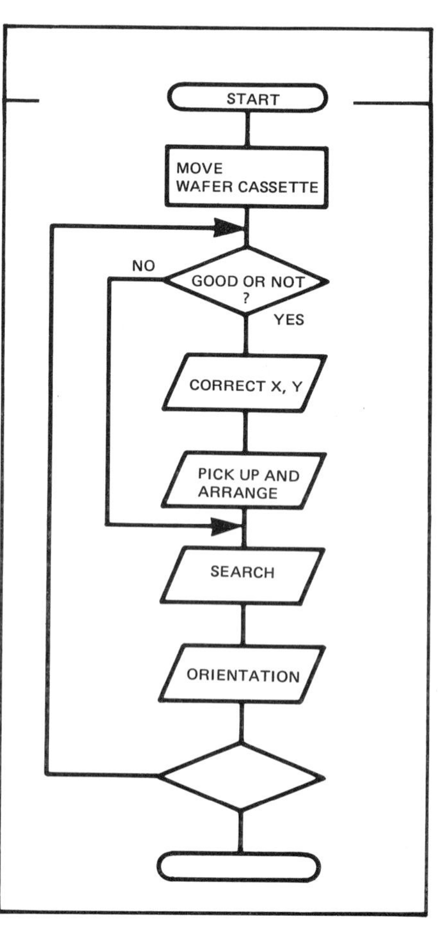

Fig. 3. Flowchart of the recognition algorithm.

*Based on the text of a paper presented at the 11th International Symposium on Industrial Robots, Tokyo, 7-9 October, 1981. Reproduced with the kind permission of J.I.R.A.

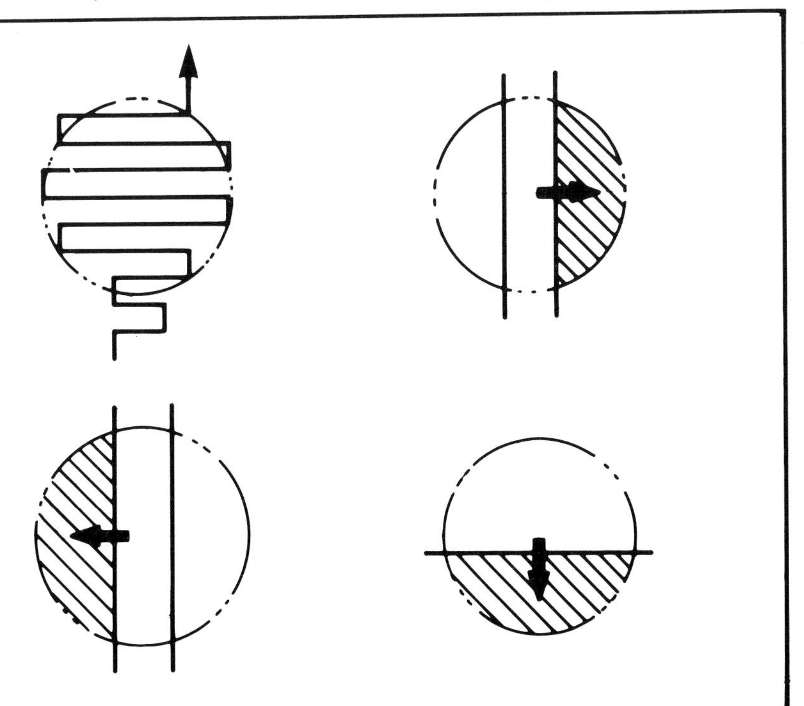

Fig. 4. *Searching algorithm for first and subsequent chips.*

where A_{max} and A_{min} are maximum and minimum areas of transistor chips respectively, the image processor judges the chip to be first. The image processor can estimate the next chip position from the present one.

The second step is to search for the end chip in X-direction. The image processor commands the X-Y table to move to the right until the condition (1) is not satisfied during three steps in the shadowed portion (see Fig. 4b).

The processor commands the X-Y table to move one step in Y-direction, and then in the leftward direction. The X-Y table is controlled at the left end in the same manner as the right end (see Fig. 4c).

The last step is to search for the end of the wafer cassette. If the condition (1) is not satisfied throughout a horizontal line in the shadowed portion (see Fig. 4d), the image processor is considered to be at the end of the wafer cassette.

The search process described above is done for the whole circle of the wafer cassette (see Fig. 4a).

This algorithm applies successfully to various chip sizes and various wafer sizes. Moreover, it is adaptable even for the processing of various eccentric wafers.

Inspection. The image processor above performs an inspection only on the pattern images which passed the check described above. The inspection algorithm consists of four checks: the chip area check, the bad mark check, the collinearity check, and the squareness check.

The area of the chip projection is checked again with narrow tolerance. The chip area check can detect a chip with a broken corner and chips stuck together.

Bad mark

In the bad mark check the scanning raster moves down, and if the chip pattern splits, the image processor begins to count up the area of black domain between the split patterns, until the split patterns meet together again. If A_g, the area of black domain enclosed with white pattern, is larger than G_{min}, the domain is regarded as a bad mark and the chip is judged as no good. G_{min} is the minimum area of bad mark and necessary in order to distinguish bad marks from defects in the TV camera.

The collinearity check is performed in order to extract the four sides of the chip and detect its cracks[4].

For the squareness check, Fig. 5 shows four terminal points of sides of the chip, which are extracted in the collinearity check. Deviation from square is defined as follows:

$$D_{sp} = |\vec{OP} - \vec{OQ} - \vec{f}(\vec{QR})| \qquad (2a)$$

$$D_{ss} = |\vec{OS} - \vec{OR} - \vec{f}(\vec{QR})| \qquad (2b)$$

where

$$\vec{f}\binom{X}{Y} = \binom{-Y}{X} \qquad (3)$$

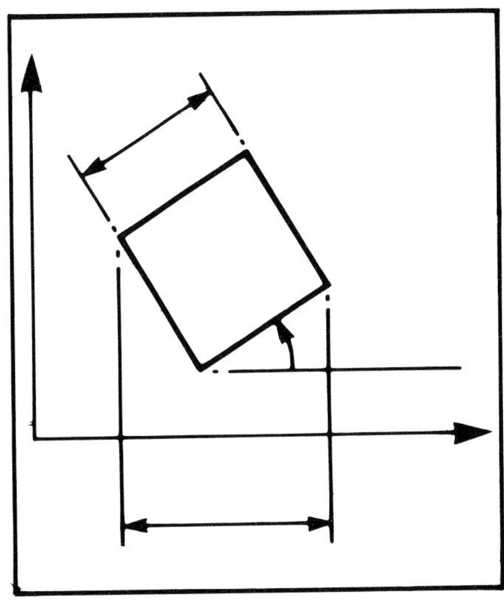

Fig. 5. *Inspection method for squareness of chips.*

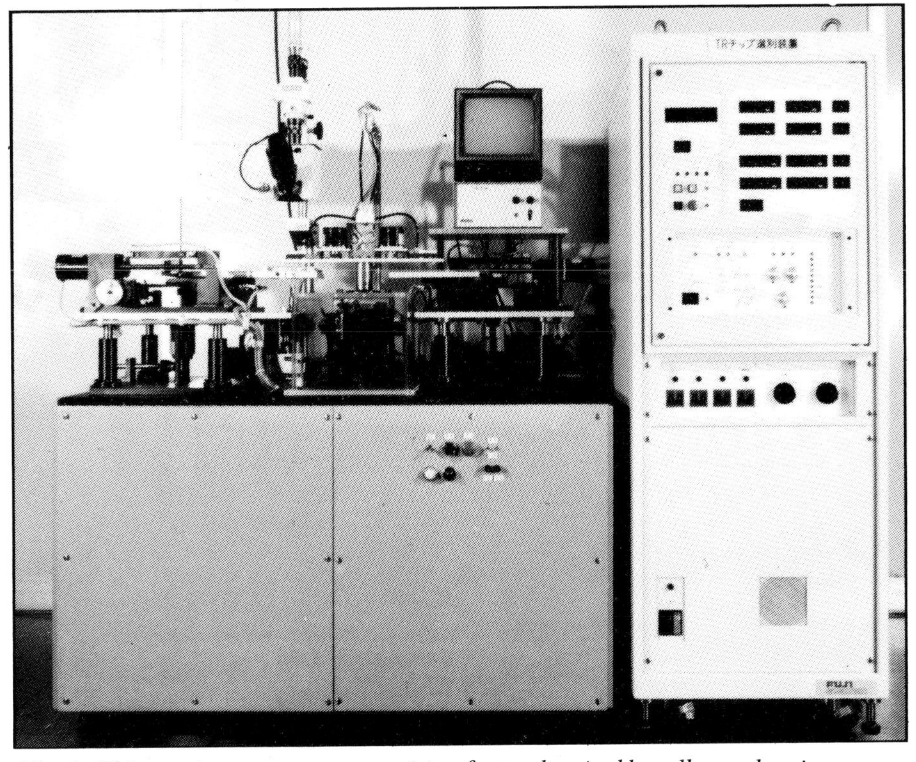

Fig. 6. *Chip sorting apparatus consists of a mechanical handler and an image processor.*

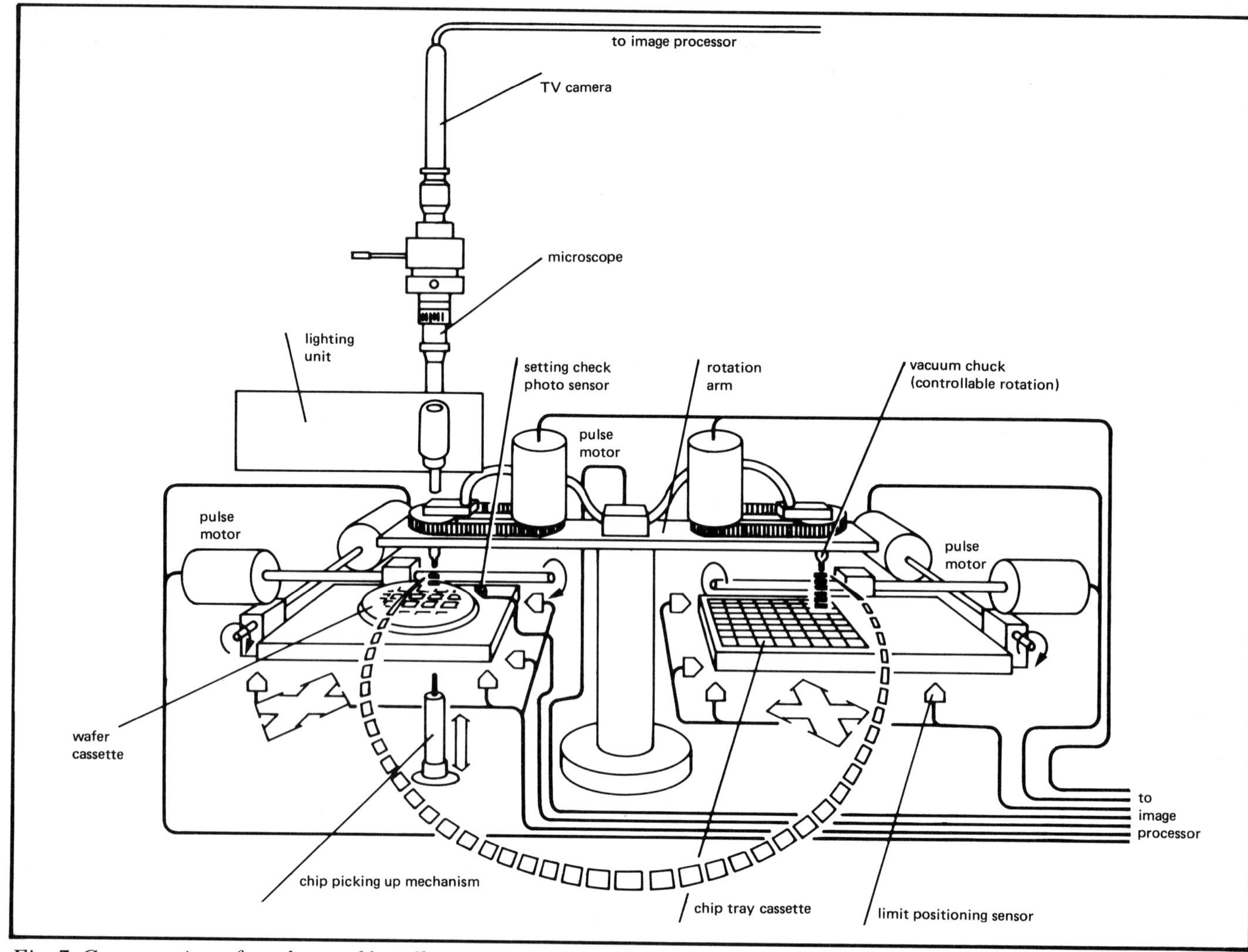

Fig. 7. Construction of mechanical handler.

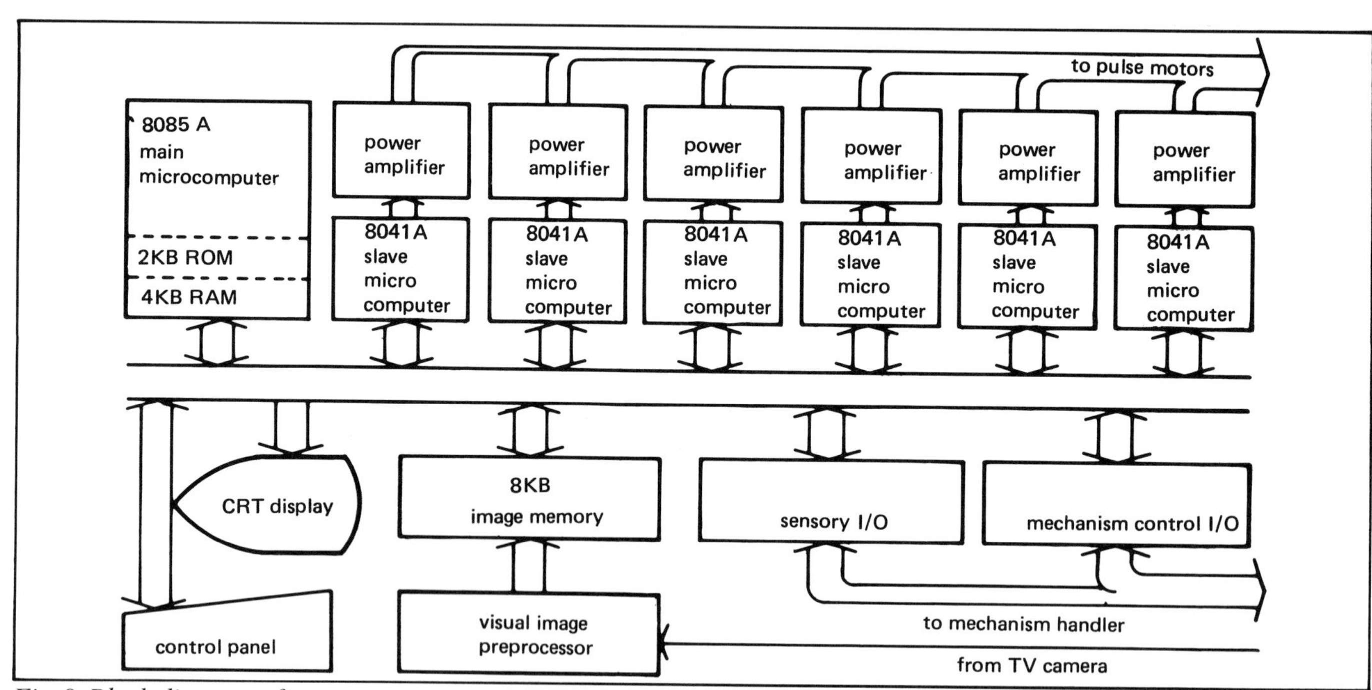

Fig. 8. Block diagram of image processor with 8-bit microcomputer.

The chip passes the squareness check when the following condition is satisfied:

$$D_{sp} \leqslant D_{smax} \text{ and } D_{ss} \leqslant D_{smax} \qquad (4)$$

where D_{smax} is the maximum deviation of a good chip. A chip with too small a corner can be detected by means of the squareness check.

Positioning

Positioning consists of rectangular positioning and orientation.

Rectangular positioning: Centre coordinates of the inspected chip are computed as the centre of its projection on each X, Y axis. This method is not affected by orientation of the chip because a power transistor chip is a regular square.

Orientation of the inspected chip from X axis, Ω, has the following relation to B and L, because a chip is a regular squre.

$$L = B(\sin|\Omega| + \cos|\Omega|) \quad (5)$$

where B is length of a side of a chip, and L is length of its projection on X axis. The equation (5) gives the absolute value of Ω. Sign of Ω is known from a comparison of X coordinates of point S and point Q.

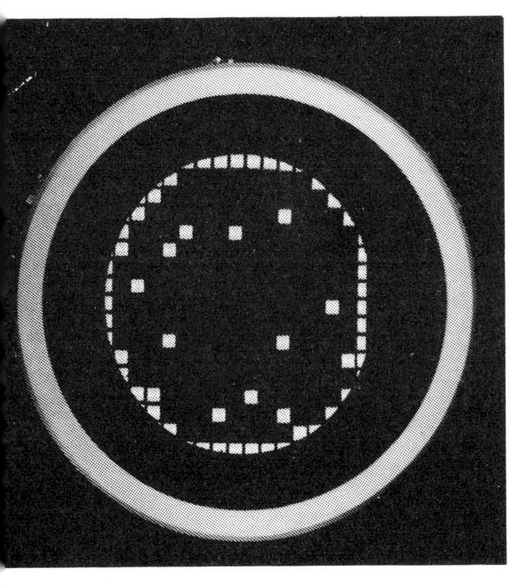

Fig. 9. Rejected transistor chips left on a wafer cassette.

Sorting apparatus

The sorting apparatus is shown in Fig. 6. The apparatus consists of a mechanical handler and an image processor. The mechanical handler is illustrated in Fig. 7. It consists of the following components:

- a rotation arm with chucks,
- a pair of vacuum chucks which are rotated by pulse motors (with one degree accuracy),
- two programmable X-Y tables whose frames are driven by two pulse motors (0.01 mm in either X or Y direction),
- a thrust rod which is driven by a cam mechanism,
- two air valves,
- a vidicon TV camera.

A wafer cassette is placed on the X-Y table. An initial calibration procedure establishes the parameters of transformation between the image coordinates of the TV camera and the axis of the X-Y table. The image processor commands the X-Y table to move the wafer cassette, searching for the first transistor chip by means of the algorithm described in paragraph 3.1. The TV camera takes a picture of the first transistor chip. The image processor analyses its image and performs pass-or-fail judgement. If the chip is no good the image processor computes the next position and commands the X-Y table. If the chip is good, the image processor computes its position and orientation, and commands the X-Y table to move the chip to a fixed thrust rod position at the centre of the TV camera's field of view. The rotation arm with vacuum chucks picks up the chip. The vacuum chuck corrects the angle. The arm rotates and places it on the chip tray cassette. Then the image processor commands the X-Y table to move to the next chip position.

Fig. 8 shows a block diagram of the image processor. It consists of an 8-bit microcomputer, six slave microcomputers with power amplifiers, an image pre-processor and memory, interface devices, a control panel, and a CRT display. The six slave micro computers operate under the control of the 8-bit microcomputer and control the X-Y tables to drive the pulse motors. The image processor accepts the video signal from the TV camera and extracts features.

References

1. S. Kashioka et al. A Transistor Wire-Bonding System Utilising Multiple Local Pattern Matching Techniques. *IEEE Transactions on Systems, Man, and Cybernetics,* Vol. SMC-6, No. 8, pp. 562-570 (August 1976).
2. D. Nitzan, Robotic Automation at SRI. *Proceedings of MIDCON/79,* Chicago, IL, pp. 1-10 (6-8 November 1979).
3. J. W. Butler et al. Automatic Classification of Chromosomes. *Proc. Conf. on Data Acquisition and Processing in Biology and Medicine* (1963).
4. K. Edamatsu et al. Capsule Checker. *Fuji Electric Review,* Vol. 26, No. 1, pp. 25-30 (1980).

Vision system aligns leads for automatic component insertion

Toshio Asano, Shunji Maeda and Toshiaki Murai, Production Engineering Research Laboratory, Hitachi Ltd, Yokohama, Japan

Two solid state cameras and a newly developed image processor have been used to measure the positional deviation of the two leads of a component and control adjustment of the X-Y table. The system allows the insertion machine to operate without mechanical guides for the leads, allowing insertion of components with different lead pitches and at high density. Application results are reported in this article.

MANY automatic machines have been developed for printed circuit board assembly. Among them are automatic inserters for electronic components. These machines insert such electronic components as resistors, capacitors, and diodes on to printed circuit boards. These components are generally taped for easy handling.

Conventional inserters have mechanical guides to keep the leads of an electronic component in position, and by using guides, the component can be inserted correctly. However, the use of mechanical guides creates the following problems. First, due to the space required for the guide, components cannot be inserted closely to one another, which impedes the improvement of insertion density. Second, due to the fixed pitch of the guide, components with different lead pitches can not be handled, which restricts the flexibility of the machine.

To be free from these restrictions, a new insertion machine without mechanical guides has been developed. The new inserter has a vision system which measures positions of two leads before insertion. The new inserter inserts radial and axial types of components with different lead pitches between 5.0mm and 25.4mm.

The vision system used for the inserter includes a new image processor called PPI-1 (Programmable Picture Input-1), which has also been developed by our laboratory. PPI-1 is specially designed for economical industrial applications, and a wide variety of applications can be expected in production automation and robotics.

This article covers the following aspects:

- □ Image processing hardware system;
- □ Application of PPI-1 to an automatic inserter;
- □ Experimental results.

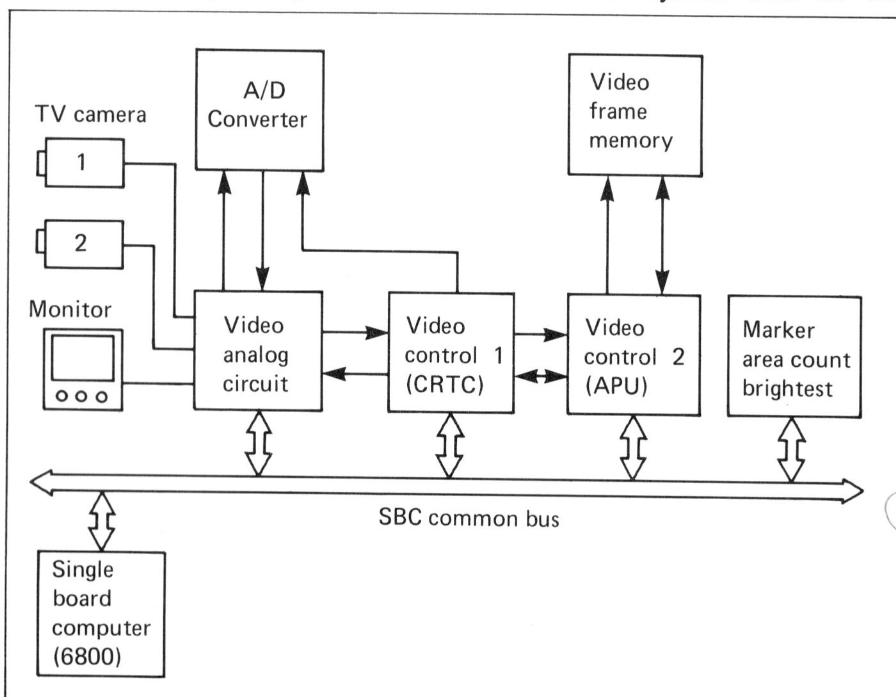

Fig. 1. Block-diagram of image processor (PPI-1).

Hardware

The need for a vision system has been increasing in automatic machines[1, 2] and robots.[3, 4] Vision systems expand the limits of automation. In this trend, one of the most important needs in a vision system is the development of a low cost and general purpose image processor. In this laboratory, several automated machines with vision systems have been developed. Through the analysis of these previous applications of vision, the specifications of the new processor were determined.

The basic ideas of the image processor PPI-1 are as follows:

- ■ Recognition algorithms are processed by software.
- ■ The vertical and horizontal resolutions of a picture are independently changeable according to the purpose of an application.
- ■ To save the image memories, PPI-1 can take a partial picture by window processing.
- ■ By computer commands, PPI-1 memorises a picture in binary mode

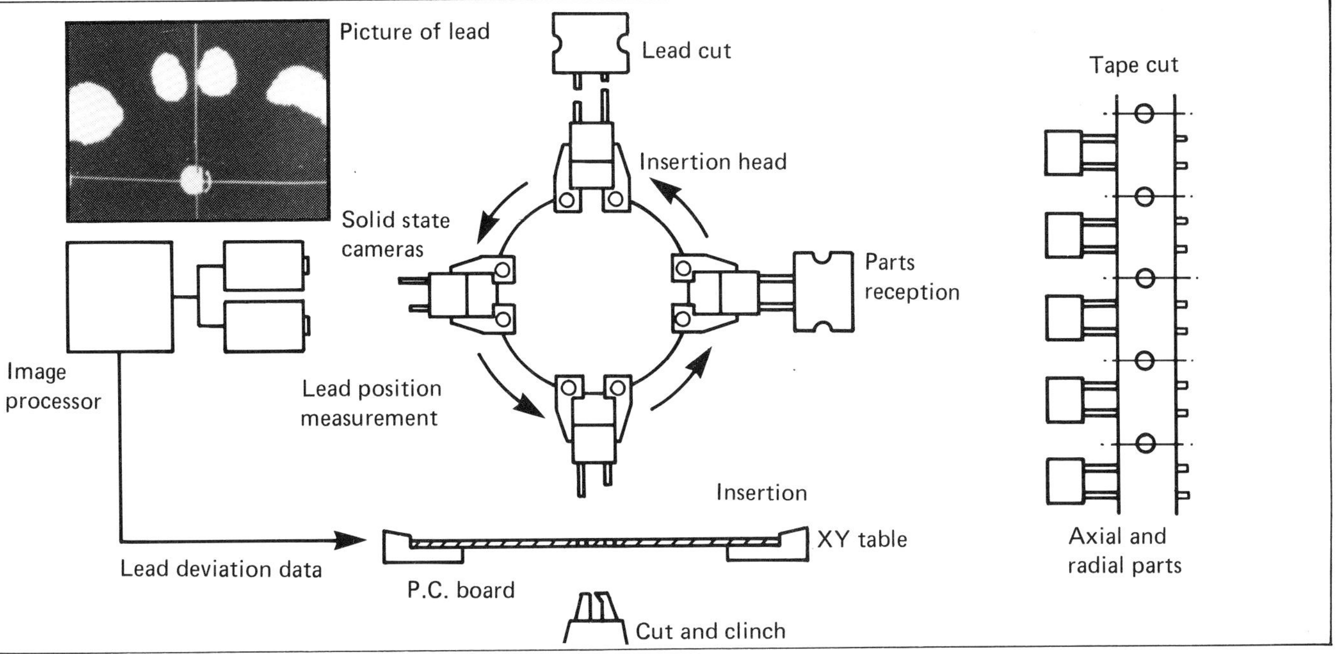

Fig. 2. Insertion process of the new automatic inserter.

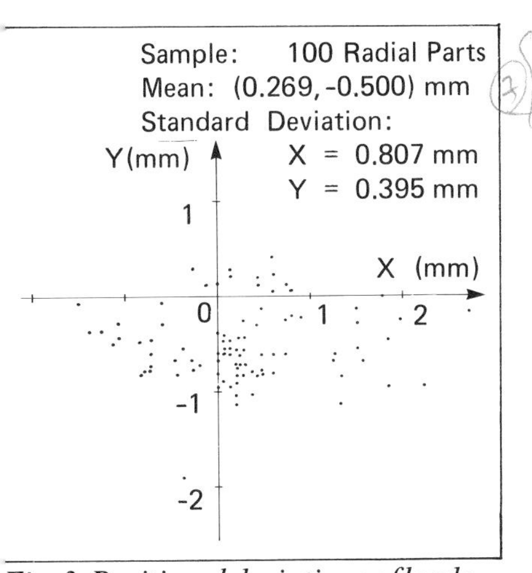

Fig. 3. Positional deviations of leads. The vision system compensates the deviations.

or grey scale mode (8 bits). This switching is performed in real time.

- PPI-1 has an arithmetic LSI (APU) which facilitates high calculation rate.

Signal

The specification and block diagram of PPI-1 are shown in Table 1 and Fig. 1. The composite video signal from a television camera is converted into a binary video signal by a comparator. If the mode is set to grey scale, the composite analog video signal is converted into 8-bit digital video signal. The video control cards control addresses of the video frame memory, and CRT controller LSI (HD46505) refreshes the memory. Using the HD46505 and multiplexing circuits, the resolution of the image can be switched in real time. The contents of the video frame memory are displayed on a television monitor. The video frame memory is mapped in microcomputer memory address space, which is easily accessible. High speed static memory ICs (HM6147) are used, and the video memory capacity is 32 K-Bytes.

Application

Fig. 2 shows the insertion process of the inserter.

- An electronic component is cut out of the bandolier at the parts feeding unit, and is sent to the insertion head.
- The component is fed to one of four insertion heads. The head holds the body of the component.
- The rotary index table rotates by 90°, and the two leads of the component are cut to different lengths.
- This image processor, PPI-1, measures the position of the two leads, and sends the position data to the XY table.
- The leads are inserted into the holes in a PCB in two steps. First, the XY table aligns the hole correctly to the longer lead using the measurement data, and the insertion head goes down until the longer lead is inserted by 1mm into the PCB. Second, the position of the XY table aligns the other hole to insert the shorter lead, and the insertion head goes down completely.
- The leads which are inserted are cut to their proper length and clinched beneath the PCB.

Positional deviations of leads are shown in Fig. 3. The standard deviation was 0.8mm in the X-direction and 0.4mm in the Y-direction. This indicates that the insertion of such components without visual correction is impossible. The allowable correction error is under ±0.13mm. The diameter of leads is between 0.5mm and 0.8mm, and the lead pitch is between 5 and 25.4mm.

In this application, accurate measurement of the position is

Table 1. Specification of image processor (PPI-1)

Items		*Specification*
Resolution	II	640, 320, 160, 80
	V	240, 120, 60
Window	II	512, 256, 128, 64
	V	Under 240
Grey level		Binary or 8 bits
Video frame memory		Max. 32K Bytes
TV camera		Max. 2 sets
Marker		V: 2 lines, II: 2 lines
Feature extraction hardware		Area count, Brightest point detection
CPU		HMCS-6800

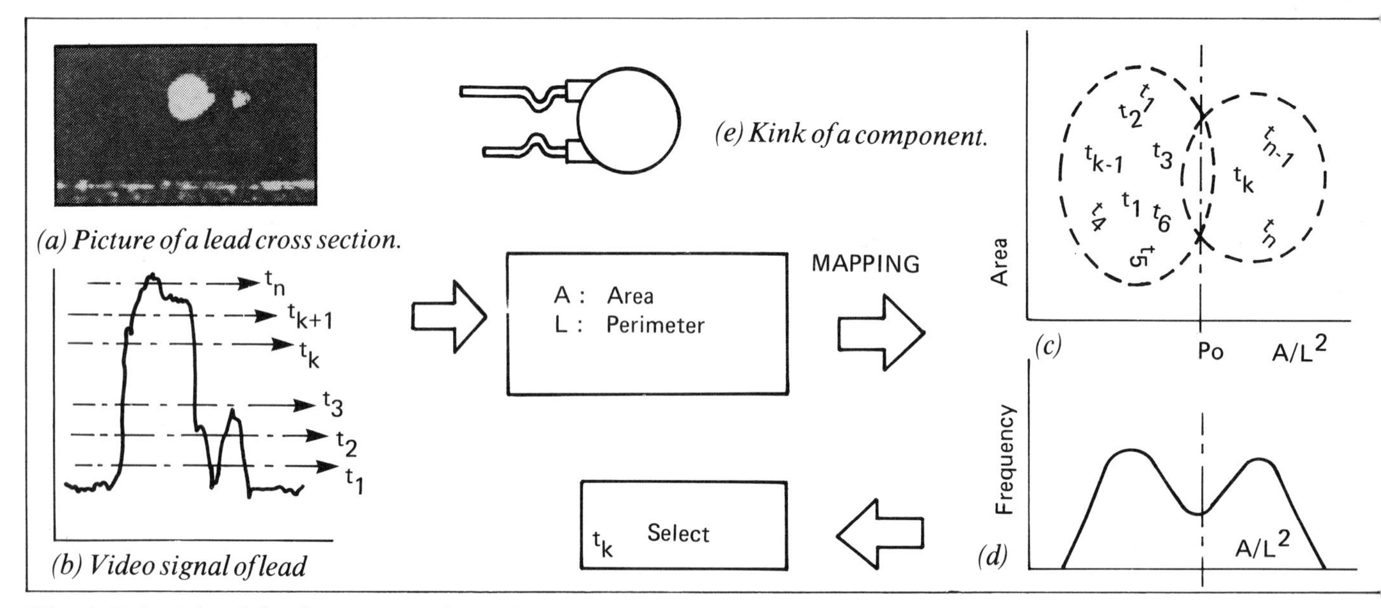

Fig. 4. Principle of the dynamic P-tile method.

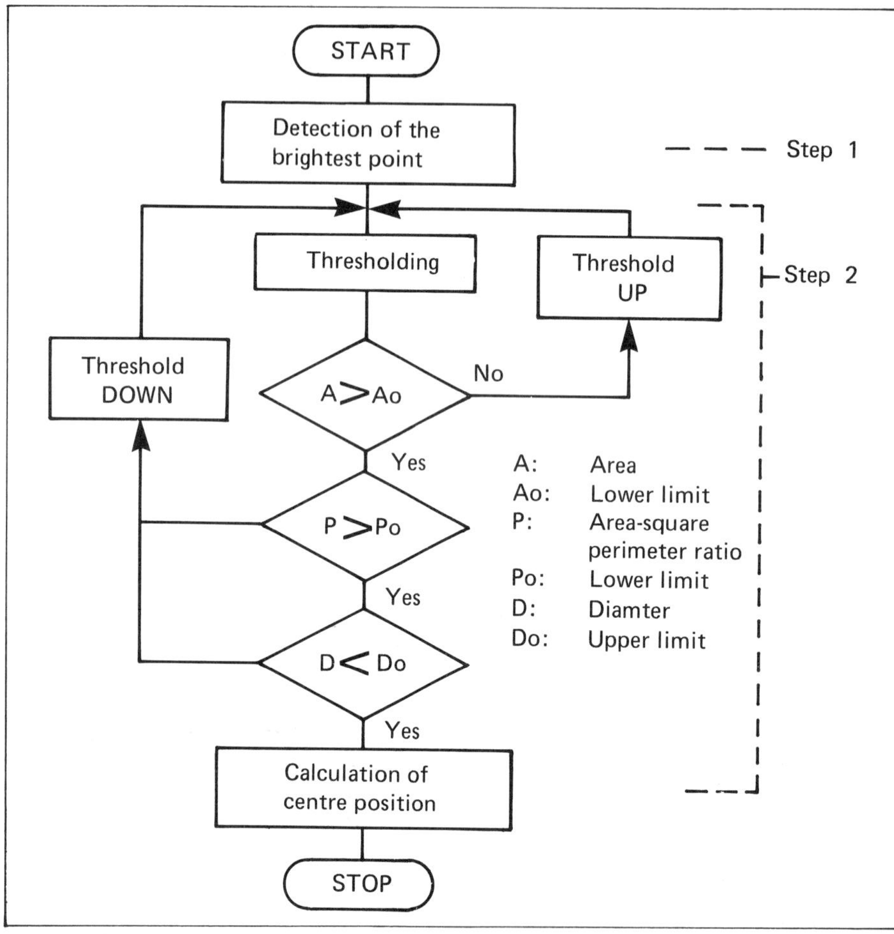

Fig. 5. Flow chart of detection software.

essential. Solid state MOS type cameras (Hitachi KP-120) were thus employed as they create no image distortion. Two cameras were used for the following reasons. Since two leads of a component are cut at different lengths the focus points are different. The resolution of the picture is also improved by the doubled use of cameras. The size of a pixel is 0.06mm × 0.06mm, and horizontal and vertical resolutions of a picture are 320 × 240 respectively.

Lighting is another difficult point in the optical systems. Several lighting methods were tested and a fluorescent light was found the best way to obtain the uniform image of lead cross section.

The selection of proper threshold level is important in extracting the cross section of a lead from the background. Sometimes a lead has a kink, and as it reflects the light well, the real lead tip must be distinguished from the kink. The shape of a kink is shown in Fig. 4 (e). Moreover, the bodies of some electronic components are made of aluminium which also reflects the light well. The distinction of the lead image from the component body is also important.

The shape of cross section of a lead is nearly circular. A circle is the most compact figure and the area-square perimeter ratio P is the biggest.

$$P = A/L^2 \quad (1)$$

where A and L represent area and perimeter of a figure respectively. Even if diameters of leads are different area-square perimeter ratio does not change. Therefore, the threshold level is determined by the ratio P, and the background or noise image can be eliminated from the lead cross section image. This method of thresholding is a P-tile method of area-square perimeter ratio and as the threshold changes automatically according to the leads, this method is called dynamic P-tile.

The principle of the dynamic P-tile method is shown in Fig. 4. Fig. 4 (a) represents a cross section of a lead. A small spot seen to the right is an image of a kink stated above. As the threshold is lowered to t_1 or t_2, the area-square perimeter ratio decreases. When the threshold is set between t_k and t_n, the area-square perimeter ratio increases and a picture of lead without background image can be obtained. Fig. 4 (c) illustrates the relation between the area and the area-square perimeter ratio under various threshold levels. The frequency distribution of Fig. 4 (c), which is given in Fig. 4 (d) with the abscissa being the area-square perimeter ratio, shows two peaks. The left peak represents a class of thresholded pictures in which leads and kinks images are seen. The right peak means a class in which only lead images are seen. P_0 is the area-square perimeter ratio with which the

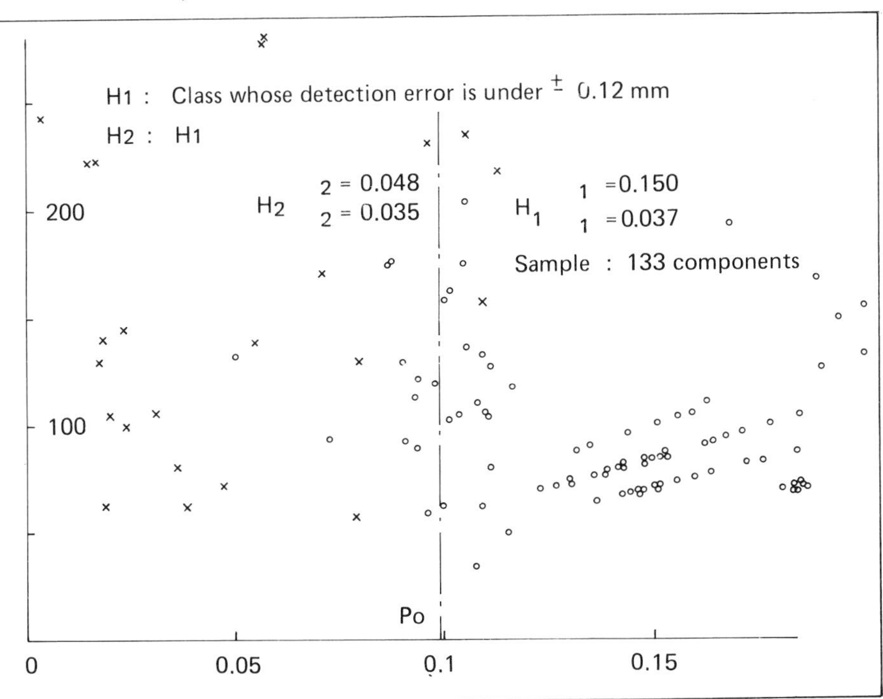

Fig. 6. Relation between area and area-square perimeter ratio.

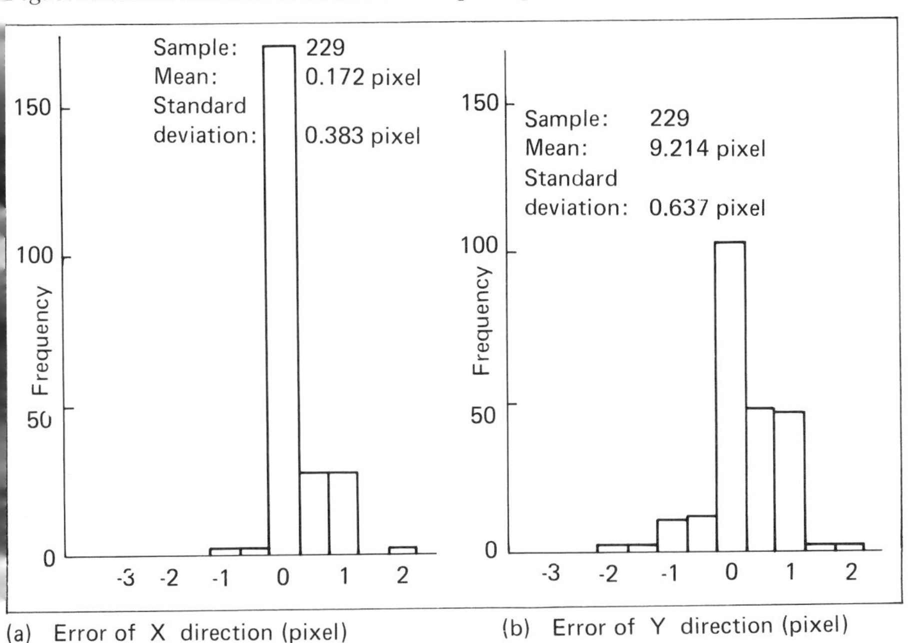

Fig. 7. Accuracy of lead position recognition.

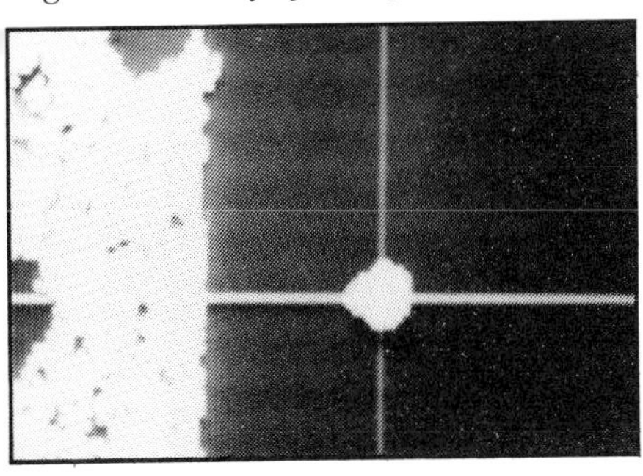

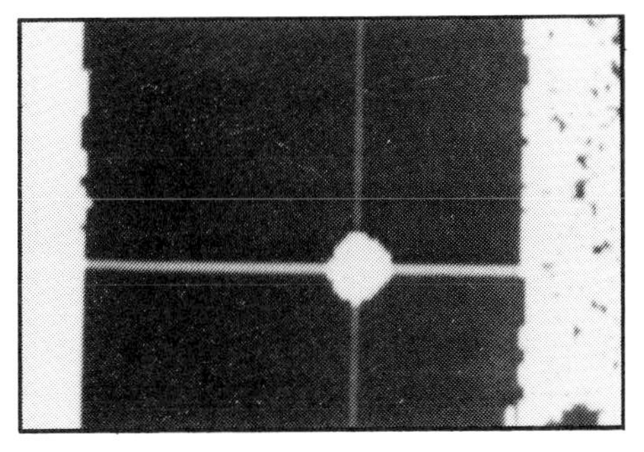

Fig. 8. Recognition results.

frequency is minimum.

By selecting such threshold t_k that the area-square perimeter ratio is greater than P_0, the accurate centre of the lead can be obtained.

Recognition

A flow chart of lead recognition is shown in Fig. 5. The procedure is divided into two stages. In the first stage, the image processor PPI-1 detects a part of lead cross section. At this stage, the grey level picture storage function (160 × 120) and the brightest point detection circuit of PPI-1 are utilised. Preliminary experiments indicated that the brightest point of the picture is always found in the lead cross section even with the presence of bright background. This has led to the use of the circuit for rough positioning of the leads. In the second stage, the picture is thresholded, and the resolution is changed to 320 × 240. The threshold is automatically determined by three properties – area, area-square perimeter ratio and diameter of a lead. In this stage, the threshold is optimised and the centre position of the lead cross section is calculated.

An experiment has been performed on 133 electronic components to determine the value of P_0 giving the lower limit of the threshold. In this particular experiment various lead images with complex backgrounds were obtained. At each recognition of the components, the area, the area-square perimeter ratio and the value of detection error were measured.

Relations

The tolerable error of lead recognition in this application of the vision system to the inserter is ±0.13mm, which derives from the dimensional relations of the lead and the hole. In Fig. 6, the detection error data are divided into two classes by this criterion. The detection errors under ±0.12mm are represented by circles (○), and the detection errors over ±0.12mm by crosses (×). The values of mean and standard deviation of area-square perimeter ratio of both classes were calculated.

The value of P_0 limiting the recognition error under ±0.12mm was calculated using Bayes decision theory and was found at 0.0988.

The experimental results of lead position recognition for 229 samples were obtained. The average recognition time was 0.122s per component, and the recognition accuracy was ±0.115mm (3σ). Fig. 7 represents the detail of accuracy data. Pictures of recognition results are given in Fig. 8 and the cross marker indicates the centre of the lead.

References

1. K. Edamatsu and Y. Nitta, 'Automated Capsule Inspection Method', *Pattern Recognition,* vol. 14, No. 1-6, pp. 365-374 (1981).
2. A. J. Cronshaw, 'Software Developments for the Visual Recognition of Chocolates', Proc. 2nd International Conference on Assembly Automation, pp. 191-199 (1981).
3. L. Rossol, 'Vision and adaptive robots in General Motors', Proc. 1st International Conference on Robot Vision and Sensory Controls, pp. 277-287 (1981).
4. A. G. Makhilin, 'Westinghouse visual inspection and industrial robot control system', Proc. 1st International Conference on Robot Vision and Sensory Controls, pp. 35-46 (1981).

Omega-Toshiba's route to automation

Toshiba has installed two robot lines for different assemblies.

PROJECTS and project teams, or task forces, are a major feature of engineering life in Japan. It is realised that the formation of teams to tackle new projects are good for morale and motivation, and get the most out of limited human resources.

So, when Toshiba decided it had to do some catching up in flexible automation, it followed the task force pattern. First, potential areas for increased automation were established. Second, the Manufacturing Engineering Laboratory was charged with the task of developing a new assembly robot, and of course, some engineers from Toshiba Seiki, which would build the production versions of the robots were involved. Thirdly, a team was built up around the nucleus of the plant production engineers to pursue the projects.

First came the Tosman TSR-701H robot, which although it looks as if it is an articulated arm device, is in fact closer in concept to the SCARA. The inclined arm can rotate on a pillar on which it is mounted, while the elbow joint also pivots on a vertical axis. The tool holder can also rotate and move vertically – just like a SCARA. Toshiba claims a positional accuracy of ± 0.05mm, and a speed of 2.2m/s.

Between 1975 and 1980, Toshiba spent £4.5 million on equipment to insert small regular-shaped parts, such as resistors and capacitors in printed circuit boards (pcbs). Most of these machines were supplied by Universal of the USA. Therefore, to increase automation, it opted for the insertion of large components of irregular shape as a major project.

However, before doing so, it decided to install one robot line for simple assembly work on a 'mature' product – one that was not likely to be subject to technological change. The domestic fan line at the Nagoya factory was chosen. These fans, designed to blow air around the room in hot weather, are sold in large numbers in Japan, but are being used less as more and more people install air conditioners.

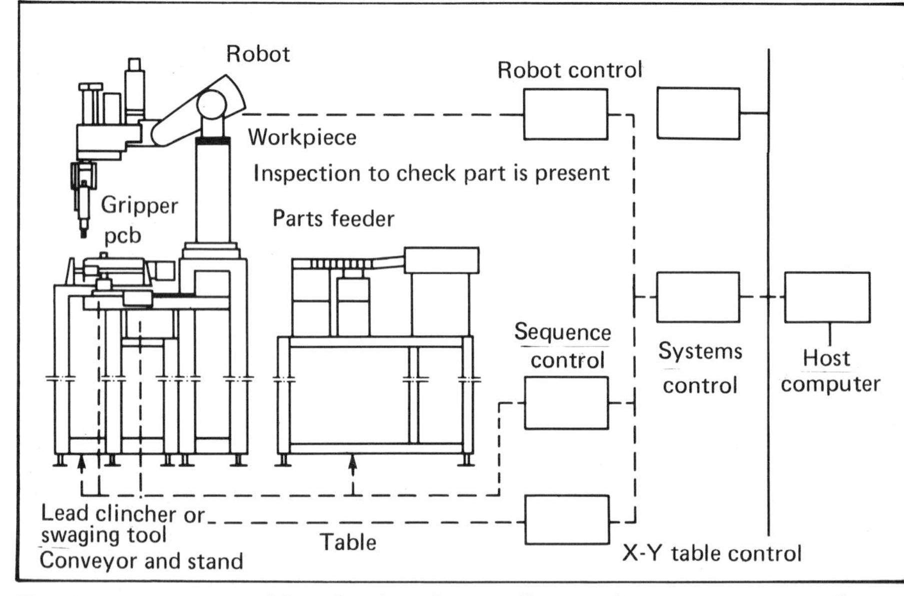

Fig. 1. Arrangement of the robot line for installation of components into pcbs at Toshiba, with the parts feeders behind the line.

The fan assembly consists of a conical plastics body with a steel stem, on top of which is the motor and fan assembly. The motor oscillates through an arc as it drives the fan. Toshiba carried out a design rationalisation programme, before the automated line was installed. Masahide Nanbu, senior managing director, said that standardising design is always essential before automation is introduced. Thus, the support tube, which used to comprise of 15 parts, now consists of 11, and it is supplied as a sub-assembly to the line – as is the gear case.

Free-flow conveyor

Altogether, 24 types of fan are produced on one line, and it takes three minutes to change from one type to another. With the reduction in manning from 29 to five people on the line, Nanbu said that the profit margin on the fans had been doubled.

Toshiba installed a row of 11 robots in a line 61m long. There is a free-flow conveyor, the assembly travelling on a pallet, which is stopped and located at each station. At the beginning of the line, the body is mounted upright on the pallet, and the first robot picks up a plastics cap from a pallet, going from position to position in sequence, and places it over the top of the stem of the body. Two screws are then inserted and tightened. When the pallet is empty a new one is presented to the robot.

There is a transfer press adjacent to the line which forms an ornamental panel, and a transfer device places it directly into the body – a good way of cutting out a lot of work-in-progress, so long as the transfer press can be guaranteed to work with a high efficiency.

Next, the body is turned over by a turnover fixture, and replaced in the pallet. A robot bends over the seven legs of the ornamental panel, and at the next station a robot inserts the support tube, four screws, and tightens them.

Installation of the timer and selector switch follow, at two robot stations. To fit the switch, two stages are needed; first, the robot picks up the switch and deposits it in a jig where it is orientated precisely. Then the robot picks it up again, and presses it on to its spindle. Another robot inserts the condenser.

Fig. 2. The robot line for insertion.

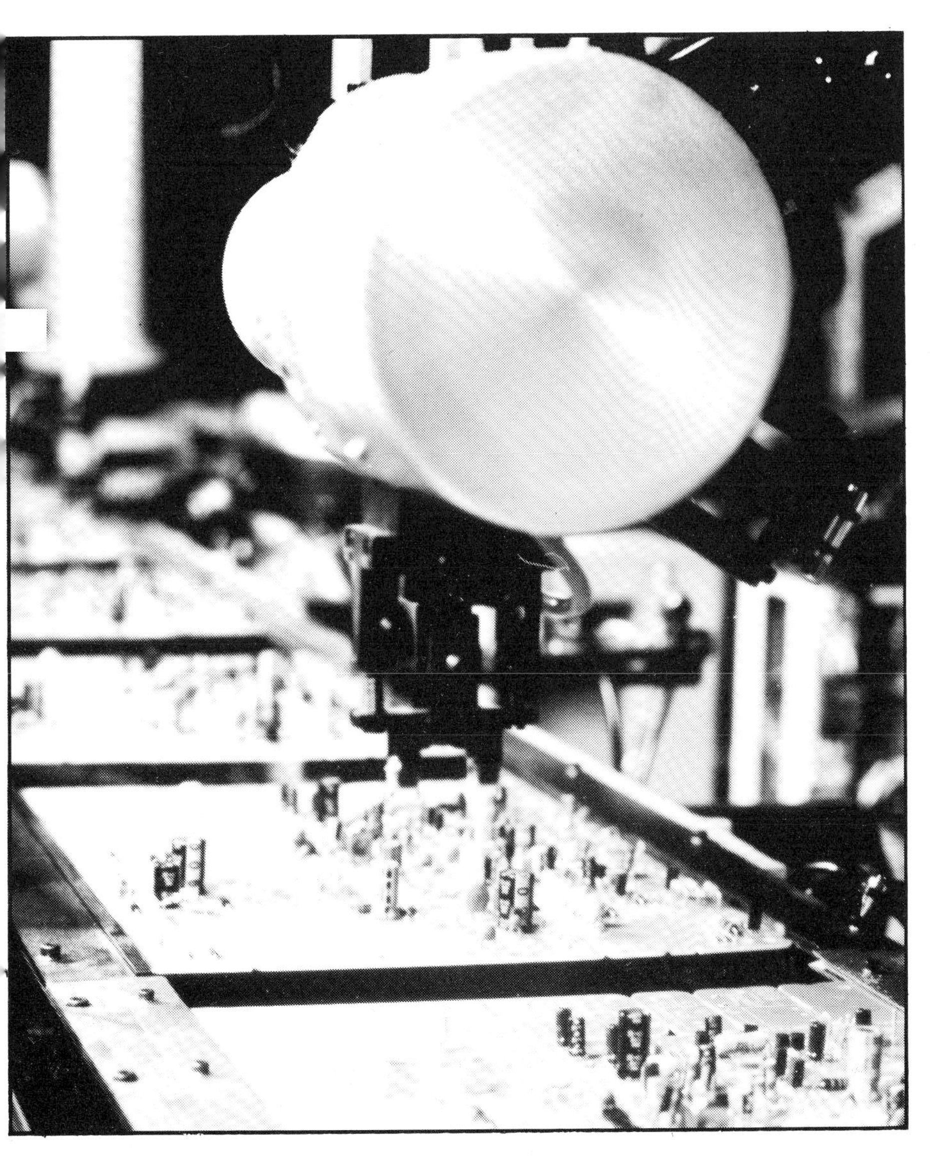

Fig. 3. Each robot carries a drum with multiple grippers.

However, the cables are too awkward for the robot to handle, so a man puts them in the correct positions at the next station, and arranges the ends of the cables on a 'separator', so that the robot will be able to pick them up individually, without touching two at a time.

Soldering gun

Nevertheless, the ends of the cables are soldered by robot. At this station, there is a robot with a soldering gun alongside the gripper, and the gripper holds the wires in position while the ends are soldered.

Once this end of the fan body is complete, the bottom plate is dropped on by robot, and then a piece of hard automation in the form of a multiple head screwdriver with automatic feed of screws, inserts and tightens the screws.

The body is inverted again, so that it is the right way up, and a robot fits a sliding knob and neck support. The next operation, however, requires manual operation – the feeding of the wires through the support tube for the motor, and the fitting of a stopper ring. Next, the neck support is screwed in place, and a timer knob is fitted. Afterwards, a man mounts the motor on the support tube, and runs the nuts on to their studs. Then, he fits the intermediate cables to the motor terminal.

Again, though, once the operator has fixed the wires in position, a machine takes over. In this case, it is a laser soldering machine, which directs lasers through a prism and along fibre optic cables to solder the wires at seven points.

Motor positioned

At the next station, the motor is positioned correctly – both the angle and height – prior to the fitting of the motor cover and its retaining peg by robot. To fit the cover, the robot has to slide it over the motor at an angle of about 5° to the horizontal. Finally, the fan itself is fitted manually, and the assembly goes on to test.

With that experience under its belt, Toshiba embarked on the Omega project at its Fukaya television factory, 70km north of Tokyo. Omega is the sort of acronym beloved of the Japanese, and it stands for 'Onward Mechatronics for Gigantic Automation System', which is such a mouthful that even Toshiba apologises for the 'Japanese English'. At Fukaya, Toshiba makes television tubes and assembles television sets, mainly 19in, although other sizes are produced. Altogether, Toshiba produces about 600,000 television sets a month, and it claims 14% of the world's market.

Fig. 4. Parts feeders, mainly in the form of vibratory bowls, are behind the robots.

At Fukaya, output is about 220,000 a month, and many operations are automated. Although the insertion of small parts into pcbs was automated some time ago, the insertion of large parts and final assembly have proved sticking points. The object with Omega was to automate some of the critical areas such as:

☐ insertion of larger components into pcbs;
☐ alignment of the set;
☐ assembly of the cabinet to the chassis;
☐ packing.

In addition, the aim is to develop an all-embracing computer control system. This way, it is expected that manning levels will be cut by 70% on assembly.

In the television tube plant, virtually all operations are automated, although the production of the pressed steel screens on Aida presses looks quite devoid of any systems approach. Since volumes are large, hard automation is the rule, although there are many handling devices that look like x-y robots; but these are all fixed sequence machines. Said Eiichi Sano, senior specialist, industrial engineering development centre: 'With these big volumes there is no need for robots; in any case, they are too expensive.' Among the recent innovations in the line, though, is a machine to fit the deflection yoke to the tube.

But for the insertion of large components into pcbs it was a different matter. The Universal machines are only capable of inserting small parts, so Toshiba decided that robots were the answer. The result is a row of nine robots in one of the six lines that go through the plant. Each is designed to process 1,000 sets/day, on a cycle of 20s, implying an operating efficiency of 75%.

In effect, there are three parallel lines in the robot system: first is the continuously-moving conveyor which carries the pcbs; next is the line of robots; and behind them are the parts feeders. The controllers are in front of the line. Since the parts are of varying shapes and sizes, vibratory bowls, chutes and pallets are all used at different stations. The parts include large condensers, transformers, terminals, a rheostat, and an integrated circuit in a standard multi-pin package. There are about 1,000 components in each parts-feeding system, or enough for unmanned operation for one shift.

Cylindrical drum

So that the robots can pick up these components, each has its own grippers. However, to speed up the job, most robots have more than one gripper. The robot carries a cylindrical drum from which the grippers project radially. The drum rotates on a horizontal axis to bring the grippers, in turn, to the vertical position for insertion. Typically, therefore, the robot rotates on its base to the area behind it – and the fact that the Toshiba robot can rotate through 300° on its base is a definite asset, when it comes to installing a practical line. The robot picks up one part, articulates so that the gripper is moved in the opposite direction to the line, to the second parts feeder. It also rotates the gripper, picks up the part, and then moves to repeat the process at the third chute. Then it moves the gripper over the conveyor, and inserts the parts in sequence.

In addition to the multiple grippers, the other keys to the system are a high degree of precision. First, the robot has a repeatability of ±0.05mm. The pcb is located with ±0.5mm as well, while the accuracy of the lead wires is similar. Then, the holes in the pcb are within ±0.1mm of their true position. Therefore, Toshiba claims that 99.9% of insertions are successful.

Tools synchronised

In some cases, once the parts are inserted, it is necessary to lock them in position to prevent movement during transport to the soldering bath. Therefore, the lead wires are either bent through 90° at the back of the pcb, or are crimped – squeezed with a pair of pincers so that the wire is flattened and cannot go through the hole. To perform these operations, tools are mounted on x-y tables beneath the pcb at the stations. The tools that bend the wires over must be synchronised to operate just after the component has

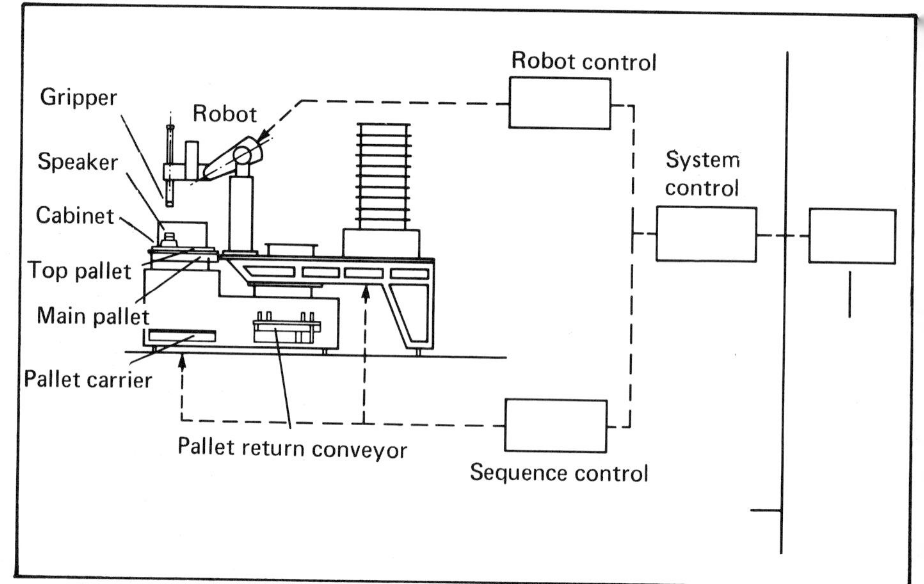

Fig. 5. Layout of the line for insertion of speakers and crts in the cabinet.

been inserted, but before the robot has released it.

Robot control

Control of the robot is exercised by a computer, which instructs the robot which program to follow, and when to start; then, it controls the x-y table, and its timing; finally, it actuates the parts feeders and synchronises their operations.

The first robot inserts small parts, and it is supplied by three vibrating bowls, arranged in a row behind it. At the end of each chute is a lift, which raises the part to a precise height for the robot to grasp. At the second station, there are two bowls, and two grippers on the chuck.

At the third station, there are two chutes fed from magazines, and one bowl. One of the chutes supplies the ics in cartridges, a finger pushing the ic out to the pick-up position.

The fourth robot is fed by bowl feeders. At the fifth robot, there are three chutes, and at the sixth there are three more bowls.

At the seventh robot, though, there are three sets of pallets, each carrying a number of parts which the robot picks up in sequence, directly from the pallets. The eighth robot mounts an angled plate, while the last one, which seemed to stop quite frequently, mounts the rheostat control, a cylinder approximately 25mm long by 25mm diameter. At the end of the line, the pcbs are loaded into racks, each rack containing about 10 pcbs.

In final assembly, there are also some robots. Two are used to insert metal components and the speaker, and there are also two automatic screw feeding/tightening machines, and a device to place the crt in the cabinet. There is a short line with a conveyor carrying double-decker pallets, which return on a separate line beneath the robots. The base pallet is a standard device that can be used for any size of cabinet; the top pallet, which is dowelled to the base, carries six locating points for the cabinet, which is an open box-shape device.

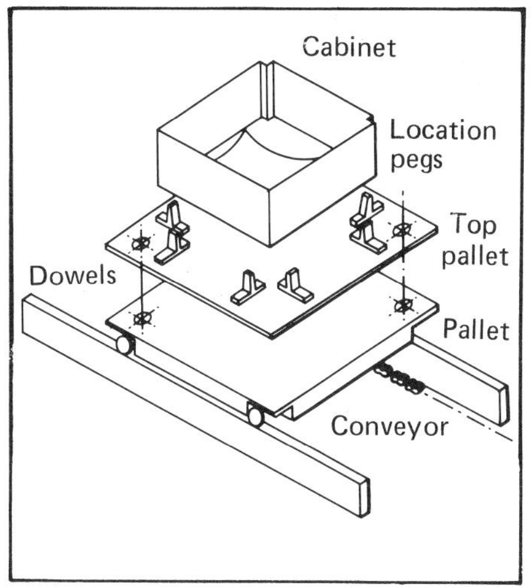

Fig. 6. The cabinet is carried on a double-decker pallet.

The first robot inserts a pressing in the cabinet, the second inserts the speaker, and then a robot inserts and tightens the screws. At the next station, a simple fixed-motion pick-and-place transfer device is used to transfer the crt to the line. Is the crt too heavy for a robot? 'No,' said Masao Teramoto, senior manager, production engineering department, 'the robot is too expensive.'

Repetitive motion

There are six lines, of course, and each line handles sets of a similar size range, so this is one repetitive motion. The transfer device picks the crt from a moving monorail conveyor, transfers it across a gantry and lowers it into the cabinet. Afterwards, more screws are inserted and tightened automatically.

The final area of partial automation is at the packaging stage, and again, robots are not evident. Open cardboard boxes are passed down a gravity conveyor to a stop, from an area beyond the end of the line. In this position, two labels with the serial numbers of the set and other data are simultaneously printed and applied to the box; the information coming from a previous station where similar numbers are printed on the set itself. As one plate pushes the label on to the lid of the box, so another plate slides down to provide a reaction pad. Since the other label is applied near the base of the box, a reaction pad is not necessary.

Next, the stopper moves aside to let the box pass on to a horizontal conveyor that transfers the box to the loading station. As the box enters the loading station, the two-piece lid is open, almost vertically. There are two slim arms at the loading station, and as the box enters the station, they move outwards, to hold the lid out of the way. A simple transfer device picks the set up and deposits it in the box. There are a couple of people in this area, but packing, usually a manual operation in Japan for assemblies of this size, is thus virtually unmanned.

In this Omega program, automation

Fig. 7. A robot inserting a speaker in the cabinet.

has been applied to assembly of the deflector coil to the television tube; mounting of large parts in the pcbs; soldering of pcbs; cabinet assembly; and packaging. The equipment cost £1.25 million, and took just one year to plan and install. According to Sano the aim was to design a system that could be installed quickly, but Teramoto pointed out that four months of concept work were done before the real planning started. The installations resulted in a reduction in the manning level of 103 people, implying a payback period, excluding the savings in overheads such as social security payments, of around two years.

Since the robot line was installed only in March, it is operating on only one shift at present. 'We intend to operate it on three shifts,' said Teramoto. There is one man on the line, and his job is to load the pallets and bowls, and to overcome minor stoppages. In due course, the one man could look after several lines.

Although that side of the equation is satisfactory, in other areas, the robot lines are clearly only a beginning. For example, the Universal or Matsushita CNC inserters for small parts operate at a rate of about 60 insertions/min, whereas the robots operate at the rate of no more than 10/min, despite those special chucks. However, there is clearly scope for an increase in speed of operation, since the robot arms could move faster. Thus, it might be possible to use only three robot lines for the work being done by six manual lines, in which case the finances would be much more attractive.

The need for special grippers, which must be changed to suit a new component, also reduces the flexibility of the line. As it is, the line can handle two or three different pcbs without any change in tooling.

As for the control system, at present, the connection of the robot line to the Toshiba host computer is not complete. 'Now, we transfer data by tape or floppy disc,' admitted Teramoto, who added that the integrated control system was taking a lot of time to develop. For communications between computers, Toshiba is using COBOL and FORTRAN languages, but the robots are controlled in a language written by Toshiba in Assembler.

Although the Omega system is not due to be finished until the end of the year, it is already paying dividends. It is another case of robots being used where their dexterity rather than flexibility is being exploited. But, of course, for a high volume manufacturer such as Toshiba, flexibility is not a question of being able to make anything, but of being able to make a small range of similar assemblies on one line, of being able to cope with seasonal demand, and of still being able to use the same plant when the parts are redesigned to suit changes in technology or fashion. In that sense, the robots are flexible, and of course, are making a significant reduction to labour costs.

4. Flexible Manufacturing Systems

To the Japanese, flexible manufacturing system (FMS) means flexible machining system, as is made clear in Jack Hollingum's review of the situation at the end of 1982. At that time, several Japanese companies were using their own interpretation of FMS in machine shops, and several relied on quite a number of operators, while there was little in the way of computer control.

Since then, as in other areas of flexible automation, the pace has quickened, with more companies installing bigger systems, often with very low manning levels. Mori Seiki's Iga plant is interesting in that a comprehensive system of unmanned vehicles was installed to link all departments together, not just to serve the FMS.

Fanuc has taken a practical approach, in that people are required to load machines, the theory being that since people are needed to keep an eye on things, they may as well be busy. Yamazaki's FMS at Minokamo is amazing for its boldness, breadth and showmanship. But that showmanship should not be allowed to hide the scale of this 'mammoth FMS.' It is very impressive not just for its size, but also because despite that high capital investment, it can break even at a low utilisation level. Indeed, it is in FMS for machining, as much as anywhere else, that the Japanese are really taking a big lead, and its machine tool makers are already reaping the dividends.

Japan's industry puts its money into FMS applications

Jack Hollingum, Editor of the *FMS Magazine*

During the past few months two groups from Britain and other European countries visited Japan to study flexible manufacturing systems in that country. This article summarises some of the findings and opinions of people from their encounter with Japanese progress in flexible automation.

WHILE MOST of the world's manufacturing companies are still shivering on the brink of flexible manufacturing, the Japanese are investing heavily in such systems. The two study tours of Japanese industry organised by IFS (Conferences) in May and November of last year covered between them only a small number of companies, but they were sufficient to show the fast-growing commitment in Japan to flexible automation.

Comments on the first study tour are based on notes from Professor Brian Hundy of Cranfield Institute of Tech nology and Dr Jeffrey Knight of the University of Nottingham. The notes accompanying this article relate to companies operating flexible machining systems. Other companies were also seen on both tours where the interest was in flexible assembly, material handling and robotics.

Dr Tkayoshi Ohmi, a senior research officer at the Mechanical Engineering Laboratory of the Ministry of International Trade and Industry, recently carried out a study of flexible manufacturing systems in Japan, and he talked to the tour visitors about his findings. He collected data from 53 systems, 20 of which were visited personally.

The list included a number of machines and cells which would generally not be classified as 'systems'. Five of the factories had stand-alone machining centres – in one case turning machines – equipped with pallet-changing facilities and equipped for unattended operation over a period of several hours. These he described as Flexible Manufacturing Cells (FMCs).

There were also 11 flexible transfer lines (FTLs) consisting NC or head-changing machine tools with automatic work transfer for a family of parts, but without flexible routing of workpieces.

Ohmi found no systems, up to early 1982, in which work and tool handling were completely automated, and the study tours found this degree of automation only in Yamazaki's 'Flexible Machining Factory' near Nagoya.

Apart from the five plants with stand-alone machines, all the systems covered by Ohmi's study had automated workpiece transfer between machines, but only six used an automated system to take tools to the machines. Most of them used conveyors for transport, but the more recently installed systems also used pallet shuttle carts on rails, and some had wire-guided carts. Tour members gained the firm impression that wire-guided carts are now the most popular transport system.

Qualification

Automatic loading of workpieces to the machine would seem to be an essential qualification for a flexible machining system but four of the systems reviewed by Ohmi did not have this facility.

As an indication of flexibility, most of the systems ran with average batch quantities between 11 and 100, but three were producing parts in random sequence with an effect batch quantity of 1, and three more had average batches of 2-10. The number of different parts produced in systems varied widely but more than half of reported cases produced 5-30 different parts.

Ohmi's picture of flexible machining in Japan tallied quite well with the observations of the two FMS tours. Of the 11 systems seen in operation, two were shops with flexible machining centres and lathes equipped for unattended operation, two were flexible transfer lines and the remaining seven were flexible manufacturing systems of various types, but only one of these – in the Fanuc motor factory – included turning machines.

All but the Toshiba Tungaloy plant had pallet shuttles or robot carts to take work to the machines, but only four had automated transport of tools to the machines. Only the Yamazaki 'Flexible Manufacturing Factory' automated the entire tool handling cycle from the tool setting department to the cutting spindle.

It is not easy to obtain information on the economic justification of these systems in terms which are comparable to European or North American conditions, though companies generally were quite willing to talk about the cost and productivity of their systems.

Several companies reported that they were able to adopt 24h working while at the same time reducing the number of people needed to run the machines. Toshiba, for example, was able to run its two-machine system round the clock with two people, whereas previously it needed six machines and 12 people on two shifts. Okuma's system with seven machining centres was running 24h daily with four operators, whereas the same number of stand-alone machines would require 10 men on two shifts. The direct labour pay rate quoted by Toshiba was Yen 250,000 per month.

Fanuc at its new motor factory reckoned to have trebled its labour productivity, with 60 employees turning out 10,000 motors a month as compared with production at the old Hino factory of 6,000 motors monthly by 108 people.

Expected payback periods varied quite widely. Fanuc talked of seven years for its Yen 7,000 million motor factory – although faster-growing sales could shorten the period. Toshiba estimated its payback at less than two years on its two-machine system, but figures would appear to suggest payback on Okuma's Yen 400 million FMS of four years or more.

One felt that perhaps in some cases some discounting must be allowed for the publicity value of a system. One felt this particularly at Yamazaki and Fanuc, where the plant was set out not only as a production unit but also to display the possibilities of the companies' machine tools and robots. Toshiba, OKK and Okuma were machine tool companies with systems based on their own machines, and Murata's system made generous use of its own warehousing and robot cart equipment.

On the other hand this showed that the companies had confidence in their contributions to flexible manufacturing. If they did not have systems of their own they would have a weaker case in selling them to others.

In system hardware there is a quite noticeable trend towards the use of robot carts rather than railed shuttle cars for transfer of work to and between machines, on grounds of cost, adaptability to existing shop layouts and scope for future extension.

Dynamic

Tool management, except at Yamazaki and Okuma, did not appear to be as highly developed as in some European and American plants. OKK had introduced at the Osaka machine tool show a large feeder tool store to back up the dynamic tool chain of each of its machining centres, in recognition of the need for extensive tool replacement capacity for unattended operation. No companies had automatic tool changing for turning machines, though some are believed to be studying the Sandvik block tool system for this purpose.

In most cases control of scheduling through the system was done by the operator with no support from computer simulation. Except at Fanuc and Yamazaki, there was no on-line link between the FMS control computer and any higher level computer.

Some standard sensing methods were universally adopted to support unattended operation of machining centres. Tool life was monitored for each tool in terms of its running time, and at the end of its nominal life it was replaced or the operator was called to replace it. Some systems had a computer link to the tool setting department to call up new tools. Automatic correction of tool offsets was in use at Yamazaki and possibly Okuma.

Spindle current monitoring was also in use to detect tool wear and breakage, and in some cases to give a degree of adaptive control catering for variations in material hardness or depth of cut.

Small diameter drills and taps absorb too little for wear or even breakage to be detected in this way, and the usual procedure was to check for the presence of the complete tool with a contact probe after each cut.

OKK, however, has introduced an acoustic emission sensor which it claims can detect damage to small diameter tools before they break. The method is used in conjunction with spindle load and tool life to determine when tools should be changed.

All the systems could also be fitted with touch probes and control cycles for measurement of work and tools. OKK had a special automatic measuring system which could measure the diameter of a bored hole to an accuracy of $\pm 4\mu m$ and a repeatability of $\pm 2\mu m$, and could make automatic correction for positional deviation in the Z-axis, due to thermal distortion, to an accuracy of $\pm 6\mu m$ and a repeatability of $\pm 4\mu m$ with resolver feedback.

From both tours the final impression was that there was nothing remarkably advanced about the production engineering of the systems, the design of machine tools, handling systems or other equipment, or the computing hardware or software – nothing in fact which could not be matched in Europe or the USA.

The fact remained, though, that Japanese companies had launched into FMS investment faster and with more confidence than their competitors in other countries, and so were gaining much valuable experience in the operation of such systems.

Behind this lies the fact that Japan's engineering companies are investing for growth in sales while the industry in most other countries, faced with a depressed market, has virtually halted new investment. The issue is less one of the relative efficiency of flexible manufacturing systems than of confidence in the future.

Brother Industries

Maker of sewing machines, knitting machines, typewriters and electrical and electronic equipment.

Two flexible transfer lines have been installed for machining of sewing machine frames. The line seen was for cast iron industrial sewing machine frames. The newer line is for diecast aluminium domestic sewing machine frames.

The industrial machine line, installed in 1980, is in three sections. It consists of 14 NC milling machines and nine NC drilling machines all with automatic tool changers, a horizontal twin-head boring machine and a special purpose milling machine, with seven automatic loaders, and a minicomputer giving DNC command of the whole system.

All the equipment was developed and built within the company so it could be fairly described as special-purpose.

Working 12-16h a day the line produces 3,000-4,000 frames monthly. The cycle time is five minutes and it take 1h 45min for a frame to go through all operations on the line. Six different frames are produced on the same line. Batch quantities are large, though change-over time is said to be short.

The line cost Yen 290 million, but only two operators are needed compared with 24 on the previous equipment, and quality is improved.

Fanuc

Two higly automated factories are now in operation at the company's Fuji Complex. The Mechatronics Manufacturing Division, which began production late in 1980, produces 300 robots, 100 wire-cut EDMs and 100 mini-CNC machine tools monthly with a total manpower of about 100 people. Assembly is entirely manual. Some welding robots are in use for fabrication, but metal plate is brought into the factory ready cut.

It has separate automatic warehouses for materials and for parts and sub-assemblies. Three wire-guided carts take materials to the machine shop in pallets or tote boxes. The are unloaded to holding stations and manually transferred, in the case of machining centres, to carousel pallet loaders. Lathes have robot loading from carousels. In all there are about 30 such machining cells.

Work is manually loaded to the carousels during the 8-hour day shift, and the machines continue unattended overnight or until they run out of work. About 450 different parts are machined, in batches of 5 to 20. Tool changing procedures are conventional.

Cost of this plant, including land in this rather remote location, was Yen 8,000 million. A number of the machine tools appeared not to be new.

In a separate building on the same site is the new Motor Manufacturing Division, which started operation in early July 1982. This is a two-storey building with the machine shop on the ground floor and assembly upstairs. A high

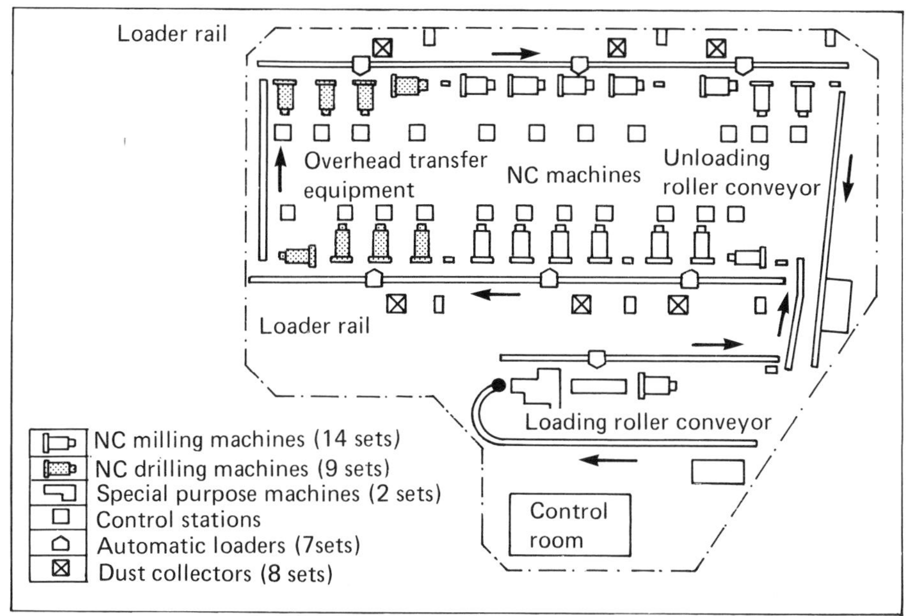

Figure 1. Layout of automated DNC machining line at Brother Industries.

automatic warehouse through the two floors receives finished machined parts and delivers them to assembly.

Members of the November FMS tour were given an opportunity to visit this factory, but they were not allowed to see it in any detail and the company was cautious in answering questions.

The factory has about 60 employees and 101 robots producing 40 different kinds of control motors at a rate of 10,000 units a month at full production.

Machining is more highly automated than in the Mechatronics Division. Wire-guided carts take work to the 60 machining cells served by 52 robots. Pallets are unloaded to stands from which robots transfer the work to each machine. Finished work is returned to the pallet which, on completion of the batch, is taken away by the robot cart. Tool settling, however, is manual and tools are taken by hand to the machines.

Upstairs there are separate assembly lines for DC servo motors, AC servo motors, AC spindle motors and AC spindle heads. These are robot transfer assembly lines to which work is delivered by wire-guided carts. There are 49 robots on this floor and the level of manning appeared to be low, but little information could be obtained from Fanuc on the operation of these lines, throughput rates, change-over times between batches and so on. Batch quantitites vary between 20 and 1,000.

Cost of this factory was Yen 7,000 million including land and building.

Murata Machinery

Maker of textile machinery, data transmission equipment, machine tools – turning machines under licence from Warner & Swasey and punch presses under licence from Weidemann, and physical distribution control systems including robot carts and high-bay storage systems. The company supplied the robot carts and warehousing for some of the other systems seen in the tours.

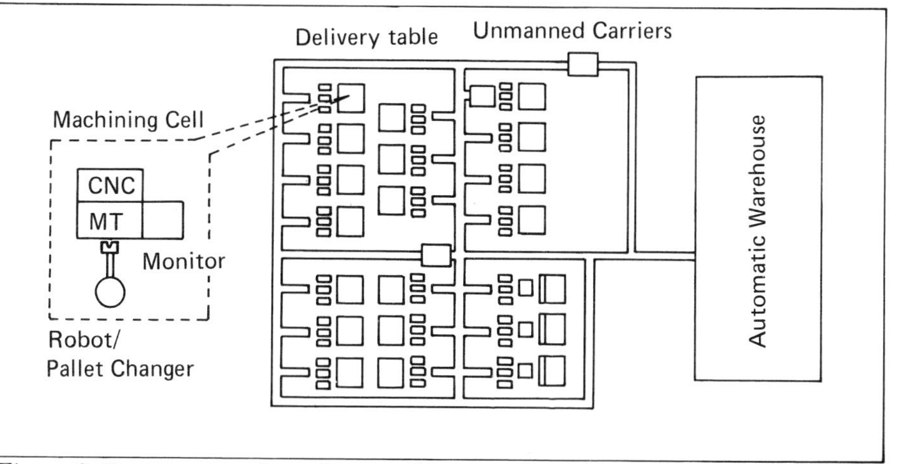

Figure 2. Fanuc system based on machining cells and assembly cells.

Murata itself has a system in its Inuyama factory linking nine machining centres and a large warehouse with load stations and robot carts. The system is unusual in that the machining centres are of different makes and types and were not all new when installed in the system. Work on the system began two years ago and production started one year ago.

It is producing up to 1,500 different cast iron and cast aluminium parts for machine tools and textile machinery, in batches of 20-50. The system has 100 machine pallets and fixtures, with different pallets for the vertical and the horizontal machining centres. The six vertical machining centres are from OKK and the three horizontal machines from Okuma and Yasda. The fixtures are semi-universal with mounting holes and clamps for a variety of different parts.

The high-bay warehouse has 288 storage racks and serves to store raw materials on wooden pallets, machine pallets with fixtures both unloaded and loaded with work before and after machining, and finished, unloaded work on wooden pallets, as well as 500 racks of cutting tools – 3,000 in all.

Incoming and outgoing work is handled by a stacker crane between the two bays of the warehouse, which also delivers work and pallets to a line of transfer stations from which they are moved across by a traverser to load stations.

Since the warehouse is used both to store raw and finished parts and as a buffer to hold loaded pallets before and after machining for unattended operation of the plant, the load arrangements are rather complicated. There are delivery stations for unfixtured work on wooden pallets as well as a row of delivery stations for pallets. All but two of these are simply stands providing a buffer between the warehouse stacker crane and the production side. The other two are the load/unload stations, which are served also by the robot carts taking work to the machines.

There is no on-line link between the system control computer and the higher level computers. The host computer in the company's head office handles order processing, inventory management and other functions and is on-line to the factory computer which deals with inventory control and production programming. Output from this is a diskette which is used by the control computer for production scheduling. This company and Yamazaki were the only ones visited where it was said that computer optimising of the production schedule was used.

Cost of the system, excluding the machine tools and pallets, was Yen 140 million. A separate estimate was obtained of Yen 200 million plus an average of Yen 10 million for each machining centre. It was built, according to the company, because in the region where the factory was employees were not prepared to work outside the eight hour day shift, while the company wanted the plant to run round the clock. If a manned rather than an unmanned system had been adopted it would have required 19-20 people instead of the present six, of whom two are in tool setting. In the near future Murata plans to reduce the number of people to four as a result of efficiency gains.

Osaka Kiko

Builder of machine tools – OKK machining centres and milling machines – special purpose machines, spinning machinery and other products. The company showed its first flexible machining system, with a vertaical and a horizontal machining centre, at the Osaka International Machine Tool Fair in October, and is in process of installing in its own factory an extended system with two horizontal machining centres and one vertical machining centre, due to start operation in February.

At the time of the November study tour the foundations were being prepared and the machines returned from the exhibition were awaiting installation.

The work handling side of the system will be fairly typical, with a shuttle car on rails connecting the three machines and a load/unload station. On the opposite side of the track from the machine tools will be 20 storage stations for the 500mm square pallets, for use during unattended operation.

Most obviously distinctive about the system are the tool storage racks, of which one was demonstrated at the Osaka show. The machining centres themselves have tool drums carrying 30 or 36 tools, linked to the spindle by an automatic toolchanger giving a chip-to-chip time of 12s.

At the back of the drum another tool change mechanism transports tools to and from a large tool rack with a capacity of 180 tools. Other sizes of tool rack are available – the one at the exhibition took 116 tools. Tools are placed horizontally into a matrix of pockets, making for a store which is fairly compact in terms of floor area. A two-axis cartesian pick-and-place device takes tools from the rack for transfer to the drum, a cycle of 16s.

The new system will be replacing 13 machine tools, of which four are turret head milling and drilling machines. It will require two operators to run it instead of the present six, working a single shift.

At the first stage, before the third machining centre is added, the system will be producing 69 different parts in batches of five or more.

Control of scheduling – input manually by the operator – and of the shuttle car and load/unload stations is through a 16-bit microcomputer, with separate Mitsubishi CNC controls for the machine tools and the tool stores.

Besides the usual supplementary functions for unattended operation, OKK has introduced an acoustic emission sensor which, the company says, can detect damage on a drill down to 3mm diameter or a tap down to M4 and stop the operation before the tool breaks. In the complete system tool replacement can be determined by simultaneous monitoring of the cumulative cutting time, the spindle load and the acoustic emission.

The company also has an automatic measuring system with a compensation facility for the effect of thermal expansion on hole centre distances and diameters.

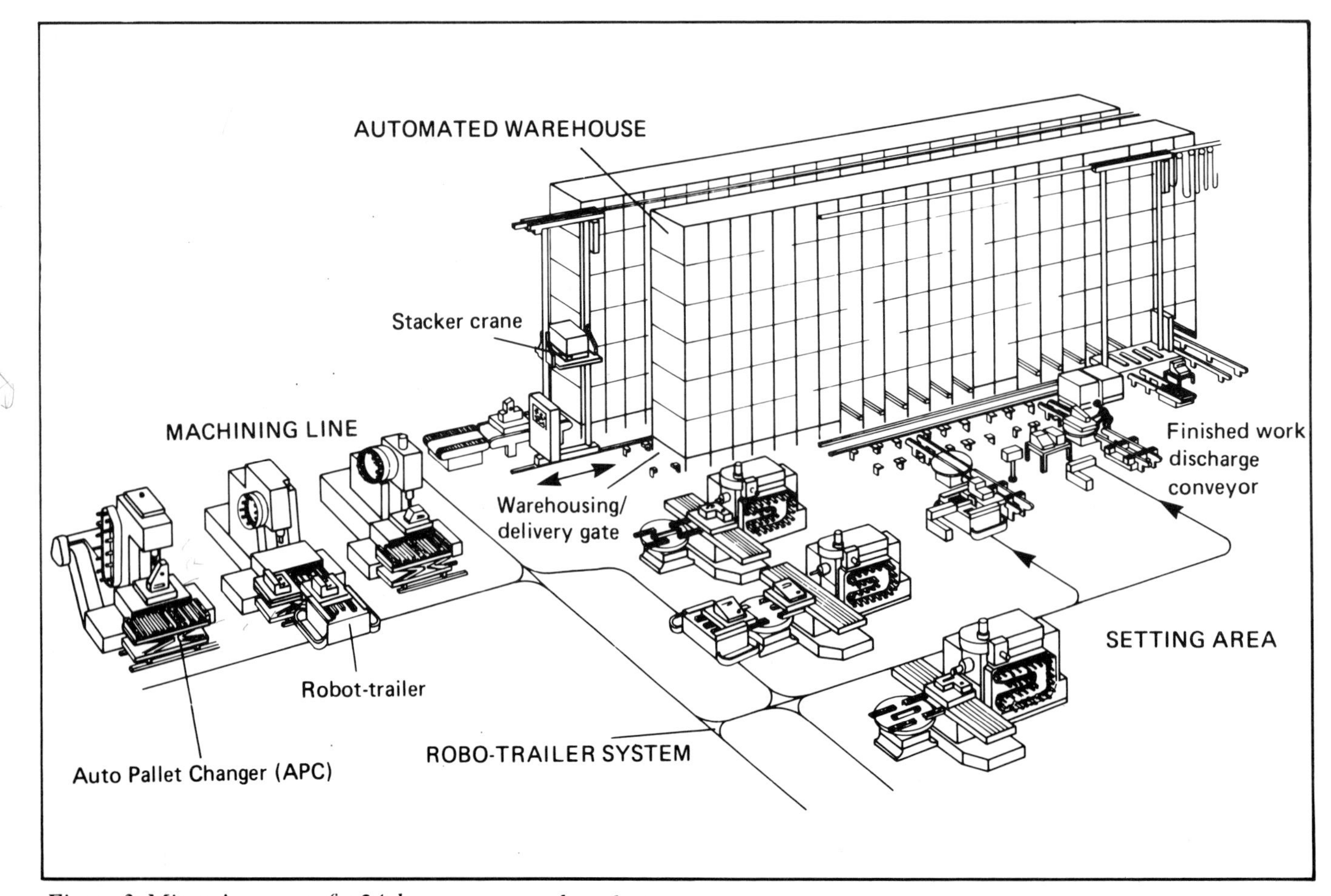

Figure 3. Mirata's concept for 24-hour unmanned production.

Figure 4. OKK's two-machine system at the Osaka show.

Okuma Machinery Works
Machine tool builder and supplier of flexible manufacturing systems. A system shown at the Osaka Machine Tool Show is described in John Hartley's review of the exhibition on page 199. Nearing completion in the company's Oguchi plant the November FMS Tour saw a flexible transfer turning line nearing completion for Canon Camera, with five LC10 CNC chucking lathes linked by conveyors and the company's own NC robots. Ten different parts will be machined in random sequence. They are mounted automatically on holding mandrels for operations which include some transverse machining such as milling. Refixturing on different diameters is also handled automatically. Parts are washed between each operation.

In the Oguchi plant is an FMS installed about a year ago consisting of seven of the company's horizontal machining centres linked by a Murata wire-guided cart, which also handles tool changing functions. The line produces 95 different workpieces in batch quantities of 10-20, mostly lathe headstocks, machining centre spindle heads, grinder tailstocks and lathe saddles.

Four of the machining centres are MC-5H models and three MC-6H. Loading and unloading of parts and fixtures is done with the help of manipulators away from the machining area and the cart takes the loaded pallets to the machines. The system has 21 pallets. There are also 17 stands in front of the machines for intermediate storage of parts for overnight unattended running.

Also away from the machining area but connected to it by the wire-guided carts is a tool control room where tools are stored, ground and preset, and from which kits of tools are delivered to magazine stations beside the 70-tool chains of the machines, for manual loading.

The control system is Okuma's own, both the individual CNC controls of the machines and the overall FMS controller which transmits data through optical fibre cables to the machine controllers and the carrier controller.

The production schedule is input manually from data supplied from Production Control, though Okuma plans a future link with the company's production control computer.

The control computer sends machining data to the machining centres, and fixturing and loading instructions to the CRT at the handling station. At appropriate times it also sends instructions to the tool control room CRT on tool kits to be prepared and preset.

The complete system cost Yen 400 million, and Okuma said the price to a customer would be about the same. The company offered the following comparison with the previous method:

	FMS	Previous
No. of machines	7	7
Running hours	24	16
Operators	3+1+0	5+5
Machine utilisation	75%	55%

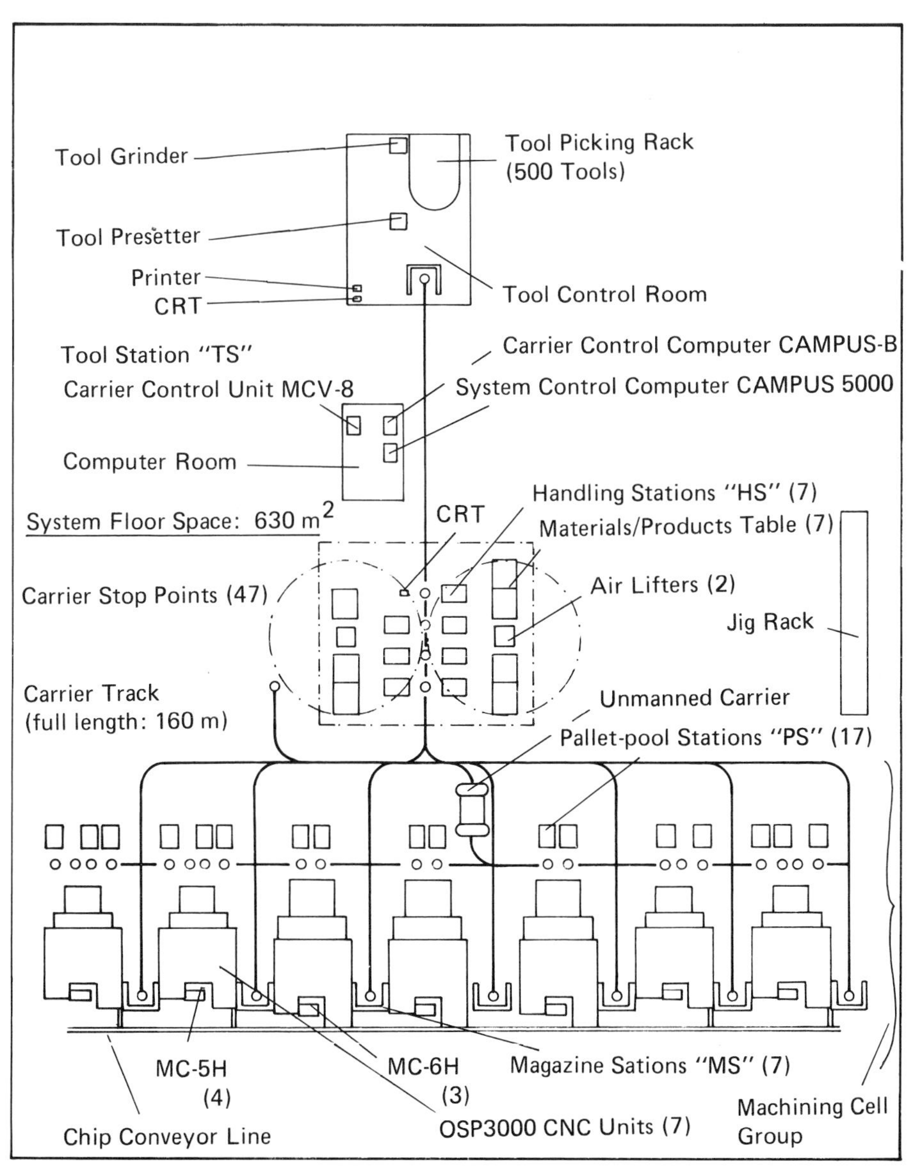

Figure 5. FMS produces parts for Okuma machine tools.

Toshiba Machine
Machine tool company with 'only' two flexible machining systems in its own plant – one making parts for injection moulding machines and the other, a new system, being proved out, for machine tool parts. Each consists of two horizontal machining centres with a single 20-station pallet carousel serving the two machines via 180° automatic pallet changers.

Alongside the new cell is an automated warehouse for storing workpieces and fixtures. The operator uses overhead tackle to transfer work to the load/unload station. Each tool magazine has 60 tools but chains can be supplied for 90 people or more tools.

Each CNC controller can store 40 programs, maximum 600m of tape equivalent. The system controller is a microcomputer which allows the 22 pallets to be selected in random order. Scheduling of the production sequence is input by the operator at the beginning of the shift and all subsequent control of pallets and tools is handled by the controller.

Other automatic functions include tool life monitoring, 'adaptive' constant load feed control, monitoring of abnormal cutting, and a thermal displacement compensation in the Z-axis.

Toshiba offered the following comparison of the first FMS in its factory with the previous production methods:

	New	Old
Number of Machines	2	6
Daily working hours	20	13
Shifts	3	2
Average set-ups per piece	1	4
Ratio of total process time	1	:4
Operators	1+1+0	6+6
Machining hours per piece	0.73	1

Cost of the system was Yen 220 million excluding fixtures and tooling. Payback time for the new system was claimed to be less than two years.

Toshiba describes these two-machine systems as 'Level 2'. At 'Level 3' the company has sold a system to IHI for machining rotational marine components, with vertical turning centres linked by shuttle cart, as well as the Toshiba Tungaloy system.

Toshiba Tungaloy
DNC system, primarily for machining steel milling cutter bodies. Equipment was supplied by Toshiba Machine Co. The system

Figure 6. Seven machining centres are cart-linked at Okuma.

consists of one CNC lathe, one vertical machining centre, three five-axis horizontal machining centres and a CNC grinder.

There is no automated handling system between machines. The lathe and the vertical machining centre have each a 10-station carousel type workpiece magazine, an auto loader and a workpiece discrimination device which checks if the piece is correct. For cutting on the horizontal machining centres and the grinder the parts are mounted by hand on standard handling arbors with discrimination rings for checking by the computer. Vertical stacking magazines hold eight components.

Tool breakage is detected from spindle motor current except for tools less than 6mm dia, which are checked for length against a limit switch after each operation. Tool wear is checked by motor current and predetermined tool life is held in the computer data base.

The system produces a large number of different workpieces – 4,000 was quoted at the FMS Conference in Brighton – in batches of 2-20.

Its capital cost was Yen 500 million three years ago the break-even is expected to be reached in less than five years. Software development accounted for 12% of the cost, and Tungaloy put in 15,000 man-hours of engineering time in developing the system.

An assessment by Toshiba Tungaloy of the performance of its

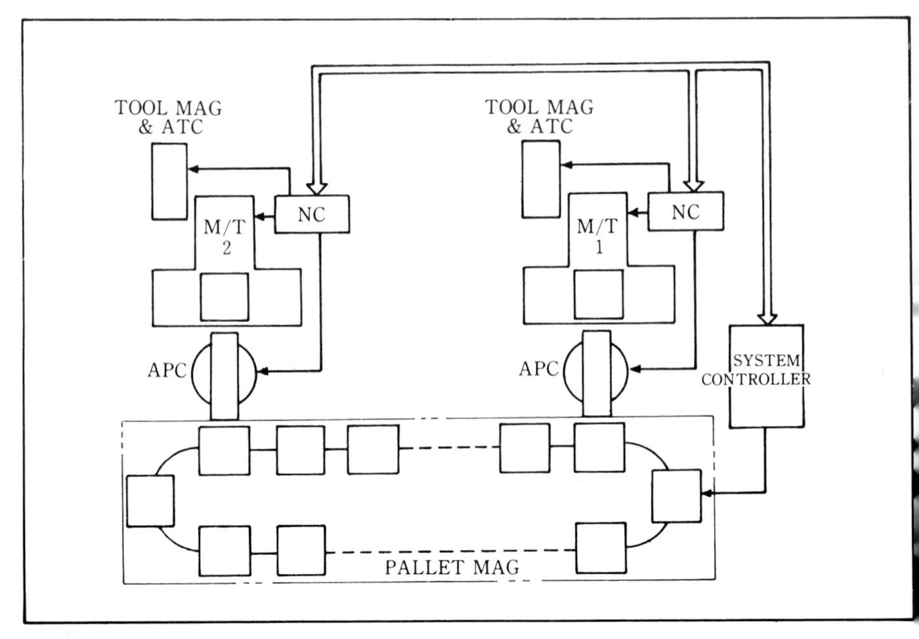

Figure 7. Configuration of Toshiba system.

'Programmable Precision Manufacturing System' gave the following projected comparison with a conventional shop:

	Conv.	PPMS
No. of machines	50	6
Labour required	70	16
Product yield rate	95%	99%
Working ratio	20%	70%
Floor area, m²	1480	350
No. of processes	15	8
Total process time	18.6 days	4.2 days

At the FMS Conference in November a paper on the Toshiba Tungaloy system reported that product yield achieved so far was 96% and working ratio 63% but that the expected performance would soon be achieved.

Yamatake Honeywell
Flexible transfer line for machining control valve bodies of 1 to 6in bore. It has been operating since 1973 and has seven special purpose machine tools and a wash station linked by a roller conveyor with pallet transfer. Datum faces are pre-machined on a conventional miller.

The line handles 400 different workpieces in batches of 20-60 with a cycle time from 3-10min. In single shift working about 2,000 bodies are produced monthly. Pallet fixtures are designed for a range of body sizes and changeover time between batches is claimed to be 2-3 minutes.

Yamazaki
Machine tool builder with a 'Flexible Manufacturing Factory' operating since November 1981 at the Oguchi headquarters near Nagoya, a second FMS starting up early this year at the Florence, Kentucky subsidiary, and a more advanced system scheduled to start operation in March at the new Minokamo plant in Japan.

The FMF has two lines of machining centres, each with pallet loading from a shuttle car. The 'A' line has eight horizontal machining centres making headstocks for CNC lathes and parts for machining centres. It produces about 800 workpieces a month of 23 different part numbers. The shuttle car taking the 1 metre square pallets to the machines has a load capacity of 3 tonnes.

Each machine has two tool drums each of 40 tools capacity, and since standardisation of product design has cut the number of tools needed to 63, each machine can produce every part and there is

Figure 8. Line A at Yamazaki machines parts up to 3 tonnes.

Figure 9. Long shuttle car on Line B takes two pallets.

Figure 10. Ten machines in Line B handle work up to 8 tonnes.

space for back-up tools in the drums. For tool changing an overhead transfer system carries a complete drum stand with two drums from the machine to the tool setting room and replaces it with another.

The 'B' line is for larger workpieces up to 8 tonnes. It has seven horizontal and three vertical machining centres producing beds, bases, saddles and columns for CNC lathes and machining centres. It can turn out 600 workpieces a month of the 51 varieties produced on this line. The interchangeable 1,600 by 3,000mm pallets can be carried two at a time on the long shuttle car for fast loading and unloading. The 30-tool drums for the B line machines are stored behind the machines and are changed manually.

An additional facility on the B line is that the pallets can be transferred on the shuttle car to a heat treatment facility for hardening of ways immediately after machining.

The FMF is designed for 24h continuous running, with two operators at the load/unload stations on each of the two lines on two eight-hour shifts, and unattended working on the third shift. The A line has 44 pallets which can be prepared for unattended work, and the B line has 29.

Sophisticated use is made of touch probes for tool breakage and wear monitoring and for checking alignment of castings, and in some cases for resetting machine zeros to match castings. The system also has the other expected functions for unattended operation, such as tool life monitoring, spindle torque monitoring, and an adaptive feed rate control based on spindle motor power.

Computer

Changing and presetting of tools is under the control of the system computer, which signals to the tool setting room when a pair of tool drums must be changed. The used tool drums are transported back to tool setting and new ones take their place. The computer's CRT display then instructs the tool setter on the names and numbers of the tools that must be reset. He puts each new tool in the presetting machine and measures the position of the tool tip, and then presses the 'measurement complete' button on his terminal. The computer stores the information, and when the drum goes back to a machine the computer automatically resets the tool offset values in the machine's controller. One tool setter is needed for each of two shifts to cover the requirements for three shifts.

The computer room likewise has one person for each of two shifts. It has a main PDP 11/23 computer with seven sub-computers used for filing, program storage for the 74 different workpieces produced, control of the handling systems, machine monitoring and so on. The main computer files all the information needed for unattended operation and it controls the scheduling of work to the line in accordance with the manufacturing programme. It can also dynamically re-schedule the line in the event of a machine failure.

The following comparison of the FMF with a conventional machine shop was offered by Yamazaki:

	Conv.	FMF
Floor space, ft^2	70,000	30,000
Machines required	68	18
Labour required	215	12
In-process time	90 days	3 days
Capital cost (incl. land and buildings)	$14m.	$18m.
Annual labour cost	$4m.	$227,000
In-process inventory	$5m.	$218,000

FMS dominant at Osaka

Many new flexible manufacturing sustems, several with robots, dominated Japan's International Machine Tool Fair, Osaka.

Okuma's FMS with its machining centre, two CNC lathes and one cylindrical grinder, unmanned trolley, pallet stations and set-up area

Layout of Yamazaki system of four machines, three with robot arms and each with automatic monitoring.

FMS, being shown by more than a dozen of Japan's machine tool makers, dominated the 11th International Machine Tool Fair at Osaka (28 October-8 November). Of course, these are really flexible machining systems rather than true FMS, but they are all aimed at the same target – flexibility with increased productivity per man.

With small lot sizes, and a lot of variety, machine tool production lends itself to a flexible approach based on unmanned cells, as Yamazaki and Fanuc have already demonstrated. In addition, a cell-system serves as an excellent showroom for the manufacturer – perhaps with some of the costs being considered to be advertising.

In all these FMS a number of CNC machines are equipped with automatic compensation for tool wear, tool condition monitors and necessary handling equipment to operate unmanned. To optimise operation, the system is computer controlled.

The latest company to install a big FMS is Mori Seiki, manufacturer of small CNC lathes and machining centres. It has invested £22 million in an FMS to increase output to £130 million p.a. True, it already has an FMS consisting of seven machining centres to machine large and medium-sized cubic workpieces, but nevertheless the return on investment will be phenomenal.

Yamazaki Machinery Works has started to build its second FMS, which will start producing machine tool parts in the spring, and the productivity per man is intended to be 10 times that of a normal Japanese machine tool maker. This is convincing evidence that where there is a lot of variety in part size, FMS is the answer.

Builders

To emphasise that point, there were many robots on show, designed to handle workpieces at machine tools – something like 12 – also some 15 Fanuc robots were demonstrated by other machine tool builders at their machines.

Of course, robots are essential for an FMS that is machining small parts with short cycle times. Most of the systems at Osaka, though, were intended to handle large parts, but some were designed to tackle the more difficult problems encountered with small parts.

But among the many FMS were some that have been proven in a factory, or are about to be installed. Among these, of course, was Yamazaki, which has been operating its first FMS for over a year, and which is starting to install its second now. In fact, its exhibit was part of the new FMS, which includes heat treatment and grinding as well as the machining of small parts.

A total of 65 machines will be used with 34 robots, and it is expected to operate for three shifts. Yamazaki was showing a system consisting of two Slant Turn CNC lathes, one V-10N vertical machining centre, and a H-12 horizontal machining centre. There was a robot arm at each lathe and the vertical machining centre. These were all drilling and tapping holes and turning discs about 100mm in diameter. Meanwhile, the horizontal machining centre was machining housings. Of course, each machine is equipped with automatic monitoring.

In concept, this system has moved on from the early Yamazaki FMS, in that the trolley is of the type that follows cables in the ground, instead of running on rails, and a CAD/CAM system is coupled up to the computer control system. But there are the same buffer stations – 12 for the four machines.

With the CAD/CAM system, Yamazaki's engineers claim that errors in design are eliminated, and that it is an easy matter to design parts

Ingredients of Yamazaki's new FMS include these workpiece carriers and the indicator panel.

Toyoda's FMS with shafts on small carriers ready for grinding.

which can be accommodated on standard fixtures and cut by standard tools. There is a Melcom 70 model 30C host computer, and data for parts are stored on floppy discs. Data is transmitted by an optical fibre data highway directly from the computer room to the FMS. The equivalent of up to 10,000m of NC tape can be stored in the system's memory.

Called System 21, the Yamazaki CAD/CAM system costs around £600,000 in Japan, whereas the FMS based on the four machines, but excluding the CAD/CAM costs around £500,000, which seems expensive.

Turrets

Mori Seiki was not actually showing its FMS – it is far too large, and it started operating in the Iga factory just before the Fair opened. The company has been operating a system, involving seven large machining centres with pallet changers, unmanned at night, to produce large components such as spindle heads, tail stocks, turrets and cross slides for five years. The pallet changers carry five to 12 pallets.

In the new system, RoboTrailers travel throughout the plant – not just between the FMS and warehouse, but also through the shops where other components are machined and assembled, and to another warehouse. The warehouse for the FMS itself is massive, with 18,000 spaces for pallets and 13 cranes. Each CNC machine in the FMS, which is used to machine beds and carriages, columns and tables, has a double pallet-changer.

There are 45 machines in the shop, of which 13 are included in the FMS; of the others, 10 are conventional, and 20 are NC machines. Outside the FMS, there are 12 machines each with an operator and 20 NC machines with seven operators. Four of these machines are housed in rooms maintained at constant temperatures. The FMS is operated by three people, since it is unmanned for two shifts. In total, therefore, 22 operators are needed for 35 machines, and 10 of these machines need one operator each.

Therefore, Mori Seiki explains, the FMS can operate for 528h/month, with three men, whereas a conventional machine shop would require 27 people for only 176h/month. These are figures that mean low costs and high profits in any language, so it is not surprising that Mori Seiki is planning to produce over 4,000 machines a year, equivalent to almost 25% of the current machine tool output of the Japanese industry.

Okuma is another company with a proven record in FMS, having gradually worked up from a single arrangement to a 'Mark III' system in its own plants. It claims to have sold 10 FMS, and was showing a system consisting of one LC40-M lathe, an LC20 lathe, an MC-6H machining centre, and a GRE6-2NF cylindrical grinding machine, designed to handle shafts, flanges and boxes in random order. There are the usual robot arms for the smaller machines, an automatic pallet changer (APC) for the large machining centre, and a set-up area.

The machines are arranged in a rectangle, one unmanned trolley following a track around the machining area. At the grinder, a camera was being used to make sure that a workpiece was in position for the robot to transfer to the machine – a rather costly solution.

Okuma claims to be using its own 16-bit mini-computer in its Campus 5000 control system, which can interact with the factory computer for production management. The system controller, Campus-A can cope with up to 10 CNC machines and their robot arms, while the sub-system, Campus-B, takes care of the transport of workpieces and tools. There is a separate fibre optic cable between the computer and each CNC, while separate computers are used to manage the fixtures and the warehouse.

ATC

Toyoda Machine Works' FMS – called AFMS TIPROS-20 – is an impressive development of its DNC systems, and a lot of effort has been put into the handling. In addition to the FVN50A and FHN50II machining centres there was a G62A cylindrical grinder with automatic load-unload arm in the cell, as well as two unmanned trolleys – one simple one for the transfer of workpieces, and a complex one to transfer tools.

Of course, each machining centre has its own automatic tool changer (ATC), but there are also two stockers, each carrying 20 tools. The Robocart-tool carrier can carry 14 tools, and has a robot arm to transfer tools between the carrier and the stocker or ATC at the machine tool. Because the robot arm is on the trolley, the handling equipment at the tool stockers and machines can be simpler than that usually found.

Voice

The system is controlled by an Eclipse MV/6000 minicomputer, while there is also a CAD/CAM system which can be linked up to integrate design and manufacture. A gimmick here is voice recognition and synthesis, so that the engineer can 'talk' with the computer. Toyoda put a price of £700,000 on the complete system, including CAD/CAM.

Hitachi Seiko was showing the System 107, it first exhibited at Chicago, built around one Model HC500 horizontal machining centre, a Model VA50 vertical machining centre, and a Hiturn II-30 CNC lathe. The Samtos arm mounted on the front of the lathe was one of the neater designs on show.

Hitachi's version differs from most others in that it measures the current needed to provide the necessary thrust at the tool to cut metal, not the current at the main spindle. If the current increases 20% above normal, the feed rate is reduced, and if the current is still too high, then a new tool is fitted.

Hitachi Seiko, not to be confused with Hitachi Seiki, was showing an FMS system that is about to be installed in its own factory. It consists of two machining centres, and is interesting in that each has a simple pallet changer, a trolley traversing between the machines, and a carousel pallet stocker/set-up area. A camera is used to identify the workpiece on the pallet prior to its being loaded on the trolley for transfer to the machine.

Ikegai layout with row of machines and simple conveyor.

Among the other FMS was one produced by Niigata Engineering, and others from Shin Nippon Koki and Osaka Kiko (OKK). The OKK set-up was based on MCH-560 and MCV-630 machining centres. There was also a large tool storage rack to hold 116 tools, and a rather small loading/unloading station with four pallet holders on a rotary table.

Shin Nippon's system is based on three large portal machining centres, and reflects a set-up installed at a Mitsubishi Heavy Industries' plant to machine V-form cylinder blocks for large industrial diesel engines.

Niigata is about to install its FMS to machine parts for machine tools, this has been developed from one built to produce cylinder heads for Niigata's big diesel engines. This is designed for large workpieces, and is based on one HN80A and one HN50A machining centre. The automatic loading/unloading system is specially designed for free disposal of swarf. A significant feature of this system is a loop-type optical fibre data highway which can serve up to 15 units, including CNC machines, terminals and trolleys.

One minicomputer can control four loops, thus providing the base for a completely computerised manufacturing system. Since a CNC machine can be plugged into the loop via an interface unit, without the need for many parallel cables running to the minicomputer, the system can be expanded easily.

Arms

Clearly, that is the data highway most suitable for FMS, and points one way ahead. But the robot arms are also now becoming essential features of machine tools, so it was not surprising that companies such as Yamazaki, Okuma, Ikegai and Hitachi Seiki have developed arms that can be fitted to their machines. Daikin Industries was one of the independent companies showing a robot arm that can be mounted on the front of a machine. The arm articulates on a horizontal spindle, and can also move longitudinally in its housing. Then, the wrist can rotate.

Toyama introduced its cylindrical co-ordinate robot for handling at machine tools. The Toyama RP-1 has a capacity of 50kg, including the hand, and has up to six degrees of freedom.

Certainly, the Osaka fair was evidence that the Japanese machine tool builders are intent on making a quantum jump in their own productivity with flexible systems – and to sell more machine tools and FMS.

Hitachi's Seiki's system includes a robot arm mounted on the front of a lathe.

The practical approach to an unmanned FMS operation

K. Kobayashi and H. Inaba, Fujitsu Fanuc, Japan

The flexible manufacturing system now in operation at Fujitsu Fanuc has attracted much attention, particularly in the aspect of use of an industrial robot combined with a CNC machine to form an unmanned machining operation.

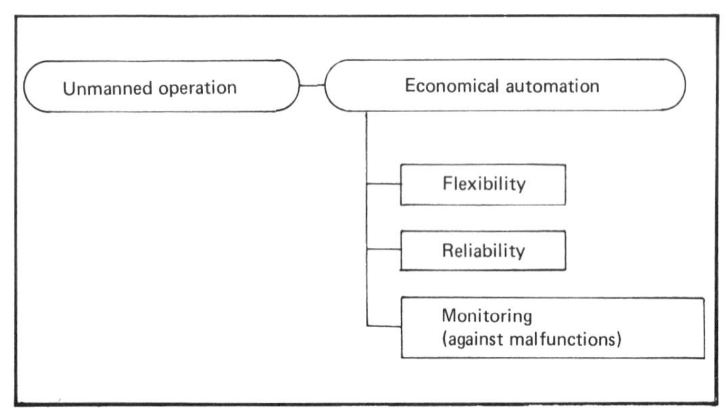

Fig. 1. Three important factors for an unmanned operation.

NC tape preparation system
Tooling
NC tape supply
Part program memory
Automatic tool change
Tool compensation
Machining
CNC machine tool
Loading/unloading
of workpieces
Robot, pallet changer
Chip disposal
Air jet, coolant
Transportation
Chip conveyor

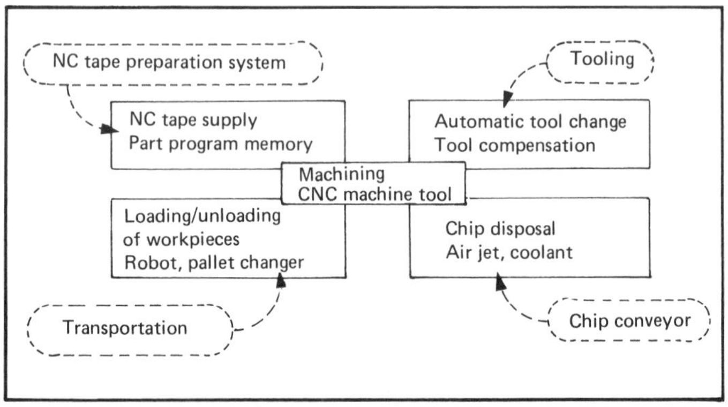

Fig. 2. Three major functions of a machining cell.

THE INSTALLATION of a wide range of CNC machine tools in machining processes requires advanced technology in order to achieve progressive and flexible production systems combining high productivity with labour savings. The loading and unloading of parts by industrial robots in combination with CNC machine tools is just one solution which can lead to automated machining cells for unmanned operations.

What is the target for making improvements to the machining process? The aim always is to improve productivity at minimum cost of investment and labour.

Also, what are the guiding principles that can help to achieve these requirements without wasting any of the investments that have been made? There are two routes: one is the flexible manufacturing system (FMS) and the other is the unmanned operation of plants. However, both FMS and the unmanned concept improve productivity but FMS provides the flexibility while the unmanned operation reduces labour costs.

Another benefit of the FMS technique is that it can provide an

This article is based on a paper presented at the 11th ISIR and is reproduced with kind permission of JIRA.

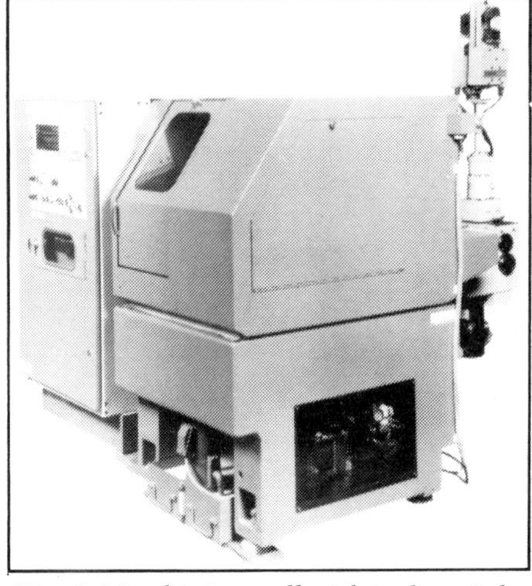

Fig. 3. Machining cell with industrial robot which handles up to 10kg.

unmanned operation with the ability to handle a wide variety of small sized parts in batch production just as if they were a continuous mass production as found in the process industry.

But the flexibility which is found in FMS is not the only target. It gives only one of the required characteristics, namely the ability to cope with variety and diversity. Others are as follows:

○ The ability to upgrade the set up,
○ The ability to expand each single element of the concept,
○ The ability to cope with a large variety of small parts.

The term unmanned operation is taken to mean the function of automatic machining with less manpower in order to achieve high productivity and flexibility using economic automation.

Understanding the features of the plant is the first step towards considering an FMS installation. The basic concept of the Fujitsu Fanuc FMS is the centralised management and distributed machining cells. The distribution of locally integrated machining cells will construct the basic core of the FMS.

The machining cell has the flexibility to become an automatic factory through step by step installations. The concept of the FMS differs greatly from DNC, which centralises both management and control. The Fujitsu Fanuc FMS guarantees flexibility and adaptability for the production of a wide variety of small sized parts as well as allowing for an unmanned operation. The change from the age of the machine tool to the age of the machining cell is now occuring.

Three major factors in an unmanned operation are:

○ Flexibility to absorb the variety of machining parts,

○ Reliability both of equipment and operations,
○ Monitoring of problem areas to guard against accidents, safety of equipment and quality of parts.

Machining cell

The machining cell is the basic element of the unmanned operation. It consists of a CNC machine tool with a part program store, an automatic components loading and unloading device, and a monitoring facility that watches for unscheduled operations.

Two types of machining cells are considered; one is the machining cell using an industrial robot, the other is the machining cell with a pallet changer. An unmanned operation is carried out as shown in Fig. 2 by three major functions, as follows: Machining using a CNC machine tool; loading/unloading of parts by the industrial robot or a pallet changer; and monitoring for malfunctions.

The machining cell can be operated without an operator, providing full preparation of the NC command data, planning of the cutting tools on the tool magazine, and the materials on the parts supplier are all carried out in advance.

The robot or the pallet changer automatically loads and unloads parts, and the CNC machine tool continuously performs its machining operation according to the NC command data stored in the bubble memory.

If any unusual or unscheduled operation is detected by the monitor during machining, the machining cell witll stop automatically.

The machining cell provides a fully automated machining system with only one operator for several machines. The operator's function relates to the preparatory jobs as follows:

○ Preparation of NC command data and to feed it into the parts program storage in the CNC machine;
○ Tooling and setting up the tooling magazines;
○ Preparation of the fixtures;
○ Teaching the robot;
○ Supplying materials to the parts supplier;
○ Inspection of the machined parts.

The operator undertakes these jobs with each machine. In the case of an overnight unmanned operation, the operator requests the preparation of parts and confirmation and resetting of tooling in the evening just before leaving the shop.

Fig. 4. Double grippers used to load and unload a lathe.

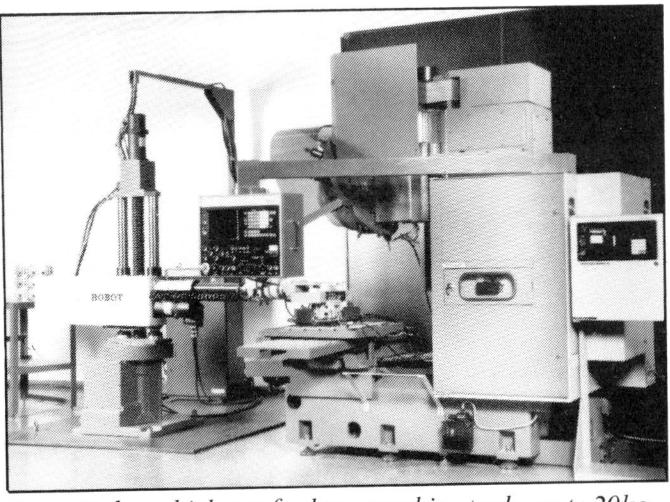

Fig. 5. Robot which can feed two machine tools, up to 20kg.

Fig. 6. The robot in this machining cell handles up to 50kg.

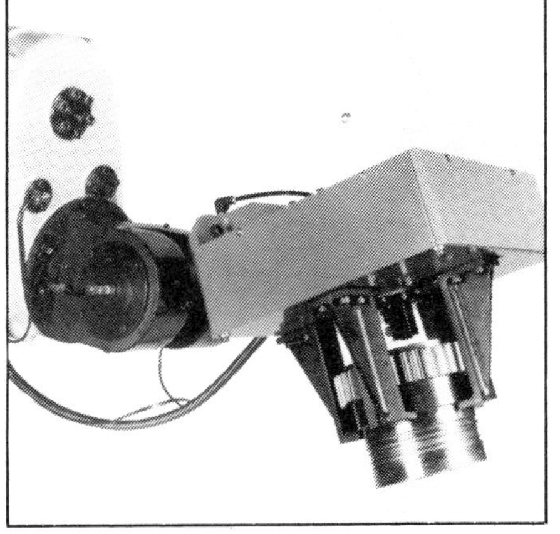

Fig. 7. A heavy component is handled by a robot.

The bubble memory used in the CNC is a small size non-volatile memory with a reliable storage capacity. It is equivalent to a tape length of 320m, enough to store the volume of NC command data required for the production of a wide variety of small sized parts. Different types of parts can be machined by calling the required NC command data from the bubble memory.

Robots

Three types of industrial robot were introduced for use with the machining cells. The industrial robot, shown in Fig. 3 is the machine-mounted type, handling up to 10kg (5kg × 2) of parts. Fig. 4 shows the loading/unloading to the lathe by twin grippers.

The industrial robot in the machining cell, Fig. 5, can handle up to 29kg of parts for each pair of machines while the industrial robot shown in Fig. 6 can handle up to 50kg of parts for each pair of machines.

The main features of the industrial robot are flexibility and reliability.

○ Flexibility in terms of the diversity of parts handled by a mechanical hand or gripper specially designed for components.
○ Flexibility of the programmable manipulator by a teach and repeat method. That is, modification of the motion and selection of alternative loading programs. For other applications, palletised parts are used.
○ Reduced waiting time at the machine.
○ Economical costs allow labour savings and improve productivity. It is more economical than a custom designed loader.

For an unmanned operation, good reliability is one of the most important features. Two degrees of reliability will be considered: one is of the hardware itself, the other is how the system operates.

The reliability of the operation is

Fig. 8. The tool monitoring system.

associated with: Preparation of a proper NC command data; uniform and stable materials and shape; long life of cutting tools; suitable tooling; and cutting conditions.

These requirements are of course necessary in the conventional production system as well as with the unmanned operation system. As a result it is possible for an efficient conventional operation to lead easily to an unmanned operation. Any unique or speically good reliability is not necessary for an unmanned operation.

Monitoring

Monitoring for faults and safety measures are essential for unmanned operations. Three major points for the monitoring are: safety of operators; maintaining the system hardware in good condition to prevent any malfunction which might have dangerous consequences; and retaining the high quality of production.

Looked at in more detail there are:

- Preventitive measures against fire and problems with power sources such as electric power cuts, changes in power, reduced voltage, faulty connections and so on. Countermeasures to these possibilities are allowed for in the control equipment by the use of sensors to detect malfunctions.
- Self-monitoring and diagnosis are used in the CNC machine tools and industrial robots to stop the system immediately, or to display a warning notice on the machine to prevent the danger or malfunction spreading. Some examples requiring detection are: out-of-control motion in the servo motor, breakdown of the feedback signal, overheating or temperature changes, power supply changes, over-loading, maximum limited speed, operation time for the interval of each function etc.
- In the machining cell, as with the robot and CNC machine tool, the motion of the robot should be halted or prevented by misfitting doors or incorrectly adjusted chuck jaws, and so on. The robot should only proceed when all the interlocks are clear.
- Inspection of quality should not be achieved by after-machining inspection, but by quality control before and during machining, to prevent imperfections caused by materials, tools, fixtures, chips and so on.
- Supplementary intermittent visual monitoring through the central monitoring system using the television monitor at the central control room is also used.

The tool monitoring system assesses the status of cutting tools in order to detect damaged or dull tools. Monitoring of the machining cell (Fig. 8) is carried out by looking for deviations in the load current on the spindle motor in an active cutting operation. The system also monitors the tool life by recording the actual cutting time on the number of drilling or tapping holes. The tool life value will be previously set before the operation and a signal is sent to stop the motor when the degree of tool wear has been achieved.

The concept of FMS in the machining process can be seen as the combination of machining cells, with an automatic warehouse and a materials transportation system.

An example of the FMS in the DC motor shop is shown in Fig. 9. In this shop 45 sets of machining cells are introduced to machine the mechanical parts for DC servo motors from the 0.15kW to 22kW.

Fig. 9. The FMS in the DC servo motor shop.

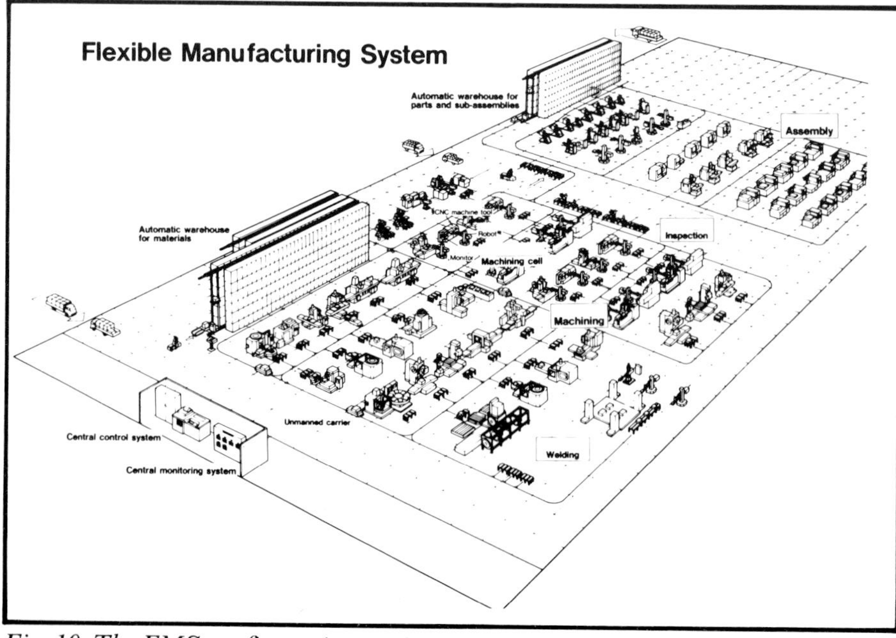

Fig. 10. The FMS configurations in Fanuc's Fuji factory.

Fuji factory

The Fuji factory (as shown in Fig. 0) was constructed for the production f industrial robots, CNC wire-cut lectric discharge machines, and small NC machine tools. The FMS with nachining cells used in the factory for nmanned operations is shown in Fig. 1.

There are automatic warehouses for naterials, another for parts and sub-ssemblies, and unmanned carriers which transport material between hese cells and the automatic ware-houses. The central control room nanages the total operation via the central control system, the central monitoring system, and the produc-tion control computer.

The automatic warehouse for materials has about 500 shelves with storage addresses and each shelf allows room for up to 1 ton of material. The material identification on the shelves, in/out control of the materials, and management of the inventories are carried out by the central computer control system.

The automatic warehouse for parts and sub-assemblies has some 360 shelves, each with an allowable weight of up to 600kg. The storage pallet for the shelf can be divided up to 12 isolated sections. So, a maximum of 4,000 kinds of parts are stored in the warehouse. Each pallet is allocated to the predetermined shelf with a fixed storage address.

Materials delivered to the factory are first loaded on the pallets which are stored in the automatic materials warehouse. Demands for materials at the machining cell are sent from the warehouse operating terminal near the machining cell by an operator. The materials required are unloaded from the automatic warehouse and delivered to the machining cell by the unmanned carrier.

The unmanned carrier, as shown in Fig. 13, passes the materials along the predetermined transportation route indicated by a magnetic inductive guide beneath the floor with the speed of either 60, 30, or 12m/min, to the machining cell. The materials are arranged on the supplying table or feeder in the machining cell by the operator.

The machined parts already at the machining cell, wait at the materials supplying station until called by the next machining cell for processing. Finished machined parts are sent to the automatic warehouse for parts and sub-assemblies by the unmanned carrier. Large parts that are not handled in the parts and sub-assemblies automatic warehouse are sent directly to the assembly floor by forklift trucks.

Such a 'demand-pull' transporta-tion system is useful for the discrete batch production of mechanical parts.

Fig. 11. View inside Fanuc's Fuji factory, showing the FMS.

Fig. 12. The automatic warehouse for materials.

Management

The total management of the FMS is achieved in the central control room by the central computer handling the inventory of materials and parts, the central monitoring system for operat-ing conditions, and the production control computer for the production schedule. The television cameras are located at various sites and are connected to the central monitoring system to observe malfunctions, as shown in Fig. 15. Abnormal operation of tools and any other abnormal operating condition are transmitted to and displayed at the central monitor-ing system.

The concept of the FMS is central-ised management with distributed processing at the job site. The planned schedule based on the production order at the headquarters is sent to the production control computer through the telecommunication line, which makes the order for the materials and parts to be purchased. Delivered materials and parts in the machining process are managed by the central control system. The production con-trol computer processes the status of these parts until the final inspection and shipment.

In the first day-time shift, the machining is fully automatic except some operators are required for pre-paratory jobs. The system required an increase of 30% in the investment for the robots; 10-20% increase in pro-ductivity in the machining cell because of the uniformness of the

Fig. 13. An unmanned carrier at the automatic warehouse.

Fig. 14. An unmanned carrier at the machining cell.

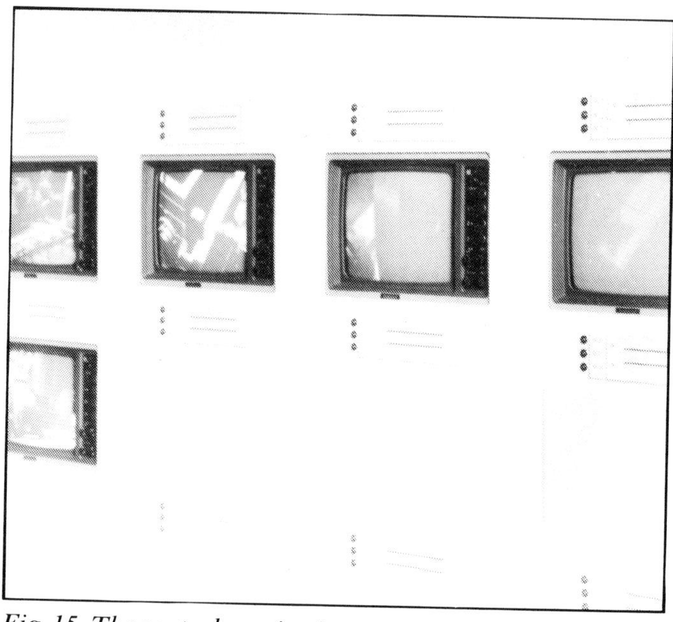

Fig. 15. The central monitoring system.

loading/unloading jobs by the robots compared with an operator; and a 50% decrease in the number of operators, depending on the characteristics of the machining job.

In practice no decrease in the number of operators occurred, because of the greater production with the increased number of machining cells.

Overnight

In the second and third shifts, the overnight operation required no operators. There was a 70-180% increase in production for each machining cell (this depends how long the machine is in operation for, based on the limitation of the parts supplier). Allowing for the absence of operators, the total productivity per machine is increased by 180-300%, and for the investment the productivity was 140-230% improved. In addition of course, the costs of parts are reduced.

The Fuji factory is operated during the day-time shift by 100 personnel, operators and supervisors, for the machining and assembly shop, and overnight by one supervisor for monitoring. In the machining shop each operator is in charge of five machining cells during the day-time shift for preparatory jobs.

Considering that the current ratio of production capacity of the day to night shift is 1:1, and that one person is needed to each two conventional NC machine tools, the productivity of the labour is reduced to one-fifth for the same amount of work.

In the near future, the improvement in production control and the central monitoring facilities could probably bring the ratio to 1:2.

The FMS based on the machining cells introduced above, has greatly contributed to improved productivity, reduced manpower and made more efficient use of investment by using unmanned machining operations in the overnight shift.

Mori Seiki moves to FMS

With the accent on modular construction to cut inventories, Mori Seiki has introduced unmanned machining systems for large and small workpieces.

AT Mori Seiki, the Japanese builder of small CNC lathes and machining centres, the accent is on self-sufficiency, flexibility and modular assembly. To pave the way to reduced costs, the company has installed three FMS in its machine shops, and each works unmanned for some of the time.

Established in 1948, Mori Seiki has its main plant and head office at Nara, the early capital of Japan some 300 miles west of Tokyo. A couple of years ago, it built a new factory at Iga, where production is to be concentrated in the future. Pride of place at Iga is taken by the new FMS, which consists of 13 CNC machining centres, the automatic warehouse, and the unmanned carriers that serve the whole plant – not just the FMS. Following the recent extension at Iga, costing £25 million, the company now has the capacity to produce 250 CNC lathes and 100 machining centres per month – with a value of £180 million.

At both Nara and Iga, there are machine shops, paint and assembly shops. Although there are many CNC machines in both factories, there are still a large number of conventional machines, from pillar drills to grinders, in use. For example, at Iga there are 300 machines of which 100 are CNC. Nevertheless, it is the CNC machines that do the bulk of the work.

Capacities

Five years ago, Mori Seiki installed its first FMS at Nara. It consists of a group of seven large machining centres with automatic pallet changers (APC). Capacities vary from five to 12 pallets according to the cycle times, which average 2-3h. Hitoshi Yabuta, director of engineering, explained: 'Here there is no need for automatic transport between the stores and machines because it would be used so infrequently.' Instead, the operators load the pallets on the APCs manually, but the machines can still operate unmanned for most of the night.

This FMS is used to machine medium size cubic workpieces, such as spindle heads, tail stocks, turrets, cross slides, and transmission housings; each machining centre is used to handle a variety of parts. A trigger on each pallet actuates a limit switch, thus indicating the type of workpiece to the controller. To allow operation for up to 24h a day, the machines are equipped with sensors that detect tool breakage.

Over at Iga, there are two automatic warehouses, separate shops for machining small parts, medium-size parts, large casings, and the FMS.

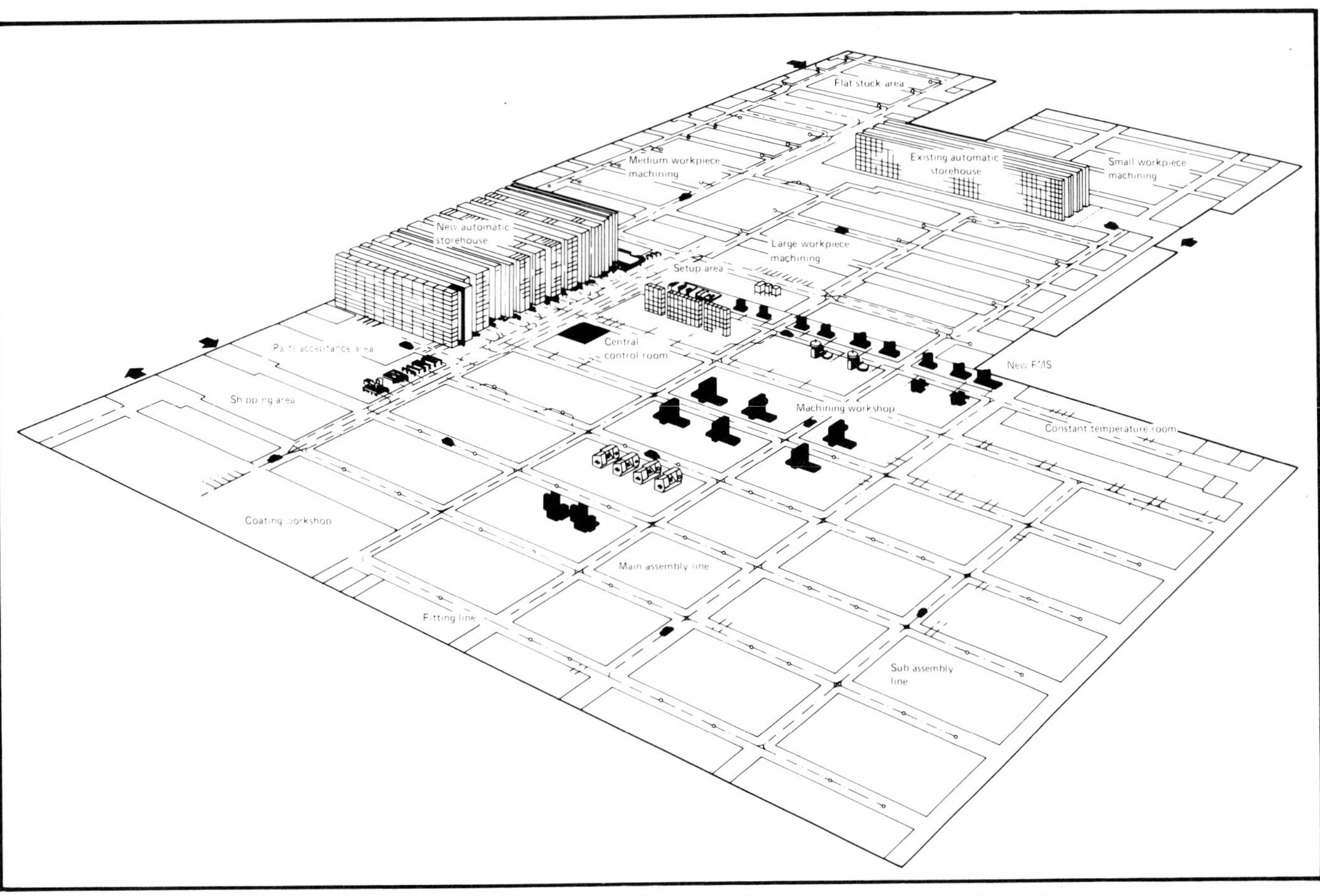

Layout of the Mori Seiki FMS at Iga, with automatic warehouse and unmanned trolleys throughout.

Many machine tools at the Nara plant are equipped with large automatic pallet changers.

One row of the machines in the new FMS, with the main automatic warehouse in the background.

Then, there is a constant temperature room, assembly, paint and fitting shops. There is also a separate building housing three grinding shops. Whereas it is common to equip an FMS with an automatic warehouse and unmanned carriers, Iga is unusual in that the warehouses and carriers serve all departments. Thus, from the first machining operation to final fitting out, there is no manual handling between sections. That in itself is worth a lot in terms of productivity.

The warehouse can hold 12,800 × 30kg pallets, and a further 7,000 pallets of 1,000-3,000kg. A total of 16 unmanned trolleys circulate in the plant, eight with a capacity of 1,000kg, four with a capacity of 2,000kg, and four more with a capacity of 3,000kg. Both the carriers and warehouses were supplied by Murata Machinery, which produces the RoboTrailers under licence from Digitron of Switzerland. The larger ones are needed for the machine beds and partly assembled machines.

Although purists might say there is only one FMS at Iga, Yabuta insists that there are two. He points out that the large machines, such as the Okuma MCM 16 and MCM 20 portal five-axis millers, of which there are about 10 installed to machine beds and slides, have such long cycles that they do not need APCs. 'Once on the machine, the workpiece is being machined for 6-7 hours,' he said, 'So if the man works an hour or so overtime, we can get two shifts' output with one shift manning.'

However, the most interesting aspect of Iga, apart from the extensive use of unmanned trolleys, is the FMS for small parts. It consists of 13 machining centres, mainly made by Mori Seiki, although some Toyoda FHN60T and FH80T machines ar[e] included. To feed this system, there i[s] the automatic warehouse, the manua[l] setting-up area, and the RoboTrailers[,] while each machine has a two-statio[n] APC and tool breakage detector[.] There is also an automatic chi[p] disposal system and, of course, [a] central computer room.

This FMS is used to process smal[l] workpieces such as levers, blocks an[d] rings, and because the workpieces ar[e] so small, it posed considerabl[e] handling problems. Mori Seiki devise[d] special fixtures to carry sufficien[t] workpieces for about 30min work. I[f] the process time is shorter, then to[o] much handling is needed. Even so, the men spend quite a lot of time setting up parts for machining, so to cut the time to the minimum, each jig has an hydraulic clamp, usually operated by a central lever. It is claimed that as a result, setting up time is reduced from 15 to 2-3min.

Machine costs

Yabuta concedes that the cost of the jigs is really too high. 'We need 24 pallets for each machine so that it can run untended for 12h,' he said. 'Each pallet/jig costs Yen 2 million (£5,500), so the set for one machine costs Yen 48 million (£130,000). Yet the NC machine costs only Yen 20 million (£55,000).

He explains that there is a trade-off between unmanned operation and tooling costs, because space is needed for the pallets, and the operators spend a lot of time unloading first thing in the morning. In fact, there are only three men in that FMS, whereas in theory, 54 machines and 27 men would be needed in a conventional machine shop to produce the same amount of work. The men load pallets with workpieces, which are then stored in an adjacent buffer store. These are transferred as required to the machine tools.

In practice, though, the FMS is not yet working for one complete shift, let alone three. Yukio Mori, senior managing director, admitted: 'We are only operating it for about half of one shift, as business is quiet. Now it is of no advantage to our company, except that we can show it as a sample system, and use it for our own research into FMS.'

Mori thinks that his FMS differs from most of the others in Japan. 'Yamazaki's and Fanuc's are like zoos,' he said, 'Customers must watch from the outside. Ours is like a Safari park – the customers can go in among the machines, and see everything.'

Mori says that he thinks the idea of FMS should be to relieve operators of heavy and dirty work. 'The operators should be retrained to use their brains,'

The unmanned trolleys transfer pallets between work stations and the stores.

Computer control room at Iga.

he said.

A Hitachi E-800 and an IBM S/I computer control inventory, warehousing/delivery, workpiece machining schedule, automatic transfer and parts supply. The software was developed jointly by Mori Seiki and Murata.

Overall, Yabuta reckons that the FMS and automatic handling, which he considers to be the most important element, are giving a gain in productivity. For example, he claims that the time the spindle is moving is 97% of working time, despite the fact that batches average about eight workpieces, and in many cases one component is made daily. 'Our total cycle of production is now one month. It used to be three months, so we have cut work in progress by 66%,' he said.

He also pointed out that in fact a normal 8-h shift, for five days a week gives about 25% plant utilisation. 'Therefore with the FMS we have the chance of increasing utilisation by four times, while reducing the cost of our investment in machines by 75%.' In practice, much of the cost saving is offset by higher capital investment costs in computers and fixtures, but the gain in productivity is nevertheless impressive.

Mammoth FMS

Yamazaki goes all out for FMS at the Minokamo plant where showmanship rivals the machining performance.

MOST FMS installations involve two to 10 machines, but at Yamazaki's brand new factory at Minokamo, near Nagoya in southern Japan, there are 43 machines. In addition, there are 17 robots, but there is a separate FMS for sheet metal parts.

Not that FMS is new to Yamazaki. Two years ago it unveiled its first two systems (consisting of 18 machines), at the headquarters plant at Oguchi. Since then, it has built systems with 28 machines at Yamazaki Seiko, and with 15 machines at its US subsidiary at Florence, Kentucky.

Boldness

Significant though those plants are, they are overshadowed by the sheer boldness and breadth of the operation at Minokamo. Built on a levelled hill-side on the outskirts of the town, the plant covers 49,000m^2 of a 330,000m^2 site. There the FMS produces flanges, spindles and small parts which are turned, milled, drilled and tapped; gearboxes and large frames are machined, flame-hardened and ground; and sheet metal parts are punched, formed and welded.

All these operations are under computer control, with unmanned trolleys transporting parts. In addition, there are some interesting developments in ancillary equipment, such as one robot to clear swarf, and another to load workpieces on to pallets. Yamazaki has further developed its principle of demountable automatic tool changers (ATCs), to provide a huge variety of tools.

Yamazaki talks of an investment of Yen 14 billion (£40 million) for the plant, excluding the land. With a staff of 240 people, it has a potential output on three shifts of 120 lathes and machining centres a month. Altogether, there are 60 CNC machine tools, 28 universal machines, 12 AGVs, and six computers in the plant. Yamazaki calls this 'System 21' after the twenty-first century, and tried to give the building a modern look to reflect that idea.

Because Yamazaki is in the business of selling machine tools there is a good deal of showmanship at Minokamo which gives the place a slightly unreal air. It hopes to be able to sell turnkey large-scale FMS too, but despite the fact that 30,000 visitors have inspected the system at Oguchi, there have been no takers yet. Either Yamazaki's systems really are for the 21st century, or they are not so flexible after all.

On arrival, visitors are supposed to press a button on a small computer to find out where they should go. In fact, one of those girls that are everywhere in reception areas in Japanese companies, soon rushes out and presses the button for you, which makes the machine unnecessary. Then, there is an unmanned trolley on which visitors can go around the plant; it was not working when I visited Minokamo. Instead, visitors can walk around the gallery, stopping to be told by cassette recorder what is happening at various stations. But at least, unlike the Fanuc motor factory, the gallery does afford a good view of what is going on.

The rectangular plant is laid out in three bays, all served by six automated guided vehicles (AGVs), with one rail-guided trolley serving each line. The first two sections of the first bay are empty, and are due to be used for assembly soon. The FMS at the other end of the bay are those for flanges and spindles. These are laid out in one long row, separated by a common area for the control computer terminals.

On the other side of the first bay are some surface grinders, a small automatic storage and retrieval (asr) warehouse, and the robot loading station for the 'box' machining line. The box line itself is in the middle bay, which it shares with the frame line. Between these two lines, which are back-to-back, is a track for the special trolley to transport tool drums to and from the tool room.

Five faces

In the third bay are a large CNC machining centre, which can operate on five faces, and a bed grinder, together forming the cell for large bases, columns of beds in the 5-25 tonne range, and the flame hardening machine. Also in this bay is the assembly area.

On the first floor there is the unit assembly area, where modules are assembled manually. Automatic lifts are used to transfer machined components from the ground to the first floor and back.

There is a main aisle for the delivery of workpieces across the plant, with spurs running longitudinally to the various areas. The six AGVs follow wires buried in the floor as they

The lines are served by individual rail-guided trolleys. Special clamps are used to hold the spindles to the disc-shape pallets.

ransfer workpieces, unmachined and nachined, between different lines and he lifts. However, these AGVs do not serve the machines themselves.

The flange line consists of seven Slant Turn 40N Mill Centres and three nachining centres – two horizontal and one vertical. At the end of the line s the control area, with two computers, and a station to which workpieces can be delivered. At the other end of the control area is the spindle line of five Slant turn 40N Mill Centres, one vertical machining centre, and a CNC cylindrical grinder.

Innovations

There are several innovations in these two lines. First, the lathe/mills are new, featuring a 15-tool turret (including milling cutters for face-milling, end-milling, drilling and tapping both on the periphery and face of the workpiece). Thus, operations normally performed on a small machining centre can be carried out on the lathe, reducing handling and setting up time.

Each machine in the line is equipped with one of Yamazaki's FLEX robots which are mounted on the front of the machine. This is complemented by a complex handling system, which is needed due to the short cycle times. First, workpieces are carried on universal disc-shape pallets, carrying 6-10 workpieces. Those used for the spindles, which are about 600mm long, incorporate clamps which pull the spindles down to the pallet.

Adjacent to each machine is one table carrying a pallet of workpieces, and then there is the track for the rail-guided trolley that serves the line – there is one for the flange lines, and another for the spindle line. On the far side of the rails are three more pallet stations arranged longitudinally for each machine. These pallet stations are therefore in between the rail-guided trolley and one of the spurs along which the AGVs can travel.

The AGVs, therefore, can bring loaded pallets from the setting-up area directly to the pallet stations, leaving the rail-guided trolley to keep the machines supplied with workpieces. The rail-guided trolley shuttles between the pallet stockers and the stations adjacent to the machines, transferring pallets between them as necessary.

Process time

The spindles are hardened outside the plant by a sub-contractor, and are returned to be finish ground. Yamazaki says that the average process time on the two lines is 20min, with the spindles taking longest – 75min, including 25min handling. The two lines can process 435 different parts – spindles, rings, housings and flanges – monthly, the potential output being 9,000 units/month. This includes 1,600 spindles of 35 types.

There are 12 horizontal machining centres in the box line, and a small asr warehouse with capacity for 228 pallets – equivalent to 348h on average for the gearboxes, headstock housings, turrets, carriages and sundry housings machined there. The average process time is put at 6h. Potential output is 1,200/month of 85 different parts.

This line, and the frame line next to it, follow the plan of the original Yamazaki FMS at Oguchi, but it has some variations. Thus, there is a rail-guided trolley for each line – chosen as elsewhere, for example, at Minokamo, for their speed – and these run between the pallet stockers and the APCs at the machines, transferring as necessary.

Two drums

Although each machine in the frame and box line has a 40-tool ATC (automatic tool changer) this is not enough to handle the variety involved and the drums are detachable. Normally there is one on the machine in use, and another on a horizontal slide that extends along the side of the

Between the spindle and flange lines is this control area.

machine. Two drums can be carried at the same time by the trolley between the machines and the tool room where tools are changed manually.

There, the operator uses a Sony instrument to measure the tools, which are reground as necessary. Before a tool is used again, the offset is measured and is fed into the control computer in preparation for use on any machine. In fact, this is one area where theory and practice have not matched up.

In the frame line, which operates on an average cycle of 12h to produce 240 parts/month – 40 varieties of beds, columns, saddles and bases – seven machining centres are installed. These are mainly YMS-H50Q horizontal machining centres, but there is also a V-20 vertical machine. The trolley can carry two workpieces, each weighing eight tonne.

Adjacent to the asr warehouse, and between the spindle line and the frame line, is the new set-up robot, evidently a Dainichi Kiko Babot. This robot is used to lift the workpieces on to the standard pallets and to control the clamps. Then, it puts the pallet/workpiece assembly in the asr warehouse. Subsequently, the assembly is transferred to the line.

Slant Turn 40N Mill Centres with robot loaders predominate in the flange line.

Special fixture

In some situations, the workpiece has to be turned over so that it can be machined on the fifth face, and in that case it returns to the set-up area where it is loaded on a special fixture that turns it over. The robot then repositions the workpiece on the pallet.

When machining is complete, workpieces are transferred to a turnover device which turns them through 180° and back again to ensure that all swarf and chips are removed.

However, for the larger frames, such a procedure is impractical, so Yamazaki has installed its 'cleaning robot'.This robot is suspended from a beam carried on a four-post structure and carries a suction pipe. Since the robot can be programmed to move in the x, y and z axes, it can traverse the periphery of the frame, sucking up the swarf from the flanges and faces as it goes. There is a central swarf disposal system which draws swarf from all the lines, removal generally being automatic.

The box line, with the frame line towards the right.

FMS shop

Yamazaki's flame hardener is a portal CNC machine which traverses the workpiece. It is in the third bay, so the workpieces are carried by AGV down the shop to a raised track, which is at the same level as the top of the AGV. The workpiece and pallet then move across the end of the line of machines, on the raised track, before sliding on to another AGV to be taken to the flame hardener.

In addition to this large FMS shop, there is a small metalworking shop where two CNC punch presses, three CNC press brakes, and two welding robots are arranged to produce fabrications. Then, there are the subassembly and assembly areas, and a paint shop. Yamazaki also produces ball screws in temperature controlled shops at Minokamo.

Modems

Yamazaki plans to put the whole system under the control of a computer at the head office, with connections by modems and the telephone lines – but no indication is given of when this will happen. Currently, each of the five computers at the Minomako plant is operated independently, with the necessary data input locally by operators, so that the system is really a collection of small FMS. The large CNC machines and the flame hardener seem to be controlled indirectly only by the transport computer arranging for deliveries to be made. Although

One of the machining centres in the frame line. Standard pallets with long :lamps are used.

from the few workpieces on the pallets, it would seem that output was still a long way from the official figures – which are for three shifts anyway. It is admitted that there have been problems in clamping the workpieces consistently in the machines.

The other major problem has been the relationship between the tools, drums, and machines. In theory, it should be possible to measure the offsets of the tools, and feed the data into the main computer so that when the tools are transferred to any drum on any machine, correct offset data should be sent to the machine. However, it was found that the data did not match actual conditions, owing to the tolerances of manufacture on the drums and machines.

Therefore, data has been collated for each drum and for each machine to compensate for variations. In theory, it is now possible to use any tool on any machine without any manual

purists may turn up their noses, in practice, such a system is adequate and has the merit of simplicity.

There is one computer to control the frame and box line; a small computer for the flange line, and another for the spindle line; there is also one computer to control the transport system, and another to control the air conditioning and energy use.

Yamazaki claims that as far as the operators are concerned, these are simple to control, since they use conversational languages – just like the CNC themselves. Since the systems are independent, a breakdown in one system has no effect on the others.

Currently, the plant is being operated 15h a day, with 240 people, and Yamazaki claims that this is solely due to the slack market. Only 39 people are needed to run the FMS, with three men/shift on the spindle line, three men/shift on the flange line, five on the box line, and four on the frame line. Three people are needed to load workpieces on to the AGVs outside the shop, and there is an operator in the computer room, and another in the tool room. The system would be virtually unmanned on the third shift, and clearly has the potential for an unmanned second shift. About 80 people work on assembly.

Teething troubles

Since the plant started to operate only in the early summer, there are obviously a number of teething troubles – mainly in the lathes and the tooling. When I visited the plant, six of the lathes in the spindle/flange line were receiving attention, and to judge

The flame hardener is controlled manually.

A robot is used to load workpieces on pallets for the box line.

checking being needed. That is one example of how much development is needed in software, and how many pitfalls are waiting in FMS development.

In several ways, this FMS breaks new ground. First, there is the extensive use of lathe/mills, which reduce setting up time while usefully extending the cycle, so that too much time is not lost in loading/unloading. Secondly, because these machines are robot loaded, they are virtually unmanned. Then the process of marking out workpieces and machining location pads or holes has been eliminated completely. The use of a robot to set workpieces on the pallets is an innovation, as is the vacuum cleaning robot, although utilisation of both robots seems to be low. Also of significance are the integration of flame hardening and surface grinding. On top of that, there is the sheer scale.

Yamazaki's comparison for Minokamo FMS with conventional plant

	FMS 21	Conventional system
Number of machine tools	*43*	*90*
Number of operators:		
Factory	36	170
Production control	3	25
Total	*39*	*195*
Machining time	3 days	35 days
Unit assembly	7 days	14 days
Overall assembly	20 days	42 days
Total process time	*30 days*	*91 days*
Floor space	6,600m²	16,500m²

What does it all mean at the bottom line? As the Table shows, Yamazaki makes considerable claims for reduced manning levels and process times. It expects to get its £40 million investment back in three and a half years – so long as it can operate on three shifts – and claims that the plant will break even at 33% of capacity, which is equivalent to one-shift operation.

In addition, the total process tim for a lathe is now put at four weeks against 12 weeks previously. What' more, Yamazaki reckons it can pu this down to seven days in due course Quite a lot of ballyhoo surrounds th plant, but it is working, and on balanc it must be well on the way to being success. If that's not an argument fo FMS, what is?

5. Research and Development

t is not easy to find good examples of advanced research in Japan. On the one hand, there are several research projects uch as the Meldog that have little practical relevance to industry. Although these are often shown publicly, they do not eem to have progressed much in the past few years.

Then, there are sound examples of development, which hardly seem to verge on research to engineers in Europe and the JSA, who are used to seeing more inventive work. Nevertheless, although the Japanese seem to be lagging in sensor esearch and development, they have produced one or two potentially effective devices, such as Mitsubishi Electric's velding sensor. One or two of the computer companies have shown a software approach to solving robot problems, while Hitachi's Production Engineering Research Laboratories have produced several examples of advanced robotics and systems, but the trend within Hitachi has been to work on research nearer the application stage – the traditional Japanese approach.

Nevertheless, the Japanese are well advanced in the use of voice recognition and are progressing with lasers, as other eatures in this section show. There is a lot of interest in vision systems and in robot languages, and although software is not an area where the Japanese have made any impact, they are now making progress towards a standard robot language. Indeed, it is likely that they will be the first to produce a standard. This will have an important bearing on the future of robot assembly, since it will allow groups of robots to be programmed off-line to work together safely and efficiently. Thus, in many different areas, the Japanese are moving ahead in applications and research into flexible automation. Up to now most of the work has had a strong practical bias, but there are some indications that the Japanese are beginning to take a more adventurous line. But even now, when it comes to flexible automation, the Japanese have made remarkable advances.

Don't look to Japan for advanced sensing techniques

The 11th International Symposium on Industrial Robots took place in Tokyo during October 1981. This report looks at some of the presentations on sensing given in the Symposium and examines some of Japan's indigenous efforts on advanced sensing techniques.

EVERY FOUR YEARS Japan's industrial robot activity comes under close examination when the International Symposium on Industrial Robots is based in that country. The 11th Symposium in the series was held in Tokyo and once again Japan's industrial robot activity came under the microscope.

Over 350 overseas delegates attended the Symposium swelling the total number to over 700. The close examination that these visitors were able to make of the industrial activity in Japan threw up some interesting facts which in general don't entirely match up with those created by the popular media image of Japan.

One image of Japan is that it is a massive user of industrial robots. This is true, although the numbers often quoted are far in excess of the true number in operation in Japanese industry today. What is a fact is: since 1977 Japanese industry has installed just over 4,000 of so called play back and NC robots. This is still a mammoth total compared with many European countries including the UK and therefore does put Japan at the head of the robot table. This number of industrial robots doesn't exactly shatter the myth about the number in use in Japan but it is certainly a significant difference from the often quoted 60-70,000 robots which includes every type of pick-and-place unit, manually controlled manipulator, hand operated lift hoist, etc. as well as the programmable type of robot.

What is not often realised is that in Japan the level of technology being used is in many respects lower than in many European and North American countries.

Shrewd businessmen

The Japanese are excellent businessmen, know their markets and have a host of excellent designers. Hence, they can produce the products needed, of high quality, and do this in the most efficient production manner possible.

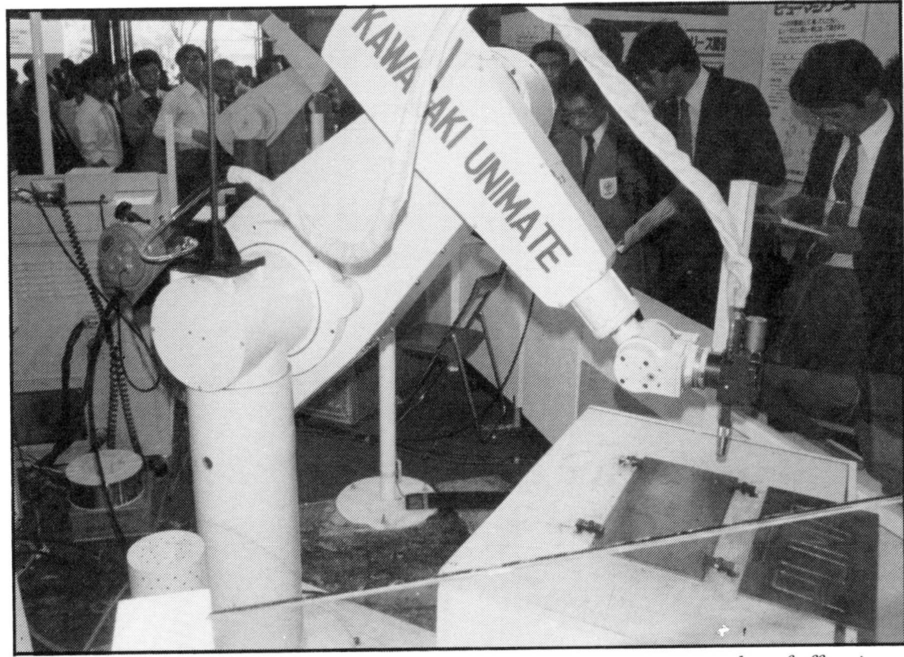

Many of the robot manufacturers such as Kawasaki are putting a lot of effort into developing weld guidance systems.

To do this does not necessarily need advanced technology. So when it comes to the use of advanced robots with sensor feedback then the Japanese are no further ahead than any other industrialised country.

The Japanese are not good innovators and tend to wait for the lead from western countries before undertaking many of the techniques that are considered to be of an advanced nature. Therefore when it comes to examining Japanese R and D and applications in advanced sensing equipment then there is not a great deal to report.

At the Symposium two sessions were devoted to vision systems with a total of ten papers being delivered. Most of these papers were delivered from research institutions. In many ways it is surprising that very little has developed over the past five or so years on this subject. Hardware has changed only in that it is more readily and more cheaply available which should allow a much wider range of activities. But as regard the techniques being used there is very little new.

Most work tends to be based on what was done at the Stamford Research Institute in the early part of the 70's. Unfortunately it hasn't gone very far into industrial application.

From the research institutes were papers on 3D geometry modelling, recognition of moving objects, identification of partially observed parts and projected image direction sensor systems. There are a variety of sensors now being contemplated for use in robots. In particular attention is being focussed on simpler types of sensors. A review of such sensors together with their tasks was given in a paper from the Tokyo Institute of Technology.

Many of these sensors can be incorporated into the gripper of a robot to produce what is often called a "smart gripper". More work is being done on grippers and there was a complete session in the Symposium devoted to this subject. Sensing featured in most of the papers of this session. Subjects covered were soft-

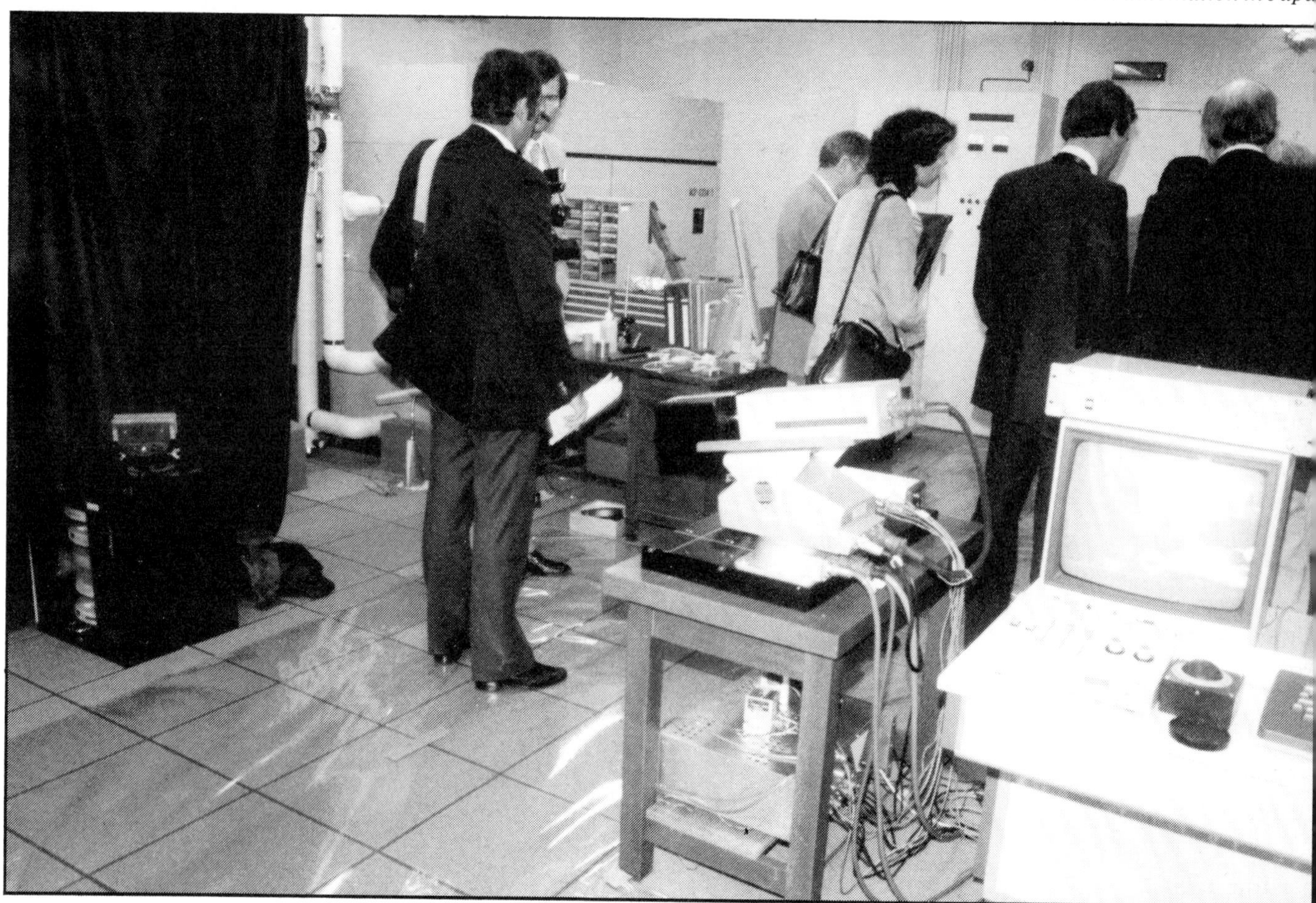

The Electrotechnical Laboratory are using a laser range finding in a project to develop direct computer drawing.

ware and hardware aspects of the sensory gripping system, bin picking using tactile sensing, automatic handling of objects and controlled force gripping.

Weld guidance

From industry, papers on sensing tended to concentrate on the area of welding guidance. There were papers from Mitsubishi Electric Corporation, Yasakawa Electric Manufacturing Company and the joint Unimation-Kawasaki team.

It is generally recognised that arc welding offers a great potential for robots. In the paper from Unimation-Kawasaki it was stated that the value of arc welding machines produced in the USA in 1980 was three times that of resistance welding machines. As the latter have revolutionised the use of robots in industry it is not surprising that the manufacturers of robots are putting efforts into sensing which can increase their application possibilities.

The guidance system being used by Unimation-Kawasaki is vision based, as is that from Mitsubishi Corporation. The concept of the Mitsubishi Corporation system is quite simple in that it uses a scanning system with a single emitter and detector and a rotating mirror. This is mounted ahead of the welding torch seeking out variations in the fit-up of the plates to be welded.

No conference on robot sensing would be complete without papers from two of the leading proponents of sensing systems, namely Westinghouse Corporation from the USA and Olivetti from Italy. The Westinghouse paper gave an update on their APAS system which is shortly to come on-line with Puma robots and vision systems.

The paper from Olivetti was a state-of-the-art evaluation and gave some of the philosophy of the Sigma robot system for use in assembly. Olivetti have always maintained that sensing is an integral part of the process and have evolved the Sigma around this philosophy. Now they have come up with a further development in the form of the Sigma 3 which is designed to integrate into a complete manufacturing system. Some examples of the use of the Sigma 3 system were given including steering box assembly, double overhead cam engine assembly and turbo charger assembly.

One surprising statistic issued by the Japan Industrial Robot Association is the number of so called intelligent robots installed. The figure installed since 1977 is given as 1,400. They define an intelligent robot as "a robot that can determine its own actions through its sensing and recognitive abilities".

Intelligent robots

On closer examination it is found that most intelligent robots are employed in the manufacture of integrated circuits. For this type of application vision sensing is used quite widely both to inspect the silicon waifers and masks and to control the placement of elements.

The large electronic product manufacturers such as Hitachi, Mitsubishi, Toshiba, etc. are users of this type of equipment.

A leading manufacturer of this equipment is Fuji Electric Company. They gave a paper at the symposium and also were one of the only two stands to exhibit any form of sensing at the nearby Robots 81 exhibition.

The Fuji Video Sensor has been applied to a wide range of applications in the food, pharmaceutical and bottling industries as well as in the electronics industries. For example it can be used for the sorting into class and grade of various types of food including tomatoes, melons, apples, peaches, etc. It can grade and select according to size, shape and colour. All the information is recorded and printed and also stored or can be passed to a host computer for auditing purposes.

Toshiba Seiki (Seiki means precision) produce a series of machines that all within this classification of intelli-gent robots. The machines are for die bonding and wire bonding of inte-grated circuits. They are fully auto-matic and make extensive use of vision sensing for fault recognition and for position servoing.

The die bonding machine takes the silicon chip from the waifer, places it into the die and meters the exact amount of epoxy to be applied for glueing the chip. The chip is correctly aligned based on visual information from a TV camera which can recog-nise the chip's circuit. It will do this at the rate of one die bond every 0.9 second.

The wire-bonding machine attaches the wires connecting the silicon chip to the metal leads and welds them in place. Wires are attached at the rate of one every 0.35 or 0.5 second depend-ing on the type of bond.

Visual insertion

Hitachi build a variety of machines for electronic assembly. They have recently completed the Mark III version of their Hisert, which is a machine to insert into PCB's a range of electronic components including integrated circuits, capacitors, resistors, etc.

The machine is flexible in that it can take different components in any order. The position of the leads on the components to be inserted are visually examined and visual servoing carried out to ensure that these leads are accurately inserted into the appro-priate holes on the board. The leads can be variable in length and pitch but the machine can cater for them all.

Nearly all advanced development work carried out in Japan is done at the research establishments of large companies such as Hitachi, Mitsu-bishi, Toshiba, etc. Whilst research is done at the Universities it is rather surprising to find that this is receiving very little support either financially or morally from Japanese industry.

Typical of the efforts going on in the academic field are the projects at the Electrotechnical and Mechanical Engineering Laboratories which have now been sited on a new Science Park some 2-2½ hours drive from Tokyo, in the town of Tsukaba. These are MITI laboratories and receive all their support from govern-ment with no input from industry.

The Electrotechnical Laboratory has done work on hand-eye co-ordination, laser range finding and mobile robots. Its current work includes imagery direct computer drawing, with input from a laser range finder tracing the shape of the objects, and a real time distributed computer system for robot control.

Various manipulative tasks have been studied and a three finger gripper is being used, which many people think is the optimum number of fingers for any gripping application. The three fingers are actuated by a wire-pulley arrangement. The gripper is fitted with photo sensors so that it can align and follow parts being handled.

The Mechanical Engineering Laboratories is less concerned with control and sensing techniques although visual navigation of mobile robots and car guidance systems is being studied. Their main efforts are on a six legged robot, a patient care robot for manipulating hospital patients and Meldog, which is a mechanical equivalent of a guide dog for blind people.

Whilst many way-out projects are under way in Japan academia this free-dom of thought and action doesn't appear to be producing much in the way of innovative designs for use in industrial processes and products. This is left to Japanese industry and it still has to "import" innovation from overseas.

It can be argued that if Japan's current economic boom is to continue innovation will have to take place within the country. On the surface there is not much sign of this happen-ing at the moment.

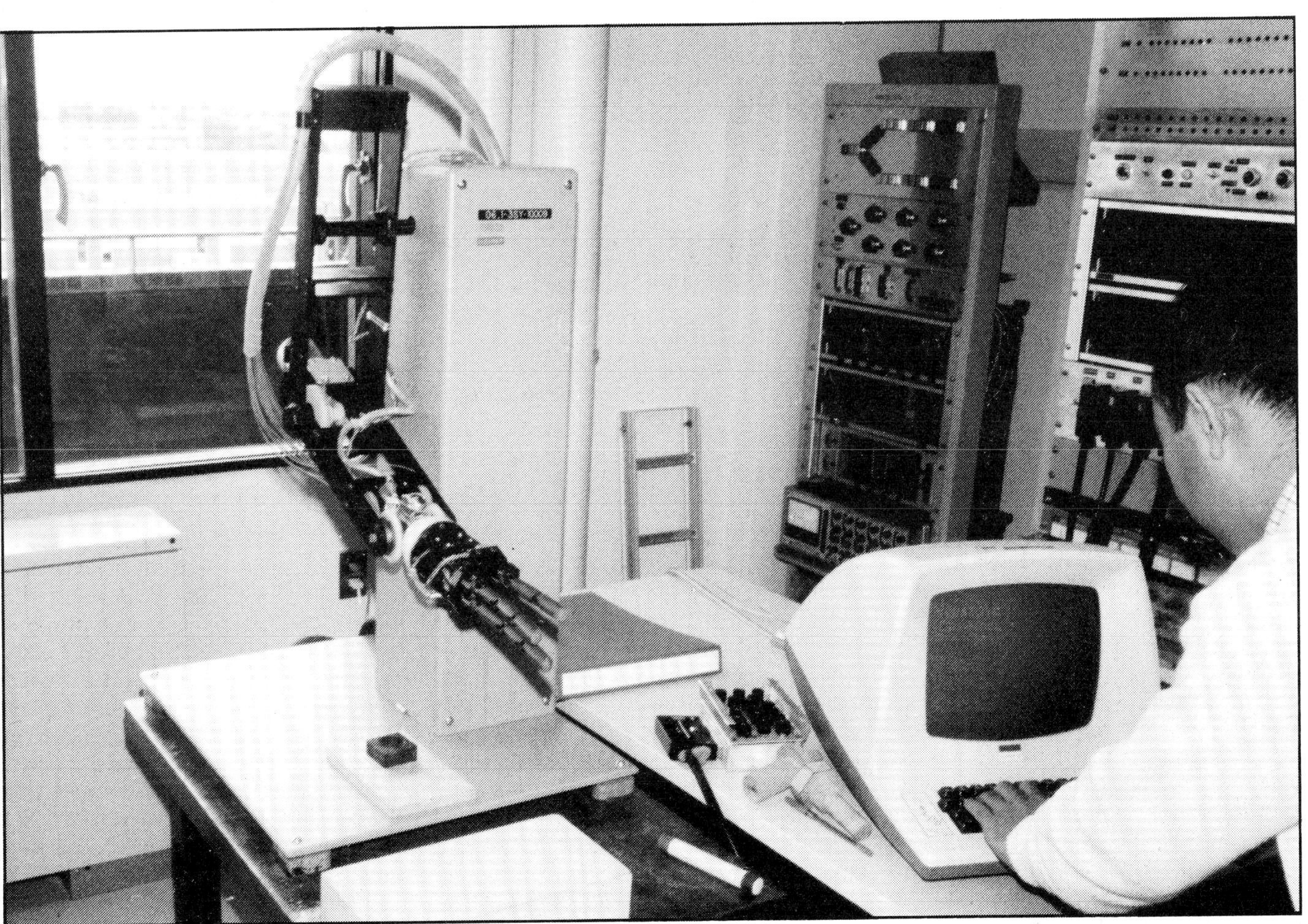

The "smart gripper" with sensors built into the hand is a goal many researchers are trying to attain.

Increasing the reliability and the flexibility of assembly automation

Hitachi's Production Engineering Research Laboratory is working on projects which vary from computer aided design for assembly through to the development of future assembly automation systems. In this special report their work on assembly automation is described.

THE NAME of Hitachi is associated with a wide range of products varying on the one hand from hi-fi and domestic appliances through to electrical traction and nuclear power stations on the other. More recently its name has become associated with the development of robots. Founded in 1910 Hitachi Limited is one of the large Japanese corporates that have been mainly responsible for taking Japan from devastation in 1946 to its present position as arguably *the* leading manufacturing country.

A significant contributor to this "economic miracle" have been the research laboratories of these companies. These laboratories have been doing work which although not necessarily at the outer boundaries of science is advancing the production and manufacturing facilities and the overall capabilities of their companies. Tremendous effort is put into research and development, for example Hitachi put 6% of all its turnover into this area.

Typical of these laboratories is the Production Engineering Research Laboratory (P.E.R.L.) of Hitachi. Situated just outside the centre of Yokohama City it is one of several Hitachi research laboratories.

P.E.R.L's activities cover the general fields of production systems, machining, forming etc., circuit manufacturing processes, automation and reliability. It has expertise both on the mechanical and electrical side as well as in-depth capabilities on software. Also it has been involved in the development of some of Hitachi's robots although the Mechanical Engineering Research Laboratory and the Central Research Laboratory are also involved in robot development.

The Process robot which was developed by P.E.R.L. is now being manufactured at Hitachi's Narashino Works.

Currently P.E.R.L. employs 450 people with over 350 of these working directly in R and D and of these over 300 are graduates. 30% of the staff work in the area which includes assembly automation, robotics and pattern recognition.

Reliability

The laboratory tries to keep a good balance between short term and long term research so that 70% of their effort is directed towards problems emanating from Hitachi factories or carrying out work for Hitachi factories and the remaining 30% is more long term in that it is funded directly from Head Office.

The General Manager of the P.E.R.L. is Mr. Masayuki Okumura. "Our main aims at P.E.R.L. are to drastically reduce costs, attain higher reliability and to ensure that there is a quick introduction of the new products into Hitachi," says Okumura. P.E.R.L. has another meaning for their acronym – Practically orientated – Economic – Reliable – Leading.

One of the outstanding pieces of work carried out by P.E.R.L. was the development of the Hitachi Process robot. This is now in commerial production at the Narashino Works and is being exported to both the United States and to Europe. Although primarily being used for arc welding it has other uses and can be applied, for example, to assembly work.

This robot follows the now well

stablished design of rotating joints out has some interesting features. It uses light alloy materials so that its otal weight is only 200 kgs. and this makes it attractive for fast precision work. Harmonic drives are used on all axes but they are unique in that rotary ransmission is used throughout. Thus he angular displacement of the arm is proportional to the amount of rotation of the motor. This has brought about a much simpler calculation of co-ordinate transformation in comparison with more conventional systems such as ASEA's, which use ball-screws for arm motion.

Robotics is just one of the technologies that P.E.R.L. is now using to attack the problem of automatic assembly. It has been evolving some innovative ideas on this subject for years.

One of the first systems ever to be devised using robots and vision systems came from P.E.R.L. This was a pilot scheme intended to achieve flexible assembly of small mechanical products.

Portable tape recorders was a typical example and work on this was reported at the 10th ISIR and the 1st ICAA Conferences in 1980. P.E.R.L. used conventional feeding equipment such as bowl feeders but then used the vision system for part recognition and part orientation. Simple orientating devices put these parts into the correct position for pick-up by robot arms.

In the experiments a total of 13 different parts were fed and assembled. The system is suitable to applications where many parts are assembled at one station. This not only means flexibility in feeding but also in the handling of the parts. It was achieved by developing a gripper mechanism that is able to accommodate a variety of small mechanical components.

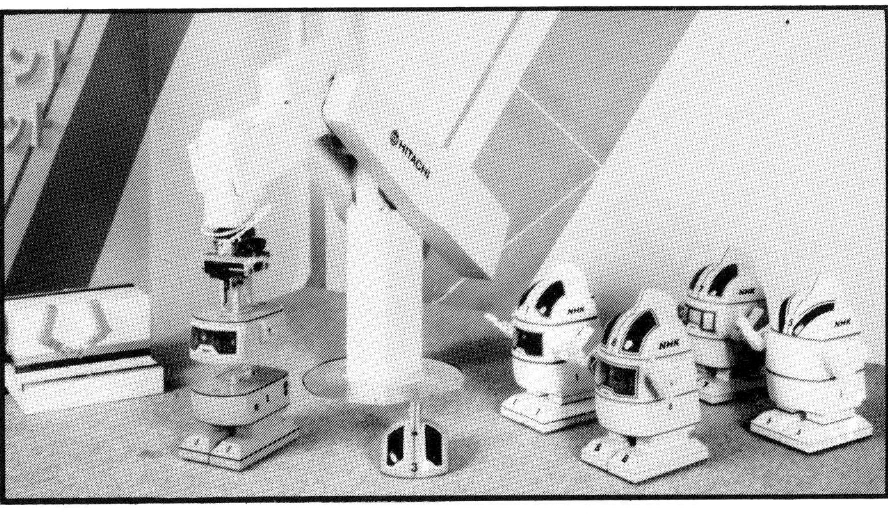

The assembly robot developed by P.E.R.L. – shown here assembling toy robots as part of a television programme produced by the Japanese Broadcasting Company NHK.

Typical of P.E.R.L's efforts to assist its factories was the development of a new concept in automatic assembly – the station module – in collaboration with Hitachi's Tokai works for assembling tape recorder mechanisms. The station module is self contained in that it has its own conveyor belt, drive, motor, positioning mechanism, control unit, etc. The module is also equipped with a guide rail to transfer work pieces directly on the conveyor belt without using a pallet. This latter method can only be used if careful consideration is given to the design of the product.

Design for manufacture is an area in which P.E.R.L. is working to try to influence Hitachi's design engineers. The advantage of the module is that being of a standard nature it is easy to construct and modify. In the event of any breakdown the station can be separated from the total line. Also having no pallet provides a high degree of flexibility.

This approach to increase flexibility is an overall philosophy that P.E.R.L. is following. Since the Laboratory's formation in 1971 work has progressed from simple repetition through to robots. For the 80s it will be working towards an intelligent type of robot with a high degree of flexibility. The Process robot could be considered to be a late 70s development of a skilful device. The assembly work is that of the 80s.

Mr. Masayuku Okumura, general manager of the Production Engineering Research Laboatory of Hitachi Limited.

Articulated robot

Thus P.E.R.L's current efforts is towards the development of an intelligent assembly robot. Some indication of the way that they think and act was demonstrated by the system shown earlier this year. It consists of a pair of six degrees-of-freedom articulated robots and a vision system. The pilot system is capable of visual recognition of the location and orientation of parts which are randomly placed and of performing co-ordinated motions of the two arms.

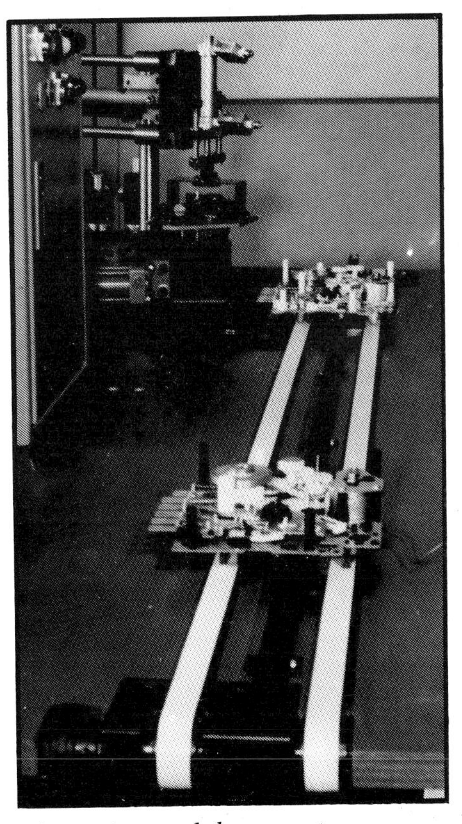
The station module concept involves a simple transfer conveyor arrangement shown here transporting the base of a tape recorder mechanism.

In order to carry out advanced control such as vision interaction and multi robot co-ordination, a control technique employing a hierarchical structure of two micro processors has been used.

Software is likely to prove the key to many successful applications in the future. P.E.R.L. have recognised this and have developed a prototype robot language for use in this system. The commands of the language are functionally classified into the three categories of teaching command, program command and monitor command.

The teaching commands are used for manual operation of the robot and for input and modification of location data of the robot. These commands are input to the control by using the teach pendant. Four different modes of manual operation have been made available and can be selected in any combination during teaching. These modes are:

- □ independent motion of joints,
- □ motion of the arm and wrist in cartesian co-ordinates relative to the arm,
- □ motion of the arm in cartesian co-ordinates relative to the wrist,
- □ motion of the arm and wrist in arbitrarily shifted cartesian co-ordinates.

The program commands are used for the description of the motin sequence of the robot. There are four categories of program command which form the main structure of the robot program. These are motion, vision interaction and triggering, program flow control and arithmetic logic.

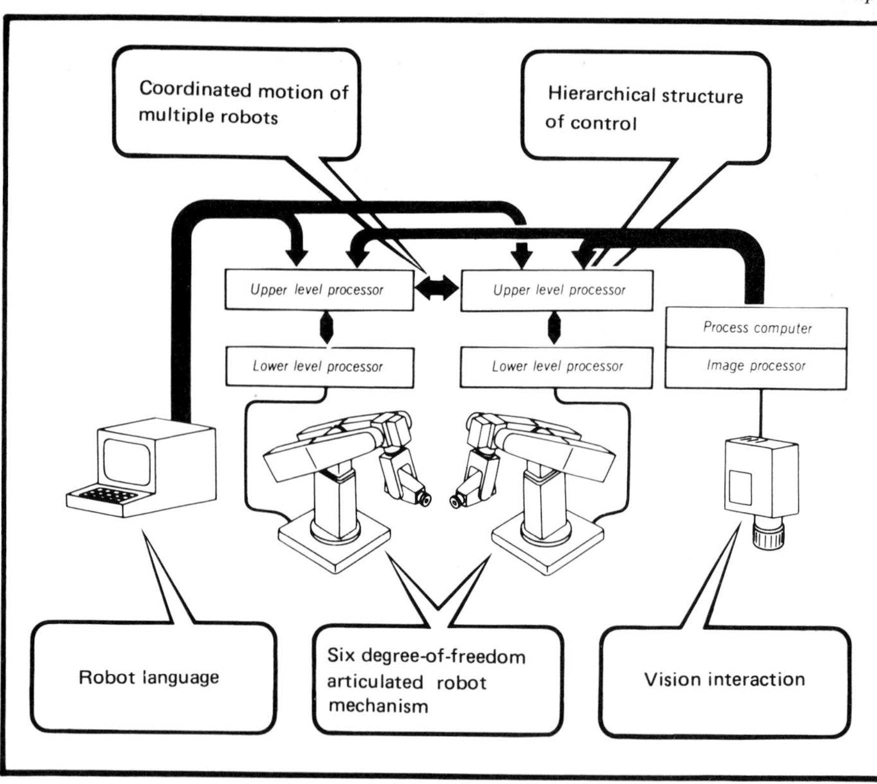

System configuration and features of the intelligent assembly robot system (Production Engineering and Research Laboratory, Hitachi Limited).

Task selection

The monitor commands are used for the selection of task control and include the following:

- □ editing and modification of the robot motion program,
- □ selection of teaching and execution modes,
- □ storing and loading of program and location data to and from magnetic tape cassettes,
- □ display of program and location data onto terminal devices,
- □ recovery from errors.

Of the functions of the languages that have been evolved the two most innovative are those concerned with the vision interaction and the other with the multi-robot co-ordination. By combining the visual information input commands and the conditional branching commands complicated operations can be easily programmed.

Thus tasks such as the following can be performed:

- □ recognition of location and orientation of randomly placed parts,
- □ selection of the assembly procedure depending on where and how the parts are lying,
- □ compensation of positional error of the robot hand for accurate part mating.

Multi-robot co-ordination has been made simple by developing a language structure in which motion flow involving multiple robots can be written in a single program. Through translation procedures this source program is automatically divided into programs for individual robots and the synchronisation commands between the robots are also automatically inserted as required in each program.

Vision

The vision system used in the pilot scheme demonstrated the feasibility of using vision which is fast, reliable and reasonably economical. For quick processing partial pattern matching techniques are employed. The vision system uses a Hidic-80 process computer with a 32k word core memory and this controls the image processor, executes various recognition procedures and communicates with the robot controllers through interfaces.

The cameras are mounted independent of the robot arms on a support frame over the operational area of the robot and look down to obtain two dimensional images of the parts placed below.

The image processor is a dedicated hardware system for fast image processing and is made up of five units linked by an image data bus. These units are the analogue processing unit, memory unit, binary image processing unit, function unit and display unit.

The binary imaging unit is equipped with four counters and a 16 × 16 pixel partial pattern matching circuit for binary imaging. In this unit various measurements such as area and width of patterns can be calculated. The function unit performs summation multiplication, logic operations and the integration of the image data.

P.E.R.L. has been able to demonstrate the feasibility of this intelligent assembly robot system but as yet have to introduce it into their factories. In fact one suspects that P.E.R.L's ideas on an ultimate commercial system may differ somewhat from that they have already shown. They say that a commercial version of a Hitachi assembly robot may be released in a year's time. Meanwhile development of assembly robots and assembly techniques is continuing.

PCB assembly

Whilst some pattern recognition is being used in the assembly work P.E.R.L. is also applying this in others particularly in the area of semiconductor manufacture and printed circuit board assembly.

In semi-conductor manufacture the Laboratory's work is to detect contaminants and flaws. Whilst in 1976 the work was directed towards looking for simple defects on flat surfaces now it is to look for contaminants in complicated patterns such as on LSI wafers, bubble memory wafers, etc.

In PC board assembly the problem is solder joints. Three dimensional inspection of solder pins is being developed to ensure that the shape of the solder is good and that the joint has been correctly made. To do this a line sectioning technique is being employed using hardware to convert this into 3D information. Using this method P.E.R.L. claim that solder joints can be inspected at the rate of 10 per second. This work will no doubt find its way into use in one of Hitachi's own factories. Much the same goes for the Hisert model 3 unit which has just been developed by P.E.R.L. This machine is capable of inserting variable pitch parts such as capacitors, resistors, etc., with variable length pins using servoing to ensure accurate and reliable insertion.

The Hisert machine will be used in-house as will the solder joint inspection machine. In fact P.E.R.L's prime responsibilities are for the development of new production technologies and machines for use in Hitachi's own factories and they do not aim to develop new Hitachi products. The robots are an exception to this principle.

Welding sensor uses structured light and 2D photocell

Takao Bamba, Hisaichi Maruyama, Eiichi Ohno and Yasunori Shiga, Mitsubishi Electric Corporation, Japan*

Vision sensor for arc weld seam tracking uses a projected line of light scan from a high power infrared LED, which is detected by a two-dimensional position-sensitive photocell. It is compact enough to be mounted on a robot hand and works well on surfaces of poor reflectivity.

IN THE WORKING environments for program controlled arc welding robots the dimensional tolerances of workpieces or fixtures are often beyond their acceptable limits. The sensing capability of path correction is needed in many welding applications. Various sensors to meet such needs have been developed, but their practical use is limited by their inadequate sensing abilities and high cost.

We have developed a new visual sensor system which satisfies the needs of many arc welding applications. Its principle is based on the optical pattern projection method which is usually used for recognition of three-dimensional objects. It can discriminate the weld area of objects by detecting the trajectory shape drawn by a scanning light spot and derive the control information for the path correction.

The sensory head is built compactly. This useful feature is made possible by using a photo sensitive position detector and a LED (Light Emitting Diode) for the main components. These components are also useful for simplification of signal processing which is done by a 8-bit microprocessor (8085A).

Experimental work with this sensor on some typical objects has shown satisfactory practical results.

Sensing mechanism

The most familiar direct way of obtaining three-dimensional information is to pass a slit beam of light to an object. Fig. 1 shows the configuration of the newly developed sensor and its sensitive area. Its main elements are a high power infrared LED, a scanning mirror and a two-dimensional photo detector with appropriate optical lens systems. A collimated light beam from

VISUAL SENSOR
COLLIMATING LENS
LED
SCANNING MIRROR
DETECTOR
SENSITIVE AREA
OPTICAL LENS
TRAJECTORY OF SPOT
WORKPIECE
WELD JOINT

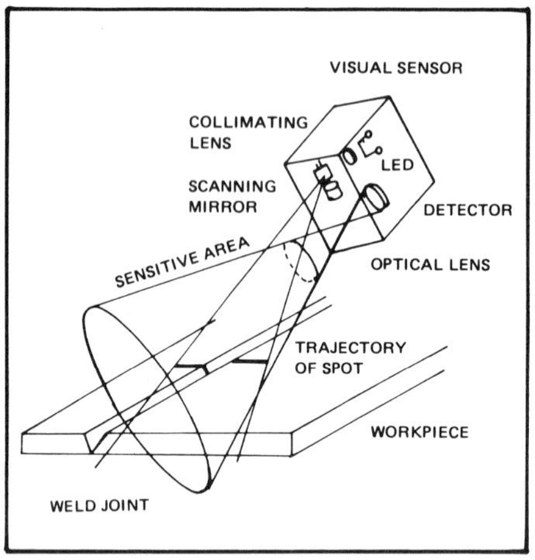

Fig. 1. Configuration of the visual sensor.

the LED is projected on to the surface of the workpiece to be welded. Before the beam reaches the surface, it is deflected by the scanning mirror which oscillates at a constant rate. Consequently the beam draws a repetitive linear trajectory on the surface, which provides three-dimensional information on the object when the trajectory is observed from a different direction.

If the weld preparation is a

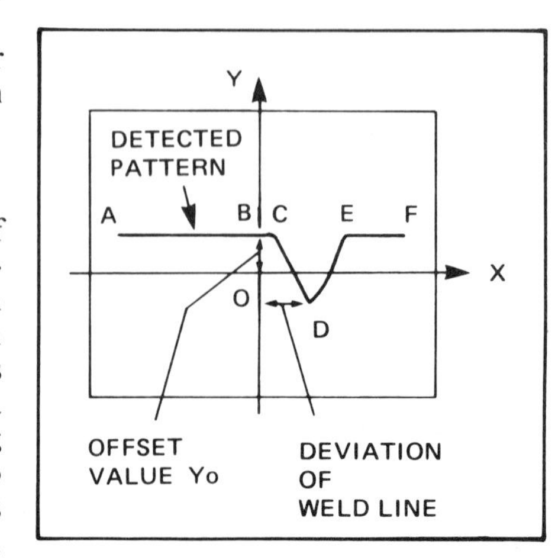

Fig. 2. Sensing image in optical pattern projection method.

V-shaped groove as illustrated in Fig 1, the sensing process will be performed as follows (see Fig. 2):

- As the pattern being picked u forms a trough (point D in Fig. 2) a the weld line, this trough in the lin trajectory is discriminated by proper means.
- The horizontal deviation of th desired welding point D from th centre of the sensing plane i calculated.
- The deviation signal is fed to a robo control system for the path cor rection.

The visual sensor also has a distanc sensing capability between the senso and the workpiece. As the observin axis is inclined to the light projectin axis, the basic line of trajectory ($\overline{AC}$ $\overline{EF}$ in Fig. 2) moves up and dow according to the relative distance Especially the Y value of the spo position B (Y_0 in Fig. 2) represents th relative distance along the observin axis. The distance (l) can be calculate by the following formula:

$$l = l_0 (1 - \alpha Y_0)^{-1} \quad (1$$

where l_0: distance between the centr of the lens and the poin where the light projectio axis crosses the observin axis

α: constant value defined by th device arrangement and len systems.

The principle of optical patter projection on the workpiece has th further advantage of its sensing ability Fig. 3 shows various types of weld lin used for arc welding and their sense images. Since each pattern has a brea point at its joint, every weld line can b easily detected by this sensor.

Fig. 4 shows a prototype of ou visual sensor. It can be compactly se up in small size by employing an LEI as a light source, which makes th sensor easily attachable to the robo wrist.

*Based on the text of a paper presented at the 11th International Symposium on Industrial Robots, Tokyo, 7-9 October, 1981. Reproduced with the kind permission of JIRA.

Parameters	*Type*	*Units*
Spectral range at 10% of peak	300 1150	nm
Peak response wavelength	900	nm
Responsivity at peak λ	0.52	A/W
Position sensitivity at peak λ	0.80	A/W/cm
Position linearity distortion from centre to a point 25% from edge 75% from edge	0.3 0.3	% %
Dark current at 10V bias	0.5	μA
Rise time at 10V bias (10%∼90%, R_c = 10 kΩ)	8.5	μsec.
Recommended mode of operation	photo-conductive	
Active area	13 × 13	mm²

Table 1 Specification of position sensitive photodetector.

Fig. 4. A prototype of the visual sensor.

In general, image sensors such as TV or CCD cameras are usually employed for two-dimensional optical sensing in robot eyes. However, they have some problems as follows:

- Appropriate illumination control is needed according to the detection situations.
- It takes a considerable time to process the image information from a large number of pixels for real-time use.
- Total system cost might be expensive for practical use because high-performance processing equipment is required for pattern recognition.

To avoid these problems, we have adopted the two-dimensional lateral photoeffect diode as the area sensing device. It is a dual-axis position sensitive detector that provides X- and Y-axis continuous position information on a light spot movement on the detector. Since the device is not divided into pixels, no internal scanning is required, so the position of the spot can be detected immediately. The peripheral circuits and image processing algorithms for the sensing system can also be simplified. Table 1 shows the main specification of this device.

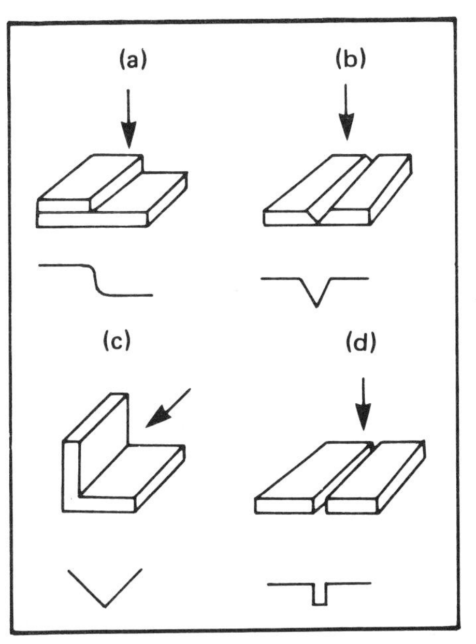

Fig. 3. Various weld joints and sensed patterns.

Signal processing

Fig. 5 shows the schematic diagram of the signal processing circuit for our sensing system. The LED is modulated by a pulse oscillating unit (f = 10 kHz). The detector picks up the reflective light from workpieces, and outputs continuous current signals which carry the position information of the moving spot image on the detector surface. Through current-to-voltage converters, the signals are fed to band-pass filters (f_c = 10 kHz) to extract effective LED signals from the background light noise. After demodulation, accurate X and Y position signals of the detecting spot are separately derived by operational amplifier circuits compensating the influences of its intensity.

An 8-bit microprocessor (8085A) is used for weld line discrimination from the sensed waveform. After the com-

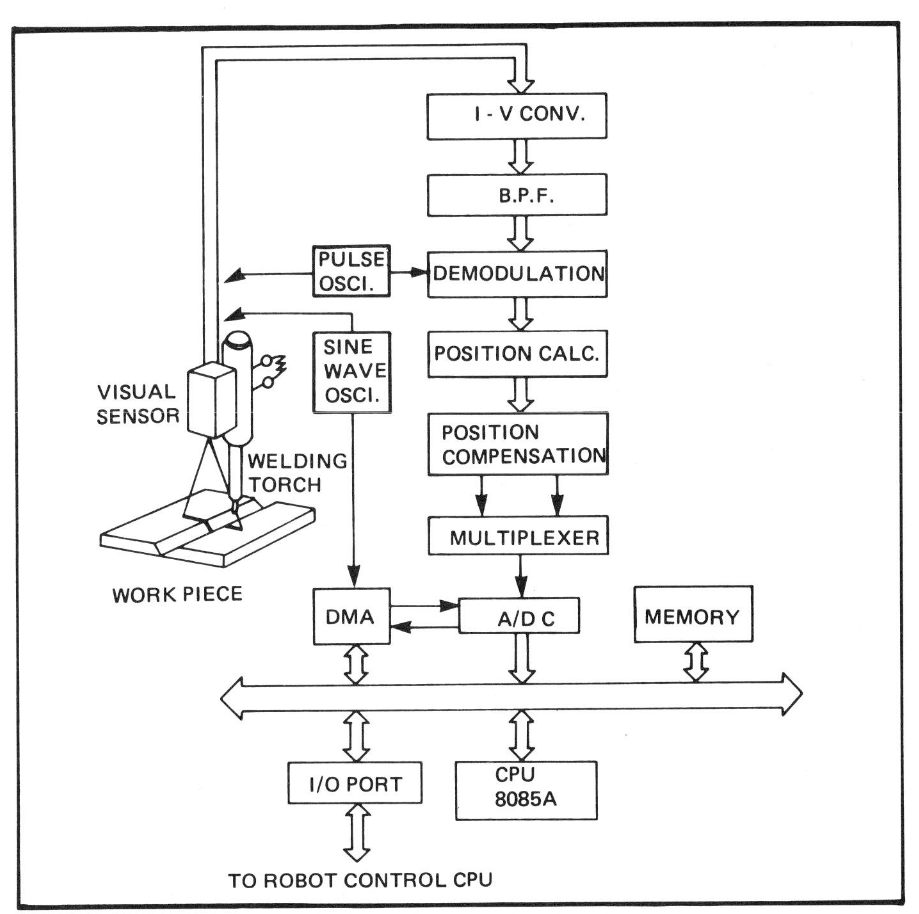

Fig. 5. Schematic diagram of signal processing circuit.

pensation described above, two-channel signals (X, Y) are digitised by an analogue-to-digital converter. A direct memory access controller (DMA controller) transfers this digital position information to random access memories (RAM) of CPU synchronously with the mirror scanning cycle. Thus one frame of spot trajectory information is stored in the memories rapidly. Refresh cycle of the spot trajectory is 10ms.

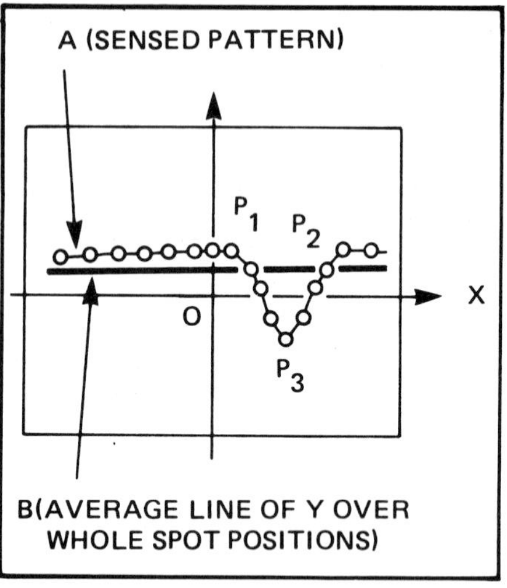

Fig. 6. Method for weld line detection.

The microprocessor perceives the deviation of weld line and the relative distance between the workpiece and the visual sensor. In the following section, algorithms on the perception will be described.

Software

Algorithms for weld line perception are simple. Assuming that the observing axis of the visual sensor is located perpendicularly to the surface plane of workpieces, the waveform can be processed easily. If the weld line is a V-shape groove, the processing procedures are as follows (see Fig. 6):

- The average value of Y coordinates over a whole spot position is calculated, which indicates line B in Fig. 6.
- Crossing points of the waveform (A) to the average line (B) are obtained. These points are P_1 and P_2 in the figure.
- The most deviated point from the average line is obtained (point P_3).
- A desired point is regarded as the most deviated point between P_1 and P_2. In this case this point is P_3.

By use of the average value, the required weld point can be detected correctly even if the waveform is disturbed by various noise. Moreover this method is applicable to other types of weld line. For example, in the case of a weld line such as (a) in Fig. 3, the weld point is regarded as the crossing point of the pattern to average line. In the case of (d) in Fig. 3, it is regarded as the midpoint of th crossing points.

The configuration of software shown in Fig. 7. At first nois reduction using the method of movin average is done. That is:

$$W_i = W_i' + (B_i' - W_i')/N \quad (i=1, \ldots, m) \qquad (2$$

where W_i : i-th spot position i present frame of image
W_i': i-th spot position i previous frame of image
B_i : i-th spot position of newl detected data
N : sampling number.

This process can remove randon noise from the sensed image. Secondl weld line discrimination is done. A proper discriminating routine i selected corresponding to the shape o weld line.

Next, the distance between th visual sensor and the workpiece i obtained by the calculation in Columr 0. In the status setting stage, the micro processor checks the sensing status For example it checks whether th light beam is correctly projected on tc the workpiece, whether the giver information is normal, and whethe the weld line is observed in the field o view. These status conditions can b used for safety control.

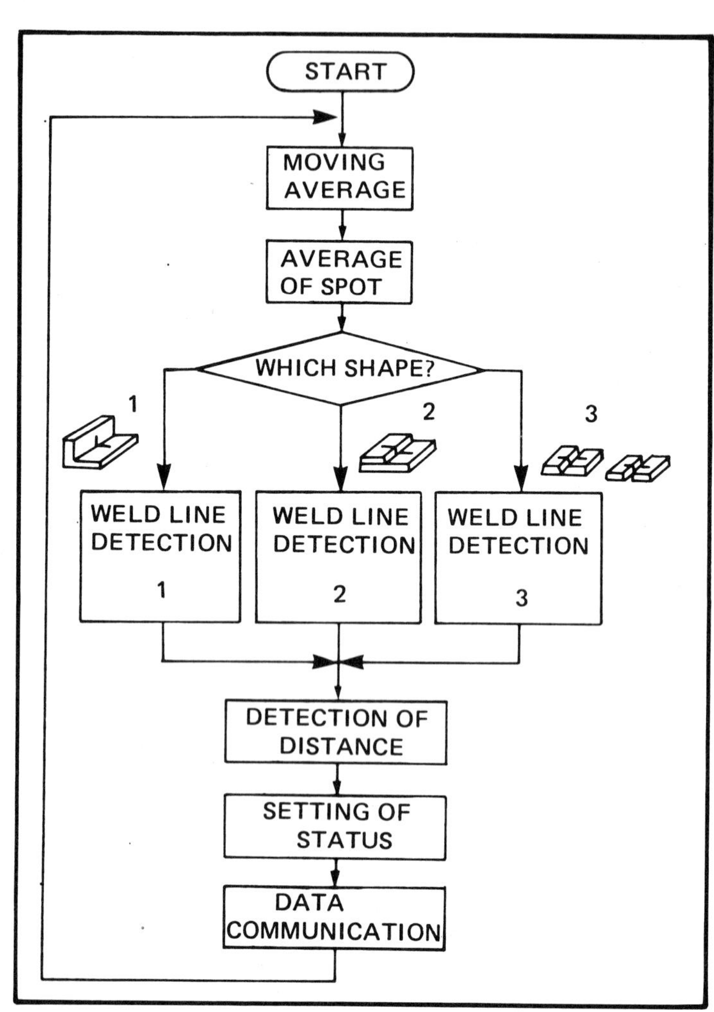

Fig. 7. Flowchart of software.

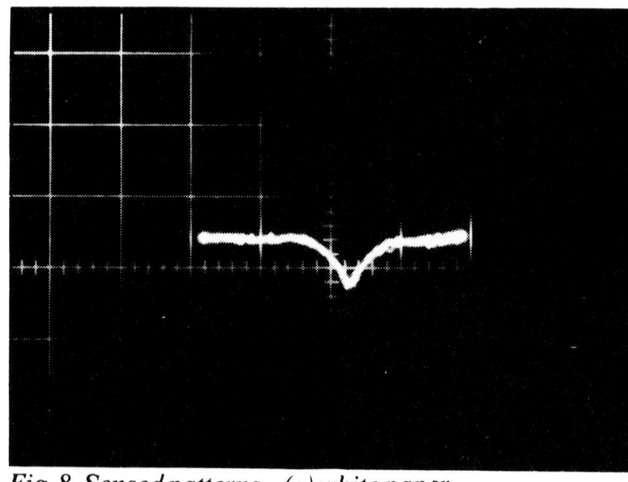

Fig. 8. Sensed patterns. (a) white paper

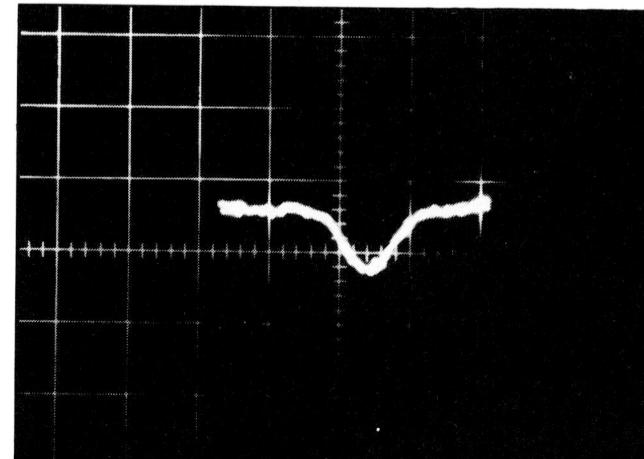

(b) rusted iron

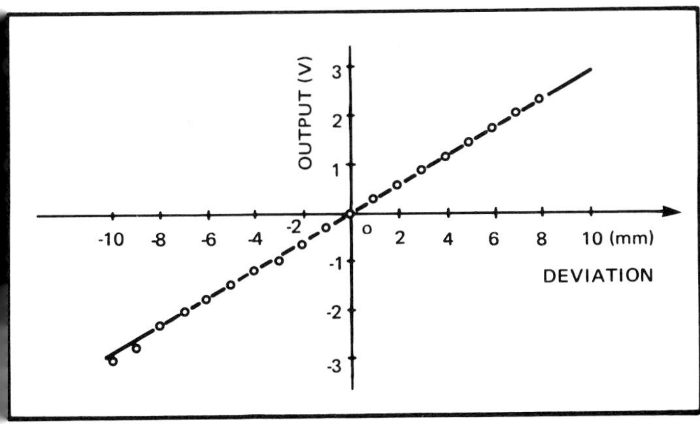

Fig. 9. Sensor output in position sensing of weld line.

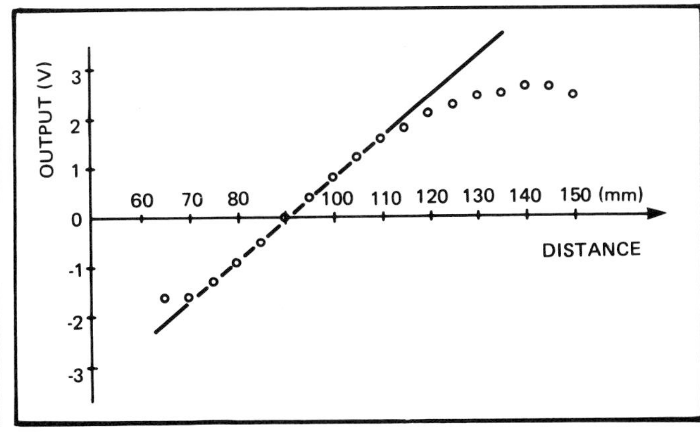

Fig. 10. Output in distance sensing.

The processing time for one cycle execution of software is less than 40ms.

Experimental results

In optical sensors, their sensing abilities depend on the optical characteristics of the workpiece surface. Unfortunately the materials used in arc welding usually have unfavourable reflective characteristics. This makes it difficult to detect the correct spot position on the workpiece. In addition, the sensing distance will be limited. We have done some experiments on such problems for our visual sensor.

Fig. 8 shows the experimental results of image detection for a *V*-shape groove using white paper (a) and rusted iron (b) respectively. Our visual sensor detects a correct image even in the worst condition such as heavy rusted iron. In further experiments, we have ensured the sensing capability with other materials such as stainless steel and aluminium.

Sensing linearity to a typical workpiece is shown in Fig. 9. The abscissa indicates the deviation of weld line from the centre of the visual sensor, and the ordinate is output from the sensor. As the field of view is approximately 20mm × 20mm at 90mm distance, it can be said that the sensed deviation of weld line is linearly proportional in almost all of the area.

The next experimental result is the sensing capability of distance between the workpiece and the sensor. Fig. 10 shows a typical response for the rusted iron. The output of the visual sensor has a good linearity in the range from 70mm to 110mm. But it gets out of ideal curve in other domains nearer than 70mm and further than 110mm. The reasons are that it goes out of the field of view at short range, and there is a lack of input power at long range.

Although these limitations on sensing abilities exist, the visual sensor is effective for almost all workpiece materials and weld line shapes with sensing resolution of less than ±0.5mm.

Universal system the target in Japan

Voice recognition research moves towards integration with knowledge base as lsi circuits reduce the cost of systems now available.

VOICE recognition research is at a crossroad in Japan, and its future is bound up with the fifth-generation computer. 'It's no longer a case of taking recognition alone', said Yasuo Kato, general manager, C&C systems research laboratories at NEC. 'For more progress we need a big dictionary, so we are talking of a knowledge base. That leads to a translating machine.'

Synthesis

Indeed, the president of NEC has called for the development of a fully-fledged voice recognition/synthesis machine that can act as an interpreting machine between different languages within 20 years. Currently, though, NEC and Fujitsu are offering voice recognition systems, while NTT, Japan's equivalent to British Telecom, has conducted field trials of a system developed by NEC. Hitachi and other Japanese companies are developing voice recognition systems, but as is often the case in Japan, they seem to be similar.

As for applications, most of the emphasis is on use by banks, so that people can telephone in and obtain information about their account. Industrial applications are mainly in warehousing.

Kato claims that NEC is the leader in voice recognition, largely as a result of the work of Hiroaki Sakoe, research manager, media technology research laboratory. About 15 years ago, Sakoe evolved his dynamic programming (DP) algorithm, on which the whole system has been based. Evidently, some Russians developed a different algorithm on the same principles at the same time, but Sakoe claims that his system gives more reliable matching.

By the end of the 1970s, voice recognition systems were available. Initially, they recognised individual words spoken by one speaker only. Then, systems to recognise individual words spoken by anyone were developed, while NEC has also developed a system to recognise words spoken continuously by one speaker.

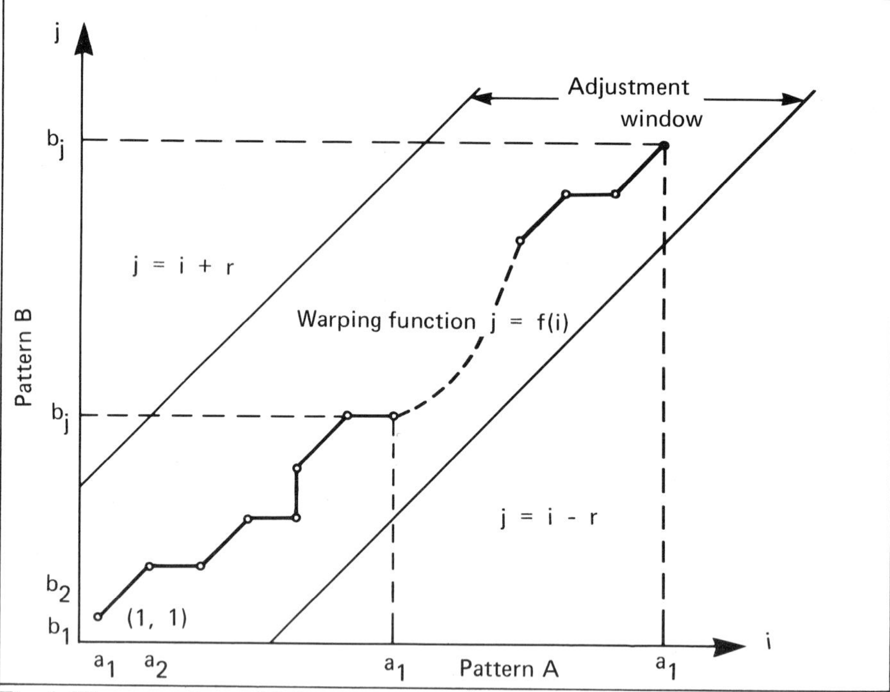

Fig. 1. Time warping, which compensates for variations in the rate of speaking, is a key feature of the NEC recognition system.

The whole recognition process starts with the phoneme, the unit of significant sound, which is usually a syllable. In Japanese, the pronunciation of vowels is always the same, so it is one of the simpler languages for recognition purposes.

Vowel sounds

Nevertheless, the speed of diction varies by 20–30% according to the speaker, most of the variation being in the vowel sound. It was in attempting to overcome this problem, so that an accurate match between the spoken word and memorised patterns could be realised, that Sakoe developed his algorithm.

He explained that speech can be expressed as a sequence of feature vectors. If A represents a pattern of a word input, and B the reference pattern, then:

$A = a_1, a_2, \ldots a_i, \ldots a_I$ and

$B^n = b_1^n, b_2^n \ldots, b_j^n, \ldots b_J^n$

The word is defined as $C(n)$ ($n = 1 \ldots N$). To compensate for variations in time – non-linear time fluctuation – the pattern B time axis is warped using the function j(i). This system was shown to work well with individual words.

For continuous speech, some other system is needed, and Sakoe discounts the segmentation method (patterns are split up at input to represent individual words), developed in the USA, because it was developed for a limited vocabulary. He considers it impractical to develop such a system for a large vocabulary. Therefore, the DP system was adapted to provide matching of the continuous sound so that segmentation errors were eliminated. In practice, two level DP-matching, of both partial patterns and distances is made. There is a process for word-matching, and another for phrase-matching.

Thus, NEC was able to develop its DP-100 series voice recognition systems, which consist essentially of the recognition unit, a terminal with crt, and output to a host computer. The recognition unit consists of a speech analyser, reference pattern

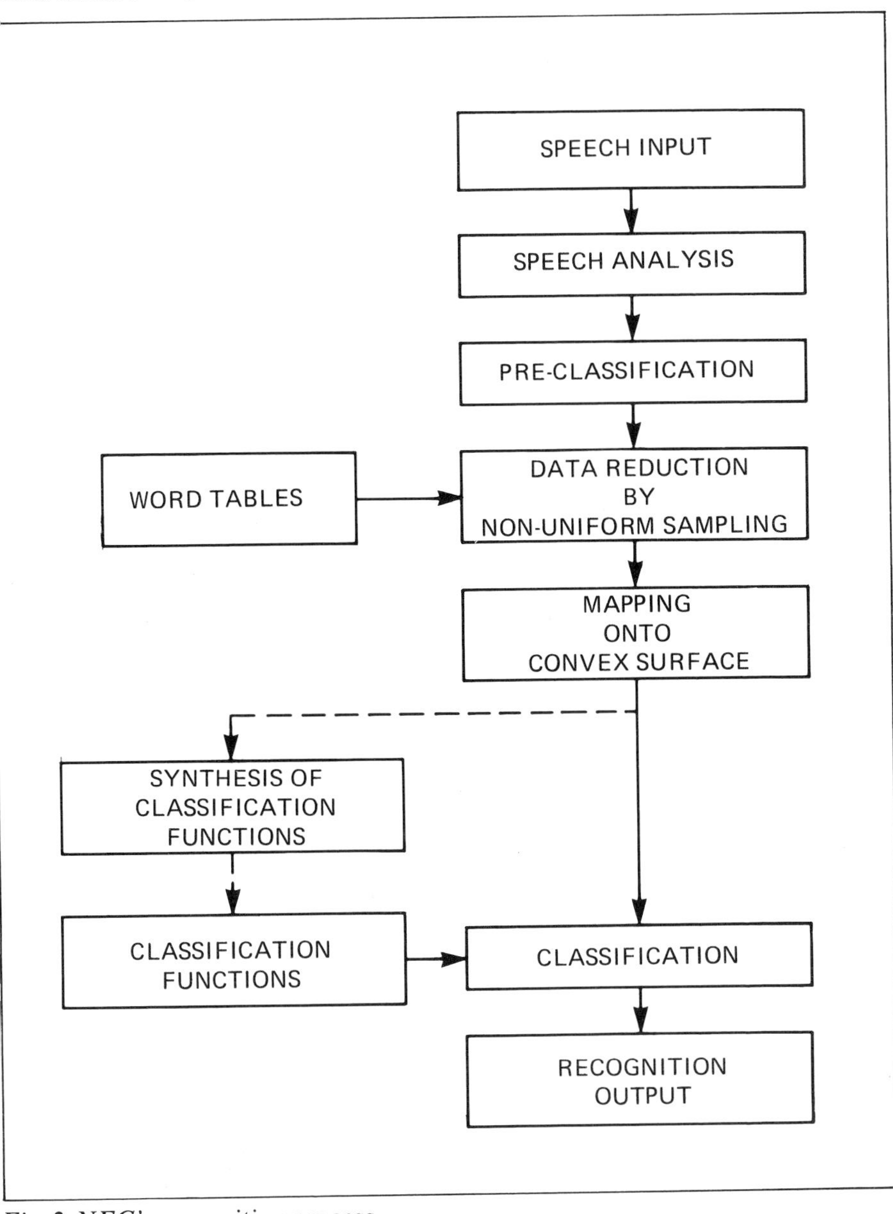

Fig. 2. NEC's recognition process.

memory, microprocessor and interface microprocessor.

Spectrum analysis is conducted on the speech input, so that the speech signal is transformed into 16-dimensional spectrum vectors at intervals of 18ms. Frequencies of up to 5,900Hz are covered. In the memory are 120 reference words, and this can be expanded to 440 words. The DP processor is specially developed hardware for DP-matching.

First, however, the operator must train the machine, and he does this by reading words individually as they appear on a plasma display. Numerals need to be spoken twice. Each sound is analysed and stored in the reference pattern memory. Words input subsequently are analysed in a similar manner, and are matched to the data stored in the memory. Sakoe claims that tests showed an accuracy of greater than 99% for words spoken by different people.

To reduce the size and cost of the voice recognition unit, NEC has now produced an lsi for DP-matching. It includes two processors, so that vector distance and DP-equation calculations can be carried out simultaneously, pipelined subtractor and adder sets, and a larger buffer memory of 416 × 8 bit words. This means that the minimum system now consists of only seven ics – the analyser, controlling microprocessor, reference pattern memory, the DP-matching lsi, input/output, latch and memory.

DP series

With the lsi, the latest DP series machines can match up to 50 words spoken continuously. In addition, there is the SR-1000 system designed to recognise individual words spoken by anyone, while Fujitsu produces the 2350 system which has similar performance.

SR-1000 models recognise four, 16 or 128 words spoken individually, and it can handle 32 or 64 calls simultaneously. It is a large device, 570mm wide by 900mm deep by 1,560mm high, and two are needed. In addition, the voice synthesiser and controls take up twice as much space. NEC claims an accuracy of 95% in field trials.

With the SR-1000, which can be used with any number of speakers, a lot of data is needed, and each work, instead of being represented by 150 parameters is represented by a cluster of those parameters. Each cluster represents the patterns for the word spoken by a large number of people. The problem is to get good separation between clusters, and obviously the more words are used, the more these problems multiply.

Banks

Already such systems, produced to different designs by NEC, Fujitsu and experimentally by NTT, are being used at some banks in Japan. Subscribers can telephone in for information; the subscriber gives his number, and can receive details of his current balance, specified transactions, interest rate, or a foreign exchange rate. Generally, the only relevant words are 'Yes', 'No', 'Please' and 'Cancel' because the synthesised voice asks questions. 'Home banking', which is sorely needed in Japan because bank tellers are so dreadfully slow, is seen as a major application for voice recognition by the computer companies.

NEC is also selling its DP8100/200/300 continuous speech recognition machines for industrial applications. One is being used at the inspection station at the end of a car assembly line. Previously, the operator had to check a number of critical items, such as tightness of wheel nuts, and then sign a form to indicate that all was well or not. That job took a long time, so now voice recognition is being used to speed the job up.

In another application, data on cars to be auctioned is input by voice in the parking area, and transferred by radio to the computer in the auction room. Input is claimed to be error-free, and again, a lot of time is saved.

Kato claims that the combination of the DP-algorithm system and the use of a noise cancelling microphone results in the DP-100 being able to operate in an ambient noise level of 85–90 dB. It is also being used in warehouses for sorting parcels and packages. Other potential applications include data entry for a CNC machine programming of robots, and at CAD/CAM and other terminals to replace the keyboard. Ultimately, of course, the applications are limitless, but they will not become widespread until the capability of both software and hardware is improved significantly.

There is a need for data to be processed much more quickly, and for a larger memory. In fact, since one word occupies about 500 bytes of memory, 500 words can be stored in 250kbytes, which is well within the bounds of a current 16 bit microcomputer. However, once memories of 5,000 words or so are considered, then the memory would present problems, especially since access must be rapid. And in the end it is the speed of the hardware that matters.

But before that stage is reached, a method of matching a large number of words is needed. Takao Watanabe, a researcher at NEC has proposed a system based on 'templates' for vowel and vowel-consonant-vowel sounds, rather than for complete words. For Japanese, 700 templates are needed, and some care is needed with enunciation. Such a system would be more difficult to apply to English.

In actual recognition, DP-matching is used, and in a test of 493 Japanese family names spoken in a quiet room, 86.9% were recognised correctly. In another test in which a 4,300-word dictionary was used, the accuracy reached 93%. This system could be a step forward, although consonant classification and the time taken to process the data need to be improved.

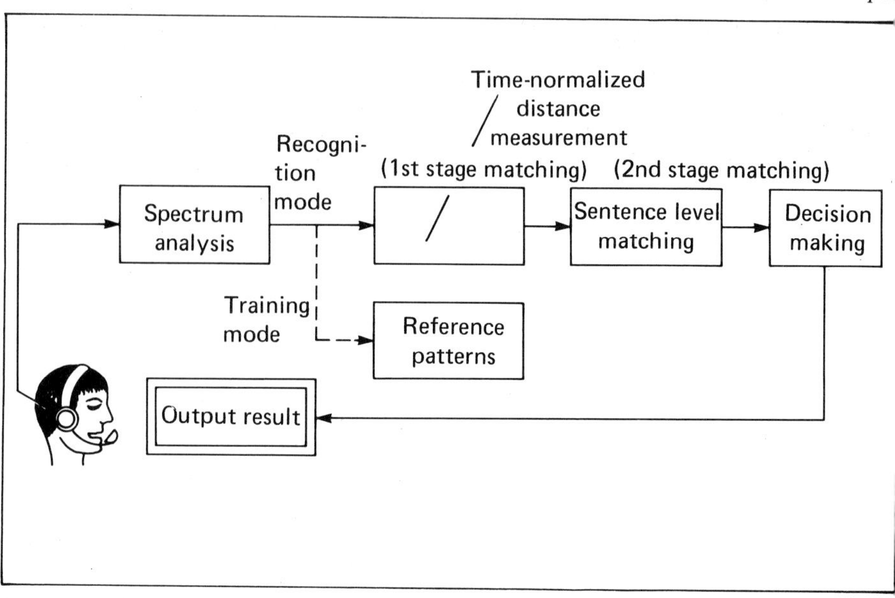

Figure. 3. Block diagram of DP-100 voice recognition unit.

But this takes us into the realms of knowledge bases, Kato argues. After all, once the dictionary has 4,000 words, we will want one with 20,000 words, and that will lead to a multi-language knowledge base. Already, though, it is practical for one person to input data verbally to a computer. Numerical data are probably the easiest, so for many people, the end of the keyboard as an essential method of input may be in sight. However, although many companies have demonstrated voice-input word processors, the complexity of providing a rapid system for such a big vocabulary is still several years away.

Bibliography

'Two-Level DP Matching – A Dynamic Programming-Based Pattern Matching Algorithm for Corrected Word Recognition', H. Sakoe, IEEE Transactions on Acoustics, Speech and Signal Processing, Vol ASSP-27, No 6 December 1979.

'A Generalised Two-Level DP-Matching Algorithm for Continuous Speech Recognition', H. Sakoe, IECE of Japan, Vol E, 16 No 11 November 1983.

How voice and laser can raise distribution efficiency

Tsutomu Abe, Kando Unso Co and Jun Suzuki, Nagasakiya

One of Japan's largest distribution centres has introduced an automatic sorting system using voice input. The workforce has been halved.

NAGASAKIYA is a bulk chain store handling mostly clothes. It has annual sales of Yen 230,000 million through 56 directly-managed branches and 90 franchised stores. It ranks seventh in Japan's chain store industry.

Goods ordered by Nagasakiya are received from suppliers by Nagasakiya Tokyo Distribution Centre and distributed by it to Nagasakiya's branch and franchised stores. The volume of goods handled by the Tokyo Centre in 1978 totalled some ,200,000 pieces weighing 100,000 tons.

In addition to the Tokyo Distribution Centre there is a similar centre in Osaka, controlled by Nippon Tsuun, acting as a base for the Kansai area and another in Hokkaido (managed by Maruwa Unyu) which services the Hokkaido area. The Tokyo Distribution Centre ships goods countrywide from Hokkaido to Kyushu and there is one shipment a day. Kanda Unso is contracted to handle all goods received and distributed for Nagasakiya.

Most of the goods are packed in corrugated cardboard boxes according to their destination. The busiest time at the Tokyo Centre is around 16.00h when 40% of the day's supply of goods arrives. The average size of each pack of goods is 3.5ft^3 or 0.1m^3.

As the volume of goods needing to be handled has increased so the Tokyo Distribution Centre has undergone changes in its goods sorting system as well as changes in its location and buildings.

The bar chart (Fig. 1) indicates the sharp reduction in the man-hours which are needed for each 1,000 pieces of cargo. However, this does not necessarily mean a correspondingly sharp drop in cost. For when the sorting was handled by men carrying goods on their shoulders or using trolleys, there was still little division of labour. The work was done chiefly by vehicle drivers and their assistants, helped also by office workers and even managers.

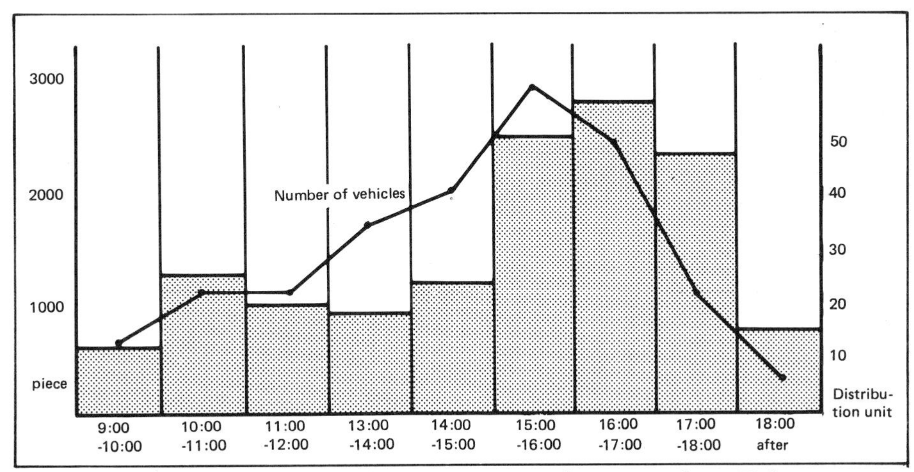

Fig. 1. Details of trucks and packages arriving at the Nagasakiya Tokyo Distribution Centre.

While in the days when belt conveyors were adopted for the sorting, students and other part-timers were employed to secure enough workforce in busy hours. And even in the case of automatic sorting machine, the investment still demands depreciation, maintenance and interest payment.

Increased efficiency

Today, the term 'logistics' is heard increasingly in Japan. This stems from the increasingly important position which physical distribution is assuming in management strategy. However, for effective development of logistics, enhancements of PDM (physical distribution management) and PDE (physical distribution engineering) are even more necessary today than in the past. Nevertheless, whilst it is important to apply new technologies to the rationalization of physical distribution this must not be done at the expense of human misery.

Nagasakiya Tokyo Distribution Centre has physical distribution system which uses a variety of machines to save manpower in the sorting works. It is linked to a computer to minimize the clerical work; an improved layout increase space utilization. The machines used in the system include a voice encoder, a laser scanner, a computer, an automatic labeller and a speaker sorter, all of which are Japanese products (Fig. 2).

The bar code is widely used to sort goods by product. At the Tokyo Distribution Centre, however it is used to identify vendors for the purposes of producing a shipping statement. Since the bar code is no longer available for goods sorting, a full keying in system is used for sorting input. Under this system, however, the machine's capacity is limited by man's ability. To rectify and improve this a voice encoder has been introduced.

Details of the sorting system at Nagasakiya Tokyo Distribution Centre and an explanation of how the system's efficiency has been increased are as follows (see also Fig. 2 and Fig. 3):

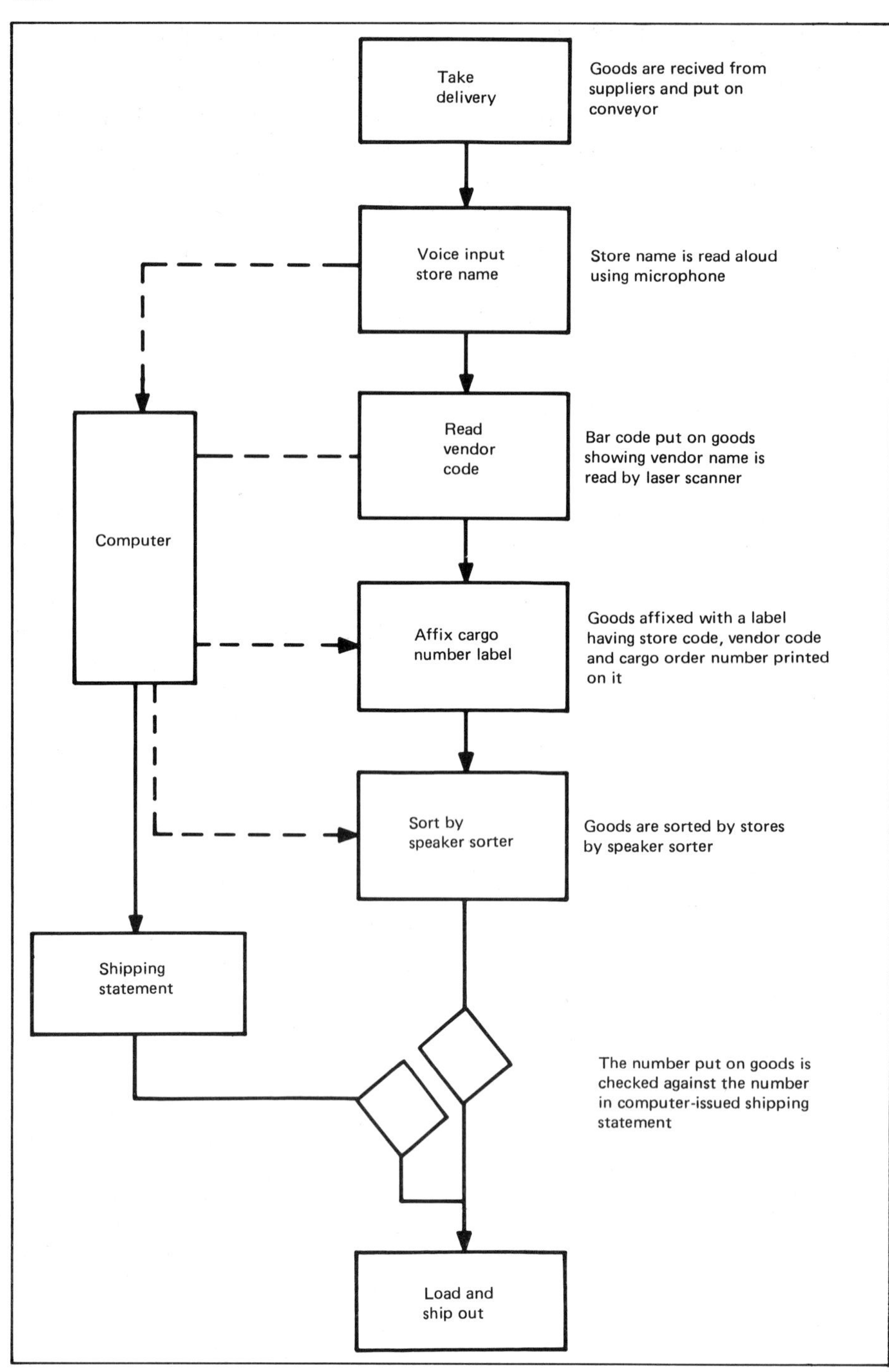

Fig. 2. The physical distribution system at the Tokyo Centre.

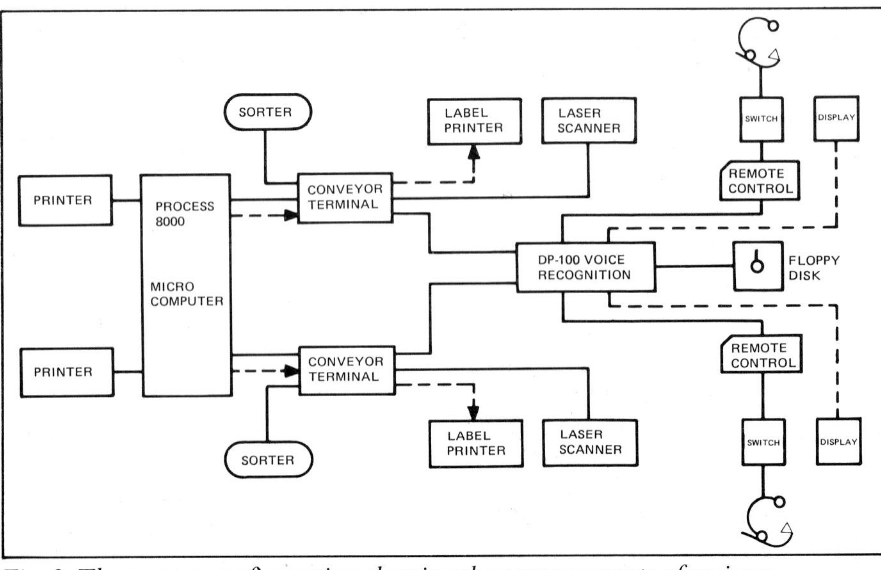

Fig. 3. The system configuration showing the arrangements of various components.

- ○ Before delivering goods to the Tokyo Centre suppliers put on their goods a special label on which the individual firm's vendor code is printed.
- ○ When received, goods are transferred by conveyor to the entrance of the automatic sorting machine (Fig. 4).
- ○ The operator reads aloud the store name while facing a parcel (Fig. 5).
- ○ The bar code on the parcel is then read by laser scanner (Fig. 8) and fed to the computer.
- ○ The store code, based on the store name, is then fed in through microphone while the vendor code from the bar code input and the cargo order number are also all fed to the automatic labeller. This unit prints the appropriate codes on a label and places it onto the parcel travelling on conveyor but before the parcel is placed on the automatic sorter tray.
- ○ Goods are placed on the high speed automatic sorter tray which channels them toward the right chute (Fig. 9).
- ○ Based on the information collected through laser scanner and voice encoder, the printer produces a shipping statement containing the store's name, arrival order and supplier's name.
- ○ Once goods have been sorted, the cargo number on the label attached to them is checked against the number on the shipping statement and the goods can be loaded onto a truck.
- ○ At their destination, the cargo number on the parcel is checked once more against the number on the shipping statement before the goods are finally delivered.

Effects of rationalization

Automatic sorting machines are used overseas to process mail orders, mail and air cargoes. These areas are also major users in Japan, but with one big difference – the extent to which they are used in the transportation industry.

Japan is unrivalled as far as the number of automatic sorting machines used by the transport industry is concerned. The number continues to increase and improvements are expected in the utilization of machines.

However, at present the wide and growing use does not seem to be justified on the basis of any calculated return on investment. Rather most companies use the machines to improve their corporate image; certainly no quantitative reports have yet been made of the economic benefits machines of this type can bring.

Most companies seem to conduct inadequate studies, both of the introduction of the machines and on the machines themselves. They are also backward in their system thinking. In contrast, Kanda Unso established clear objectives of rationalization, worked out a system of operation, selected machines accordingly and achieved the required objectives.

The first benefit stemmed from an increase in space utilization as a result of a good layout. What has struck us in looking around some of the distribution centres and terminals run by other companies is that the automatic sorting machine occupies as much as three-quarters of the loading platform, resulting in effective use of the building. In a country such as Japan, where land prices are high, the higher costs resulting from such ineffective use of space adversely affects profits.

At Nagasakiya Tokyo Distribution Centre, the horizontal circular sorter and speaker sorter are installed 4.5m above the loading platform. Also, the chutes are made to cross leaving a gap of at least 15m between the chutes. Beneath crossing chutes there is a 2m by 2m column space where a conveyor is installed to accept manually goods unsuitable to be processed by the automatic sorter.

The conveyor transporting goods to the entrance is raised to the height of the automatic sorter, making the accumulator conveyor run on a level with the sorter. There is a distance of more than 100m between the cargo receiving point and the entrance to the sorting area to provide a buffer facility for the smooth flow of goods. By this means the cargo receiving conveyor need not be stopped even if the entrance conveyor stops temporarily for whatever reason.

The real estate costs for each pack of goods is set at Yen 19. If the utilization rate rises from the present 80% to 100%, the real estate cost will go down to Yen 15 per piece. Although this is twice the warehouse storage, it should be considered inexpensive in view of the once-a-day turnaround and the facility of the distribution centre.

Fig. 4. Entrance to the automatic sorting machine.

Fig. 5. The operator feeds the store name through the microphone.

Higher throughput

The second advantage of the system is the reduction in the machine cost as a result of the throughput. Automatic sorting machines are introduced to increase the speed of working; to respond more quickly to customer needs and save manpower. But to achieve higher working speed it is important that machine characteristics are fully understood in order to make best use of the machine's capacity and to minimize the workforce.

Nagasakiya Tokyo Distribution Centre sorts 7,200 packs hourly in conditions where the maximum cargo is 90cm, the tray pitch is 1,370mm and the line speed is 110m/min. For the size of cargo, this rates the centre as having the highest handling capacity in Japan. If another preparation area is added and the line speed increased to 120m/min, the capacity can be raised to 10,000 pieces an hour.

In a situation where goods received by 17:00h must be shipped out on the same day, higher hourly processing capacity leads to reduced equipment investment. The hourly processing capacity of a linear sorting machine cannot be more than 5,000 pieces if the pitch is set at 1,300mm. If greater capacity is needed, either two machines must be used or rough sorting must be conducted involving the use of the upper and lower sides of the belt. In either case, the same number of workers will be required regardless of the volume of goods.

Using the machine layout adopted at Nagasakiya Tokyo Distribution Centre, part of preparation area can be halted according to the volume of goods being handled so helping to cut back fixed personnel cost. The high hourly processing capacity also enables work to be completed early during busy seasons, such as the year end, and minimizes overtime.

The third benefit is increased efficiency through the use of the voice encoder. Normally, the keying method is capable of inputting 40 pieces a minute. But the speaker sorter at the Tokyo Distribution Centre is capable of 40 pieces a minute per entrance. In this case, because of the lack of leeway between the human and machine capacities, the operation rate at the most is 80% or 32 pieces a minute.

Voice encoder

If the voice encoder is introduced, the input capacity increases to more than 60 pieces a minute, allowing operator freedom of action. The operator's input capacity will exceed the capacity of machine which makes it possible to raise the operating rate to over 90%. It will become possible in the future also to handle an increased volume of goods by increasing the machine speed.

In the past, a handler and a key operator have worked as a team. However, the introduction of the voice encoder eliminates the need for the handler, thus reducing the work-force by one for each entrance.

Two workers can be eliminated for two entrances, resulting in an annual saving of Yen 6 million in labour costs. Therefore, investment in the voice encoder can be paid off in only two and a half years.

The fourth benefit stems from the minimization of clerical work by the use of the laser scanner. For although distribution centres generally use laser scanners to sort cargo according to store destinations, the Tokyo Distribution Centre uses it as an information system. Special Nagasakiya labels are given to suppliers in advance with vendor or bar codes printed on them. Goods delivered by suppliers are laser scanned before being placed on the automatic sorter tray, in order to register the vendor code in computer.

By combining this input with that of the store name fed in through voice encoder, the computer can automatically print out a shipping statement.

In the past, this shipping statement used to be handwritten for each store. It was usually done during busy hours of the evening, and used to result in high personnel expenses.

Fig. 6. In the previous system the key input was used to code in the store number.

商品発送票 婦人服・スカート スラックス

送主 商品名 長崎屋

中央区東日本橋三—七—一四
(株)長崎屋本社

店

10901

仕切書 No.

発送日 · ·

個口

Fig. 7. The bar code, used to sort goods by product, is attached to the parcel.

Fig. 8. The laser scanner reads the bar code.

	1976	1979 (old system)	1979 (new system)	New/ old
Number of pieces of cargo handled	3 million	3.5 million	3.5 million	
Total number of workers (month)	682	795	594	
Number of workers	56.8	66.3	49.5	
Number of pieces of cargo per worker per month	4,398 pcs.	4,400 pcs.	5,892 pcs.	134%
Labour costs per year	Yen 145 million	Yen 205 million	Yen 153 million	134%
Labour costs per piece of cargo	Yen 48	Yen 58	Yen 44	76%
Increased costs per year			Yen 40 million	
Total costs per year	Yen 145 million	Yen 205 million	Yen 193 million	94%
Cost reduction			Yen 12 million	

Table 1 – Effects of rationalizing cargo handling work by new system

	1976	1979 (old system)	1979 (new system)	
Number of workers per month	141	164	140	
Worked hours per year	390,000	421,000	360,000	
Number of pieces of cargo per hour	7.7	7.7	9.2	
Labour costs per hour	Yen 1,050	Yen 1,377	Yen 1,377	
Total labour costs	Yen 410 million	Yen 580 million	Yen 495 million	
Saving in labour costs per year			Yen 85 million	15%
Actual saving			Yen 45 million	8%

Table 2 – Saving in total labour costs by new system
(Including costs for drivers, clerks, cargo handlers)

Fig. 9. The Tsubaki speaker sorting machine has 52 chutes and travels at 10m/min.

High-speed automatic sorter
Maker: Tsubakimoto Chain Co.
Type: Tsubaki Speaker Sorter
Length: 165m
Sorting number: 50 chutes
Speed: 110m/min.
Capacity: 7,200 pieces/h
Induction: 3
Cargo handled: Maximum: 900L x 800W x 800H
Minimum: 300L x 200W x 150H
Weight: 40 kg per piece
Control system micro-computer, pulse oscillating system
Laser
Maker: Anritsu Electric Co.
Type: LSB-2A, 3 units
Code: 2 out of 5
Reading speed: 70 m/min.
Automatic labeller
Maker: Anritsu Electric Co., 4 units
Type: PROCESS 8000
CPU: 32 KB
Conveyor terminal: 4 units
High-speed printer: 2 units
Floppy disk: 1 unit
Voice encoder
Maker: Nippon Electric Co.
Type: DP-100
Word number 120 words per set, 240 words
Word length: 0.2s to 1s
Recognition speed: 0.3s

Table 3 - Equipment at the Nagasakiya Tokyo Distribution Centre

Today some carriers adopt a system whereby no shipping statement accompanies the delivery truck. However, most transporters continue to produce such statements, either by using copying machines or by mobilizing every available clerk, simply to prevent accidents and provide much needed information about cargo.

With our system, however, the shipping statement is automatically issued based on the information collected from goods travelling on the conveyor. This not only saves manpower but increases the speed of clerical processing. In addition, the ability to check cargo number against the number on the shipping statement has reduced the possibility of accidents to near zero.

The final advantage can be found in the cost reduction compared with belt conveyor sorting. The sorting system at the Tokyo Distribution Centre, through the introduction of automatic sorting machines, has brought a 25% reduction in labour costs for the sorting work. In 1979, this resulted in a saving of Yen 12 million after deducting equipment costs. Also, because drivers' assistants handled cargo too, increased utilization of trucks and reduced overtimes led to total labour costs for clerks, drivers, being decreased by Yen 45 million in 1979.

Japanese robot languages

New languages designed to lead off-line programming of robots.

ALTHOUGH a lot of progress has been made in simplifying the teaching of robots, complete teaching on-line is inherently slow. Moreover, where 20 or more robots are used on one assembly line, the need to teach each individually is a major deterrent to the use of robots to assemble a wide variety of products in short runs – and that is where robots should be the answer.

Therefore, many researchers are working on robot languages that will overcome this problem, particularly since it will be impractical to teach robots with multiple sensors on-line. The aim is that the same language can be used at the robot, at the teachbox, and at mini and mainframe computers.

Unimation's VAL, and IBM's AML are two important languages, but at the 4th Conference on Assembly Automation in Tokyo, two new Japanese languages were discussed – STROL/STROLIC, from Tokyo University, and RVL/A from the government-operated Electrotechnical Laboratory, and there was an interesting report on work at the National University of Singapore.

Language types

There are basically three types of robot language. These are:

□ A motion orientated language, in which there is complete correspondence between instructions and the basic robot motions;

□ Task-orientated languages, in which tasks such as 'grasp' or 'release' are described by instructions, which are independent of the robot motions;

□ Job-orientated languages that decide automatically where to grasp a component and where to place it, with the aid of a database describing the job that has to be done.

In theory, all these types can be used to program off-line, but type No. 1 clearly takes more time than No. 2, which in turn takes more time than No. 3. In fact, with the motion-orientated language, only the movements can be programmed off-line normally, the actual positions being taught with the teachbox in the normal way. At present, most robot languages are motion-orientated, but for the next generation of robots, more advanced languages are required, and the aim is to find a standard language.

Hitachi's ARL is a typical nev motion-orientated language. It i Pascal-like, and various robot contro modes can be specified. The robot ca be co-ordinated with peripherals, an is suitable for use with sensors, whil the language can process on/off signal as well as 8bit analogue data.

Pascal was chosen as a base becaus it is a popular language, while th grammar is relatively easy, an Hitachi considered that the bloc concept is inevitable for robot contro However, because Pascal is a difficul language for operators rather tha programmers, though, a simplifie form is provided for end users.

The main functions include MOVE which assumes interpolation, and fo which S is used to indicate speed, an H to indicate that the hand is to b operated. PERFORM invokes move ment of a group of points, the grou being specified by use of the teachbox There is the ON condition to chec condition statements, such as I and WAIT, which are valid only afte the robot motion is complete. ON i valid for one statement only, bu ENABLE and DISABLE are paired a the beginning and end of a span, t

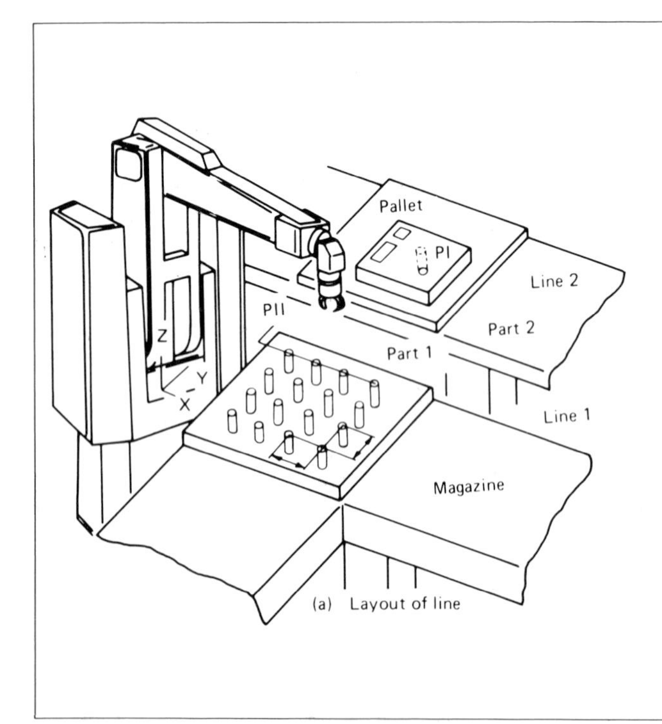

Part 1 (of which 16 pieces are arranged on a magazine) is inserted into Part 2 one piece at a time. It is assumed here that when the magazine and pallet arrive at their preset positions, limit switches for each flip to ON. The robot performs the assembly task while constantly checking that the limit switches are ON. Part 1 is taken in the order of 1 to 16. The two points that must be taught to the robot are P11 and P1.

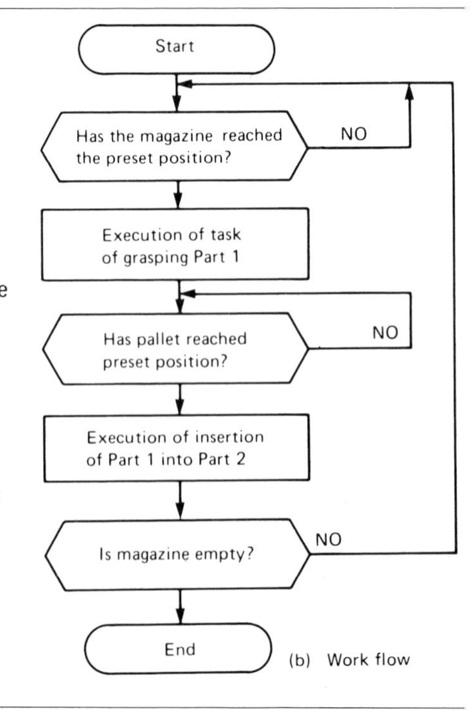

Fig. 1. Hitachi robot with palletising routine in ARL.

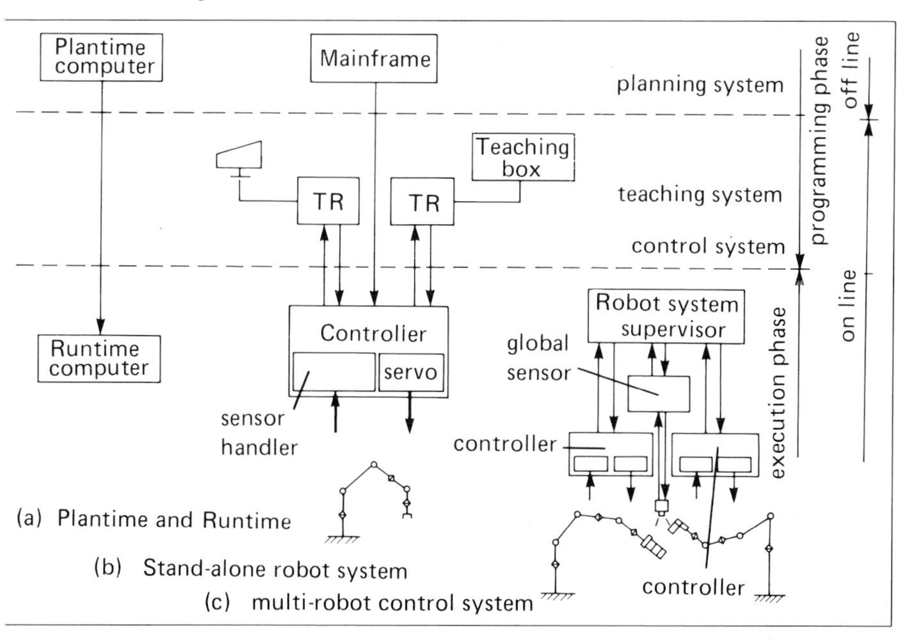

Fig. 2. System architecture of control by STROL/STROLIC.

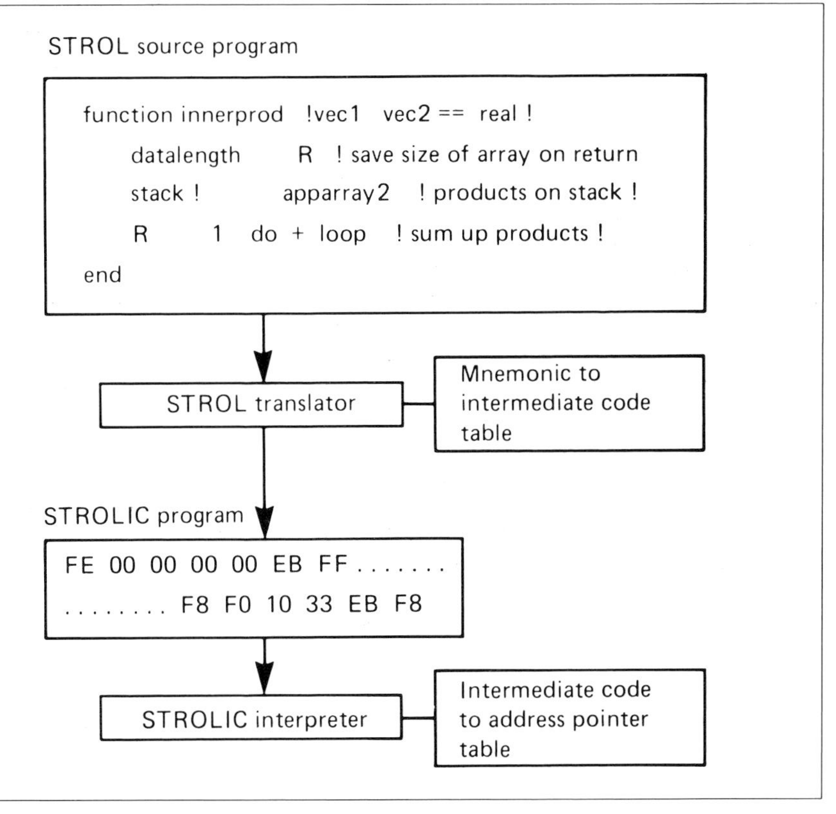

Fig. 3. An example of a STROL function of an inner product.

heck for specified conditions hroughout the span.

ARL has been designed to accommodate changes in input robot anguage, while the structure of the ntermediate language is independent of the robot. Fundamental subroutines re modular, and there is pipeline rchitecture to speed up processing. Standard Fundamental Modules SFM) are used to program the robot so hat independence from special in/out peripherals can be maintained, and to educe the time needed to make the oftware.

Basic motions

In practical use, the main advantage f ARL is that the basic motions can be pre-programmed, while operations such as palletising can be programmed with the minimum of instructions. In fact, to palletise, the motions as the component is raised from the conveyor position vertically taken horizontally then lowered to the first corner of the pallet are defined. The distances between pallet positions in x, y, and z planes are specified. Then, there are instructions to open and close the hand, to repeat motions, and to move in increments from P003 and P004 to the other positions in the pallet. In practice, 21 lines are needed for this routine, irrespective of the size of the pallet, and many are single word instructions.

The aims of STROL/STROLIC are rather broader, and it is being proposed as a standard language to the Japanese Industrial Robot Association, but is just one of several languages being discussed in committee as a possible standard.

According to Tomio Arai and Akihiro Matsumoto of Tokyo University, NC languages such as APT and EXAPT are not suitable for robots, owing to the complex movements involved. Nor are established languages such as Pascal and Fortran specialised enough for a standard robot language. In addition, they said that established languages such as AL/3 and VAL/3 cannot 'adopt sophisticated control algorithms'.

Therefore, the new language is based on the stack architecture of FORTH, and to eliminate the need for parsing, reverse Polish notation is used. It has POP-2 like syntax, and is designed so that the same instructions can be sent by mainframe computer, on-line computer, or teachbox. Since instructions will normally be sent from a computer binary codes are preferred to ASCII codes.

The following specifications were laid down for a standard language at the conceptual stage:

- ○ Cpu independent;
- ○ Robot independent;
- ○ Motion description only;
- ○ Representation of movements in Cartesian co-ordinate form;
- ○ Machine-machine interface by intermediate codes;
- ○ Data structures available;
- ○ Run time interpreter to optimise debugging;
- ○ Unified interface with sensors and servo mechanisms.

High-level

In fact, there are two languages: STROL is the high-level language in which the robot is programmed, and this is automatically converted to STROLIC, which is in machine codes for a virtual computer. STROLIC is the intermediate code language in which actual instructions would be sent from the computer to the robot controller. Arai said that the use of machine code, which of course requires far less memory than a high-level language, is essential for a standard language. It is intended to run on the robot controller, which is seen as a virtual computer with stack architecture.

STROL is claimed to be as powerful as an ordinary programming language, and five functions are prepared as system routines, including MOVE, which moves from current point to destination by PTP control; IMOVE which moves incrementally; MOVES for straight line interpolation, and

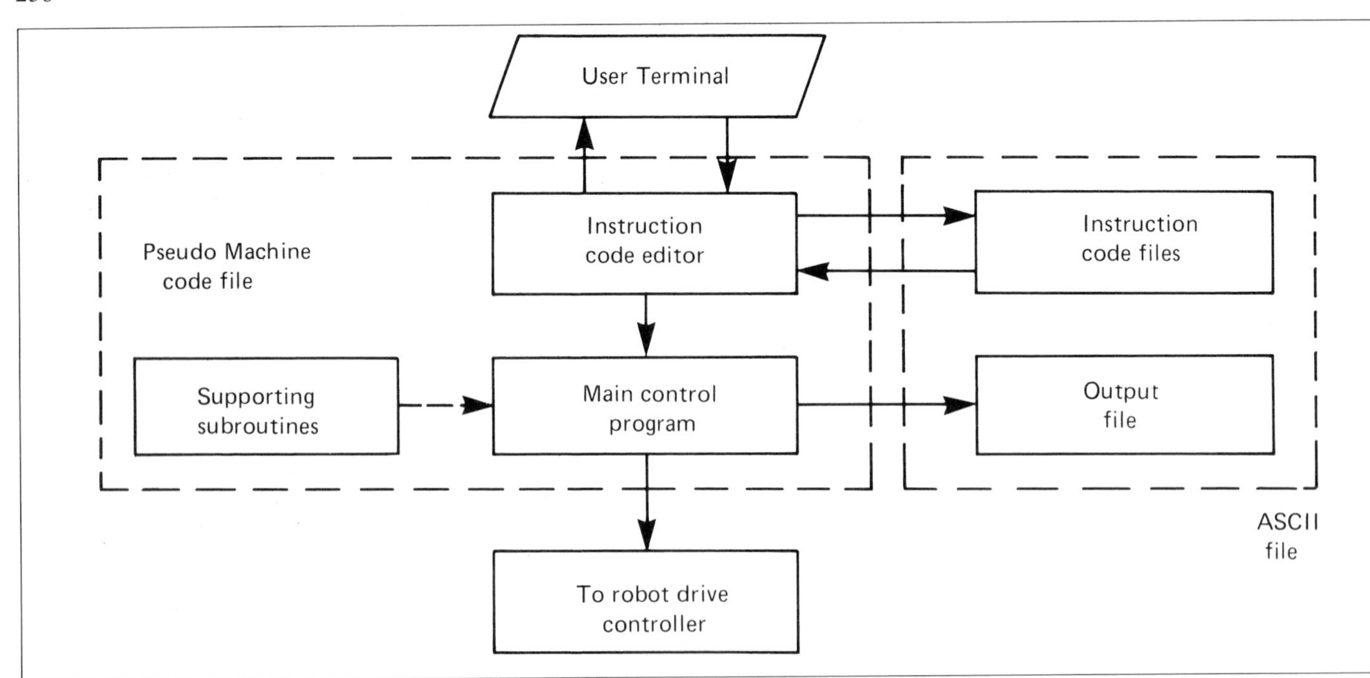

Fig. 4. System software organisation of language developed at the National University of Singapore on an Apple microcomputer.

MOVEC to move to a corner or point vector.

With its dual structure, STROL/STROLIC would seem to have some elements of a standard language, but one problem with attempts at standards, of course, is that they can be overcome by particularly innovative developments, or by the commercial success of a particular product.

Less ambitious were the aims of Wee Wang Loy and Kah Bin Lim of the National University of Singapore. They set about simplifying the teaching and manipulation of a robot with the use of a microcomputer – and Apple II with an extra 16k ram card – and have produced software so that simple and comprehensive instructions can be used to teach and simulate the movements off-line.

First, the program for the kinematic motions of the robot was written for a five-axis articulated type robot, the systematic approach of J. Denavit and R. S. Hartenberg was adopted to relate the joint angles to the links. It is possible, therefore, to teach the robot with reference to a fixed co-ordinate system, rather than by defining joint positions.

To simplify teaching while keeping the software within the limits of the small microcomputer chosen, a short list of instruction sets was selected. They are: MOVE, APRH(approach), PICK, PLAC(e), DRAW, ROTA(te), OPEN, CLOS(e) and HOME. Some of these instructions are very powerful. For example, PICK has five arguments, and causes the robot to move into an intermediate position, open the gripper, approach the object with the required orientation, close the gripper, and then move back to the intermediate position. PLAC is similar, except that the robot places the object held according to the position and orientation specified.

There are two categories of files, those in ASCII code, and those in pseudo-machine code, and the main control program interprets the instructions, and performs the various calculations with the aid of subroutines. Control codes are then sent to the robot controller and output file. The authors claim that to program the robot, the operator needs only to understand the significance and format of each instruction code, know the space co-ordinates of each critical point, and the operation to be performed.

Simple instructions

Toshio Matsushita and Tomomasa Sato of the Electrotechnical Laboratory, Ibaraki, Japan, devised a language for robot vision based on simple instructions. Called RVL/A, it has a 'level of image understanding' but unlike most other vision languages is intended for three-dimensional work. Therefore, it is based on the concept of range data processing.

It was developed so that:

- Object identification and measurement can be described directly;
- Object models which support

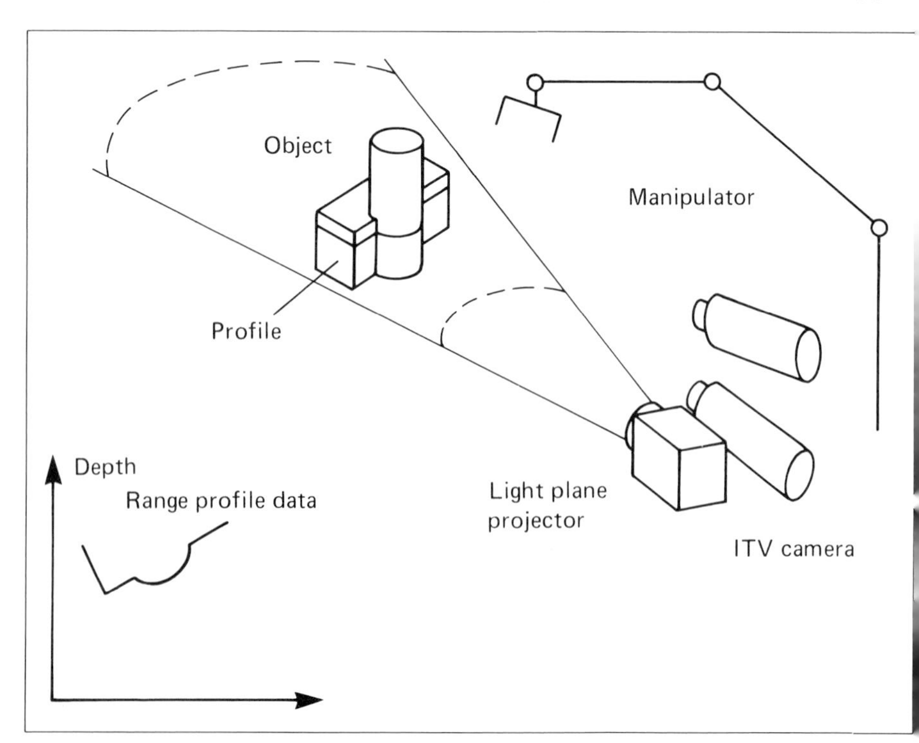

Fig. 5. Method of obtaining range profile data for RVL/A language.

functions for understanding images can be constructed;
- It can deal with three-dimensional data about objects;
- The use of simple visual data and model descriptions permits real-time processing.

It is implemented in PETL, a LISP-like interpreter language, so it can be used interactively. There are three stages in visual data: one-dimensional range profile data stored in a real number array; a profile list, which is the result of analysis of the range profile data; and visual model level expression, which describes the shape and dimensions of the object in the form of an attribute list of PETL.

Number array

In operation, the range profile data is first stored in a number array. Secondly, range profile data is analysed and divided into segments. Then, a visual model level expression is produced. The model of an object is defined by APPEARANCE and VIEW. View is defined by a viewpoint and a light plane projection, while APPEARANCE is a description of the appearance of the range profile when observed from a certain VIEW condition.

Once the model has been defined, the user can write a program to find the object quickly; reference is first made to the description of the shape from that viewpoint, and then the procedure for analysing that form is invoked. If the analysis is successful, the list pointer and item number corresponding to the target object are shown.

However, because RVL/A is an interpreter language, it is not an easy programming language. Therefore, over 120 sub-routines have been written in Fortran for this purpose. Although this language has been used for some experimental work, it is not yet complete, and the researchers say that it is necessary to attempt to model independent of viewpoint and viewpoint control.

As the use of optical and tactile sensors on robots increases, so the need for robot languages will become urgent. Whether standardisation is practical remains to be seen, but at least researchers should aim to adopt standard nomenclature for instructions and statements, so that operators can use the different languages without too much confusion. Whatever happens, as these papers show, progress is being made.

Company Index

Japanese industrial robot specifications

The data presented is aimed at supplying basic information that will act as a guide and a comparison. It cannot be used as an absolute specification as it is based on manufacturers' data and not on independent testing. The following notes will enable the reader to interpret the data presented.

Model The designated number, name or series given by the manufacturer.

Load The maximum load capacity in kilograms (unless otherwise stated) that the robot can carry at its wrist and at the designated arm velocity.

Configuration Refers to the mechanics of the main axes. It may be cylindrical with two linear and one rotational axes, spherical with one linear and two rotational axes, jointed arm which utilises rotary joints to achieve horizontal and vertical motions, and cartesian which is all-linear. Additionally the robot may be equipped with a linear lateral motion. A gantry device operates in cartesian coordinates and is suspended from an overhead structure. A portal device also is suspended but employs some rotary motions.
There are three types of jointed arm: one has joints with horizontal axes which swing in a vertical plane (designated V); the second has vertical axis joints which swing in the horizontal plane (designated H); and the third type is of the SCARA (Selective Compliance Robot Arm) design developed at Yamanashi University which swings in the horizontal plane with a high compliance, but is very stiff in the vertical plane and is well suited to assembly (designated S).

No. axes Is the total number of main and wrist axes available as standard.

Drive system Refers to the method of powering the main axes, other methods *may* be used for grippers and/or wrist.

Control Refers to the method of positional control. It is normally servo control employing encoder, resolver or similar position feedback in either point-to-point (PTP) or continuous path (CP) mode – the latter will involve velocity feedback. It may also be sequential control which is non-servo and often uses a programmable logic controller using limit switches and/or mechanical stops for position control.

Reach Indicates the maximum reach in the horizontal (H), vertical (V) and lateral (L) directions, except where indicated by (T) when just the travel is specified. For cylindrical configurations the maximum horizontal reach can be obtained at any vertical position and vice-versa. For spherical configurations the maximum vertical reach can only be obtained at maximum horizontal reach. For jointed arm types maximum horizontal cannot be achieved at maximum vertical and vice-versa.

Repeatability Is the maximum variation that occurs on repeated returns to a taught point and should not be confused with accuracy. Accuracy is the smallest increment in three-dimensional space which the robot's tool (end-effector) is capable of achieving.

Tool velocity Is the maximum speed of the robot's tool (end-effector) under the load specified. If there are different speeds in different directions these are qualified by the direction – horizontal (H), vertical (V) or lateral (L).

Memory size Indicates the number of positional points or action steps that can be stored in standard memory. Often these can be optionally extended.

Applications Indicates the principal applications of which the machine is capable. Specialist applications or those for which the robot is well qualified by virtue of a large number of applications are indicated in **bold**. Loading indicates loading and unloading of machine tools and presses. Handling is a general category and includes, for example, loading conveyors and implies handling a variety of materials.

Notes, comments These may be specific comments which have a reference number or general either to the robot model or the manufacturer.

Manufacturer	Model	Load (kg)	Configuration	No. axes	Drive system	Control	Reach (mm)	Repeatability (± mm)	Tool velocity (m/s)	Memory size	Applications	Notes, comments
Aida Engineering Ltd., 2-10 Ohyamacho, Sagamihara Kanagawa Tel: 0427 (72) 5231	A-1	3	cylindrical	5	hydraulic	servo PTP	1440H 1320V	0.25	1.5H 0.8V	544 steps	press loading	
Daido Steel Co. Ltd., Machinery Div., 7-13 Nishi Shinbashi 1-chome, Minato-ku, Tokyo Tel: (03) 501 5261	ATR- HA100	150	jointed arm (H)	6	electric dc	servo PTP	2450H 2000V	1.0	0.95V	9800 steps	**handling**	
Daikin Industries Ltd., Robot Systems Division, 700-1 Hitotsuya, Settsu-shi, Osaka 564 Tel: (06) 349-7361	Robotel M3220	15	swing arm	3	hydraulic	mechanical stops	520	0.1	0.22	50 steps ×4	**loading**	
	Robotec W-4400	–	cartesian	4	electric dc	servo CP	325H 200V 5000L	–	0.075	–	arc welding	
Dainichi Kiko Co. Ltd., Kosai Ind. Complex, Kosaimachi, Nakakoma-gun, Yamanashi Tel: 05528 (2) 5581	200H	2	jointed arm (H)	4	electric dc	servo PTP	450H 265V	0.05	1.4H 0.12V			
	200V	2	jointed arm (V)	4	electric dc	servo PTP	450H 800V	0.05	1.7H 1.7V			
	300H	5	jointed arm (H)	4	electric dc	servo PTP	700H 310V	0.1	1.6H 0.1V		assembly, handling, loading	
	300V	5	jointed arm (V)	4	electric dc	servo PTP	700H 1465V	0.1	1.6H 1.6V			
	500	7	jointed arm (V)	5	electric dc	servo PTP/CP	1165H 1045V	0.2	0.4H 0.4H			
	550	7	jointed arm (V)	5	electric dc	servo PTP/CP	1033H 1312V	0.2	1.0H 1.0V			
	Hanbot	10	cylindrical	5	pneumatic	mechanical	850H 1000V	0.2	1.0H 0.6V	–	**press** **loading,** handling	
	PT600	12	jointed arm (V)	5	electric dc	servo PTP/CP[1]	–	0.1	–	250–1000 steps[1]	welding, loading, handling	[1] 4 control systems available, from 'one-axis-at-a-time' PTP to 6 simultaneous axes CP control
	FR10	10	jointed arm (H)	3	electric dc[1]	servo PTP/CP[1]	2870H 150V (T)	0.5	–		plasma cutting, adhesive/ sealant application	
	800	25	jointed arm (V)	5	electric dc	servo PTP/CP	1350[2]	0.5	2.3		assembly, welding, handling, loading	[2] in all directions
	1000	340	jointed arm (V)	6	electric dc	servo PTP/CP	1736H 2775V	0.5	2.5			

Manufacturer	Model	Load (kg)	Configuration	No. axes	Drive system	Control	Reach (mm)	Repeatability (± mm)	Tool velocity (m/s)	Memory size	Applications	Notes, comments
Dainichi Kiko Co. Ltd. *(continued)*	Babot 1440	50	jointed arm (V)	4-6	electric dc[3]	servo PTP	1780H 1525V	0.5	0.6		handling, welding, palletising	[3]fitted with pneumatic counter balance
	Babot 2600	100	jointed arm (V)	4-6	electric dc[3]	servo PTP	2570H 1950H	1.0	0.6			
	Babot 4700	350	jointed arm (V)	4-6	electric dc[3]	servo PTP	2570H 1950V	1.0	0.4			
Daini Seikosha Co. Ltd., Electronics Equipment Div., 31-1 Kameido, 6-chome, Koto-ku, Tokyo 136 Tel: (03) 682-1111	XY-2000	0.5	cartesian	4	electric dc	servo PTP	120H 50V 200L (T)	0.005	1.2	400	assembly	
	TT-2000	1	jointed arm (H)	4	electric dc	servo PTP	330H 150V (T)	0.025	1.6			
	RT-2000	1	cylindrical	4	electric dc	servo PTP	525H 100V (T)	0.01	1.6			
	RT-3000	5	cylindrical	4	electric dc	servo PTP	700H 120V (T)	0.025	1.4			
Fanuc Ltd., 5-1 Asahigaoka 3-chome, Hino-shi, Tokyo 191 Tel: (0425) 84-1111	A0	5	cylindrical	5	electric ac	servo PTP	650H 1065V	0.05	1.2H 0.6V	300 points	**assembly**	
	A1	15	cylindrical	5	electric ac	servo PTP	1190H 1570V	0.1	1.2H 0.6V	6000 points	**assembly,** handling	
	S1	10	jointed arm (V)	5	electric ac	servo	–	0.2	–	6000 points	**welding,** sealing, deburring	
	M0	20	cylindrical[4]	3	electric dc	servo PTP	763[5]	0.5	0.5	300 points	**loading**	[4]mounts on machine tool [5]measured from mounting point
	S2	20[6]	jointed arm (V)	5[6]	electric ac	servo CP		1.0		6000 points	**arc welding,** sealing, flame cutting	[6]15kg with 6 axes, 7th axis available of 1, 2 or 3m travel
	M1	47[7]	cylindrical	3[7]	electric ac	servo PTP	2160H 1450V	1.0	1.0H 0.5V	300 points	**loading,** assembly, handling	[7]47kg load only in 3-axis form, 31kg with 5-axis robot
	M2	60	cylindrical[8]	4	electric ac	servo PTP	300V (T) 300L (T)	1.0	–	–	**loading**	[8]lateral and vertical with 180° rotation about horizontal
	M3	80	cylindrical	5	electric ac	servo PTP	2035H 1850H	1.0	1.0H 0.5V	300 points	**loading,** handling	
	S3	60[9]	jointed arm	5[9]	electric ac	servo	–	1.0	–	–	**spot welding,** handling, loading	[9]60kg load only in 5-axis form, 40kg in 6-axis form lateral travel of 1, 2 or 3m available

Manufacturer	*Model*	*Load (kg)*	*Configuration*	*No. axes*	*Drive system*	*Control*	*Reach (mm)*	*Repeatability (± mm)*	*Tool velocity (m/s)*	*Memory size*	*Applications*	*Notes, comments*
Fujitsu Ltd., Industrial Engng Div., 1015 Kamiodanaka, Nakahara-ku, Kawasaki-shi, Kanagawa-ken Tel: (044) 777-1111	Farot S4B	3	jointed arm (S)	4	electric dc	servo PTP	550H 80V (T)	0.05	0.68H	1000 steps	**assembly**	
	Farot S4C1	3	jointed arm (S)	4	electric dc	servo PTP	550H 50V (T)	0.05	0.9H	850 steps	**assembly**	
Hikawa Industry Co. Ltd., Kinomoto 49-2, Miyoshicho, Nishi Kamo-gun, Aichi Tel: 05613(4) 1611	HS-U	5	jointed arm (V)	5	electric dc	servo PTP/LP	984H 1106V	0.3	–	1200 steps	**arc welding,** handling, sealing	
Hirata Industrial Machines, 5-4 Myotaiji-machi, Kumamoto 860 Tel: 0963-44-2266	AR-250	2	jointed arm (S)	4	electric dc	servo PTP	450H 100V (T)	0.05	2.09			
	AR-300	2	jointed arm (S)	4	electric dc	servo PTP	575H 100V (T)	0.05	1.7H 0.25V	500 points	**assembly**	
	AR-450	2[10]	jointed arm (S)	4[10]	electric dc	servo PTP	695H 150V (T)	0.05	1.95	1000 steps		[10]load with 4 axes; 3 axes-4kg, 2 axes-10kg
Hitachi Ltd., Industrial Components and Equipment Div., 6 Kanda Surugadai, 4-chome Chiyoda-ku, Tokyo Tel: 03 (258) 1111	A4010	1	jointed arm (H)	4	stepper motor	servo PTP	500H 150V (T)	0.2	0.73	5 points 25 steps	assembly	
	A3020	2	horizontal jointed arm (S)	4	electric dc	servo CP	700H 200V (T)	0.05	1.5	500 steps	**assembly**	
	A6030	3	jointed arm (H)	6	electric dc	servo CP	800H	0.1	1.5	500 steps	assembly	
	SPR-6V	5	jointed arm (V)	6	hydraulic	servo PTP/CP	2000H 3145V	–	1.75	1000 steps[11]	**surface coating**	[11]bubble memory
	Mr Aros	–	cartesian	5	hydraulic	servo CP	1000H 1100V 1500L (T)	1.0	0.1 (F) 0.01 (W)	1000 steps[11]	**arc welding**	(F) fast traverse (W) weld speed
	Process robot	10	jointed arm (V)	5	electric dc dc	servo CP	1257H 1588V	0.2	1.0	2000 steps	**welding,** sealing, loading, assembly	
Ishikawajima-Harima Heavy Industries Co. Ltd., F. Project Group, Control Systems Div., 2-9-7 Yaesu, Chuo-ku, Tokyo Tel: (03) 277-4161	FE-10	10	jointed arm (V)	5	electric ac	servo CP	1306H 1609V	0.2	1.0	800 steps	**arc welding**	floor, overhead, or wall mounted

Manufacturer	Model	Load (kg)	Configuration	No. axes	Drive system	Control	Reach (mm)	Repeatability (± mm)	Tool velocity (m/s)	Memory size	Applications	Notes, comments
Kawasaki Heavy Industries Ltd., Hydraulic Machinery Div., 4-1 Hamamatsu-cho 2-chome, Minato-ku, Tokyo Tel: (03) 435-6583	9653	5	jointed arm (V)	6	hydraulic	servo PTP/CP	2116H 2900V	1.0	–	–	**surface coating**	
	3000 series Unimate[12]	50	cartesian	4	hydraulic	servo PTP	750H 300V 1500L (T)	1.0	0.5 in all axes	510 steps	**spot welding,** palletising, handling	[12]Kawasaki also builds US designed Unimates under licence
Kayaba Industry Co. Ltd., Engineering Administration Dept., 4-1 Hamamatsu-cho, 2-chome, Minato-ku, Tokyo Tel: (03) 435 3511	KMR200	40	cylindrical	4	hydraulic	servo PTP	1185H 1485V	1.0		32 steps	loading	
	KMR300	30	cylindrical	3	hydraulic	servo PTP	710H 1820V	1.0	0.8H 0.5V	32 steps	handling, palletising	
	KAR 5000	150	jointed arm (V)	3	hydraulic	servo PTP	2500H 2100V	2.0	0.8H 0.6L	500 steps	handling	
Kobe Steel Ltd., Welding Div., Daiichi-Tekkon Bld., 1-8-2 Marunouchi, Chuo-ku, Tokyo Tel: (03) 218-7013	Arcman	5	jointed arm (V)	6	hydraulic	servo CP	2560H 1220V	1.0	0.08[13]	–	**arc welding**	[13]weld speed; also agent for Trallfa
Komatsu Ltd., Research and Development Dept., 3-6 Akasaka 2-chome, Minato-ku, Tokyo Tel: (03) 584 7111	RCA70	70	cylindrical[14]	6	hydraulic	servo PTP	2250H 1400V 1500L	1.0	1.5H 0.5V 1.0L	550 steps	**forging,** machining	[14]plus linear lateral travel
Matsushita Industrial Equipment Co. Ltd., 1-1, 3-chome, Inazu-cho, Toyonaka-city, Osaka Tel: (06) 862 1121	Pana Robo AW-0400	5	jointed arm (S)	4	electric	servo CP	890H 80V (T)	0.1	–	500 steps	arc welding	
	Pana Robo AW1000	10	jointed arm (V)	5	electric dc	servo CP	1320H 1870V	0.2	–	1000 steps	**welding**	
	RW2000	30	cartesian	3	electric ac	servo PTP	500H 300V 800L (T)	2.0	0.2	253 steps	**welding**	

Manufacturer	Model	Load (kg)	Configuration	No. axes	Drive system	Control	Reach (mm)	Repeatability (± mm)	Tool velocity (m/s)	Memory size	Applications	Notes, comments
Matsushita Electric Industrial Co. Ltd., Manufacturing Equip. Div., 2-7 Matsuba-cho, Kadoma-shi, Osaka 571 Tel: (06) 901 1171	Pana Robo CM	1	cylindrical	4	electric	servo PTP	462H 50V (T)	0.1	0.8H 0.2V	–	**assembly**	
	Pana Robo HI	5	jointed arm (H)	3	electric	servo PTP	700H	0.05	1.78	–	**assembly**	
	Pana Robo HII	10	jointed arm (H)	4	electric	servo PTP	1005H 200V (T)	0.05	2.0	–	assembly	smaller versions also 5kg capacity, 100 travel, 1.6 speed
	Pana Robo VI	5	jointed arm (V)	6	electric	servo PTP/CP	900H 1500V	0.1	–	–	assembly	
	Pana Robo VII	10	jointed arm (V)	6	electric	servo PTP/CP	900H[15]	0.2	–	–	assembly, handling	[15]in both directions
Meidensha Electric Mfg. Co. Ltd., Mech-Electronics Div., 1-17, 2-chome Ohsaki, Shinagawa-ku, Tokyo 141 Tel: (03) 492-1111	GHR25S	25	jointed arm (V)	5	electric dc	servo CP	1361H 2372V	0.2	1.6	1280 steps	handling	
	GHR50S	50	jointed arm (V)	5	electric dc	servo CP	1702H 2774V	0.2	1.3	1280 steps	handling	
	GHR100S	100	jointed arm (V)	5	electric dc	servo CP	1895H 3108V	0.3	0.8	1280 steps	handling	
	GHR200S	200	jointed arm (V)	5	electric dc	servo CP	1895H 3108V	0.3	0.4	1280 steps	handling	
Mitsubishi Electric Corp., Industrial Machinery Dept., 2-3 Maranouchi 2-chome, Chiyod-ku, Tokyo Tel: (03) 218-2164	RW-1	10	jointed arm (V)	5	electric dc	servo CP	1280H 1768V	0.2	1.0	1000 steps	**welding**	
	RV-131	2	jointed arm (V)	6	electric dc	servo CP	600H	0.05	1.5	1000 points	assembly	
	RV-242	60	jointed arm (V)	6	electric dc	servo PTP	2516H 2456V	0.1	2.0	1000 points	spot welding	
Mitsubishi Heavy Industries Co. Ltd., Precision Machinery Div., 5-1 Maranouchi 2-chome, Chiyada-ku, Tokyo Tel: (03) 212-3111	Painting robot	5	jointed arm	6	hydraulic	servo PTP/CP	2200H 2230V	2.0	2.0	20 minutes	**surface coating**	
	Robitus RD	25	jointed arm (V)[16]	5/6	hydraulic	servo PTP	870H 1410V 2000L (T)[16]	1.0		460 steps	**spot welding**	[16]with lateral travel, no rotation
	Robitus RC-RH	30	cartesian	6	hydraulic	servo PTP/CP	1450H 1300V 2000L (T)	1.0	0.4	1433 steps	**spot welding**	
	Robitus RC/RL	60	cartesian	up to 6	hydraulic	servo PTP	1200H 1650H 500-1500L (T)[17]	1.0	0.4H 0.3V 0.4L	200 steps	**spot welding**	[17]lateral travel in 500 mm increments

Manufacturer	*Model*	*Load (kg)*	*Configuration*	*No. axes*	*Drive system*	*Control*	*Reach (mm)*	*Repeatability (± mm)*	*Tool velocity (m/s)*	*Memory size*	*Applications*	*Notes, comments*
Nachi-Fujikoshi Corp., Machine Tool Div., 4-1 Hamamatsu-cho 2-chome, Minato-ku, Tokyo Tel: (03) 435-5031	5000	5	jointed arm (V)	6	hydraulic	servo CP	1950H 2430V	2.0	1.75	500 points	**surface coating**	
	200	2	jointed arm (S)	4	electric dc	servo PTP	360H 60V (T)	0.03	1.27	700 points	assembly	
	400	6	jointed arm (S)	4	electric dc	servo PTP	650H 50V (T)	0.05	1.27	700	assembly	
	600	30	jointed arm (S)	4	electric dc	servo PTP	1230H 100V (T)	0.05	1.1	660 points	assembly handling	
	7000	10	jointed arm (V)	5	electric dc	servo PTP	1190H 1650V	0.3	0.78H 1.0V	1000 points	**arc welding, sealing**	
	8000	50	jointed arm (S)	6	electric dc	servo PTP	1936H 2157V	0.5	–	1000 points	spot welding, handling	
Nippon Electric Co. Ltd., Production Facilities Dev't Div., 1-17 Shibaura 2-chome, Minato-ku, Tokyo Tel: (03) 451 5131	B-10	1	jointed arm (S)	3	electric dc	servo PTP	600H 200V (T)	0.05	–	768 steps	**assembly**	
	B-50	5	jointed arm (S)	3	electric dc	servo PTP	800H 300V (T)	0.05	–	768 steps	**assembly**	
Nippon Robot Machine Co., 73 Yohge Nihougi-cho, Anjon City, Aichi Pref. Tel: 0566 (74) 1101	NR-605RH	3	cartesian	5	pneumatic	sequential	1200H	1.0	0.8	–	**die casting take out**	
Nitto Seiko Co. Ltd., Assembly Machine Div., 20Inokura, Ayabe-city, Kyoto 623 Tel: (0773) 42-3111	Picmat PRO 221	6	jointed arm (S)	4	electric dc	servo PTP	410H 100V (T)	0.05	–	–	**assembly**	
	Picmat PRO 423	10	jointed arm (S)	4	electric dc	servo PTP	650H 100V (T)	0.05	–	–	**assembly**	
	Picmat	30	jointed arm (S)	4	electric dc	servo PTP	1030H 1000V (T)	0.05	–	–	**assembly**	
Osaka Transformer Co. Ltd., 1-11 Tagawa 2-chome, Yodogawa-ku, Osaka Tel: (06) 301-1212	Thor-T	3	cartesian	5	electric dc	servo CP	100H 500V 500L (T)	0.1	0.2	1000 steps	**arc welding**	
	Thor-K	–	jointed arm (V)	5	electric dc	servo CP	1450H 1465V	0.2 0.2	–	–	**arc welding**	
Pentel Co. Ltd., Production Dept., 1-8 Yoshi-cho 4-chome, Sohka City, Saitama Pref. 340 Tel: 0489 (22) 1111	Puha 1	3	jointed arm (S)	4	electric dc	servo PTP	246H 75V (T)	0.05	0.8	152 points	**assembly**	

Manufacturer	*Model*	*Load (kg)*	*Configuration*	*No. axes*	*Drive system*	*Control*	*Reach (mm)*	*Repeatability (± mm)*	*Tool velocity (m/s)*	*Memory size*	*Applications*	*Notes, comments*
Sanki Engineering Co. Ltd., Industrial Plant and Engineering Div., 1-4 Yurakucho 1-chome, Chiyoda-ku, Tokyo Tel: (03) 501-6111	Uniarm	8[18]	jointed arm (H)	7[18]	electric dc	servo PTP	1240H 300V	0.05	0.4H 0.5V (T)	1000 steps 670 points	assembly	[18]5-axis version 20kg load
Sankyo Seiki Mfg. Co. Ltd., 1-17-2 Shinbashi 1 chome, Minato-ku, Tokyo Tel: (03) 508-1156/(03) 591 6886	Skilam SR-3	6	jointed arm (S)	4	electric dc	servo PTP	650H 75V (T)	0.05	1.0	300 points	**assembly**	also marketed by IBM
Shinko Electric Co. Ltd., Electronic and Precision Products Group, 12-2, 3-chome, Nihonbashi, Chuo-ku Tokyo Tel: (03) 274-1111	SR-5	5	cylindrical	3	pneumatic	sequential/ mechanical	1070H 1085V	0.5	1.0H 0.5V	–	**press loading**	
	SR-10	10	cylindrical	4	pneumatic	mechanical	901H 1270V	0.5	0.5H 0.3V	–	**press loading**	
	SRH-10	10	cylindrical	5	hydraulic	servo PTP	1325H 1355V	1.0	0.8	–	handling	
	SR-25	25	cylindrical	4	pneumatic	sequential[19]	924H 1470V	1.0	0.5H 0.3V	120 steps[19]	handling, loading	[19]pin board memory
Shinmeiwa Industry Co. Ltd., Machinery and Plant Div., 1-1 Shinmeiwa-cho, Takarazuka City, Hyogo Pref. Tel: (0798) 52-1234	RJ-65	10	jointed arm (V)	5	electric dc	servo PTP	1357H 1947V	0.2	–	850 steps	**arc welding**	
Taiyo Ltd., 1-1 Kita-Eguchicho, 1 chome, Higashi-Yodogawa-ku, Osaka Tel: (06) 340-1111	Toffky	–	cartesian	3	electric dc	servo PTP	500H 200V 750L (T)	0.05	0.1-0.15	254 steps	**sealant application**	
	N-25	25	cylindrical	5	hydraulic	sequential[20]	1030H 1400V	3.0	0.5	24 steps	handling	[20]pin board memory
Tokico Ltd., Equipment Export Dept., Hitachi Kamakurabashi Bldg., 1-1-14 Uchikanda, Chiyoda-ku Tokyo 101 Tel: (03) 258-0431	105	5	jointed arm (V)	5	hydraulic	servo CP	1400H 1785V	2.0	0.8	–	**surface coating**	
	155/856[21]	5	jointed arm (V)	5/6[21]	hydraulic	servo CP	2090H 2286V	2.0	1.0	–	**surface coating**	[21]155-5 axis 856-6 axis

Manufacturer	*Model*	*Load (kg)*	*Configuration*	*No. axes*	*Drive system*	*Control*	*Reach (mm)*	*Repeatability (± mm)*	*Tool velocity (m/s)*	*Memory size*	*Applications*	*Notes, comments*
Tokyo Keiki Co. Ltd., 2-16 Minami Kamata, Ohta-ku, Tokyo Tel: (03) 732-2111	Matbac Irb-10	15	cylindrical	5	hydraulic	servo PTP[22]	1530H 1067V	2.0[22]	0.8H 0.2V	1200 steps[22]	**palletising,** handling	[22]alternative sequential on-off control available, 0.75 repeatability, 28 step memory
Toshiba Seiki Co. Ltd., 14-33 Higasi Kashiwagakya 5-chome, Ebina City, Kanagawa Pref. Tel: (0462) 31-8111	TSR-701H	8	jointed arm (S)	4	electric dc	servo PTP	650H 125V (T)	0.05	1.5	512 points	**assembly**	
	TSR 500V	2.5	jointed arm (V)	6	electric ac	servo PTP	–	0.05	2.0	512 points	assembly (electric)	
	IX-15S	20	spherical	5	hydraulic	servo PTP	2103H 2350V	1.0	0.7H	512 points	**spot welding,** loading, palletising	
Toyoda Machine Works Ltd., Robotics Department, 1 Asahimachi 1-chome, Kariya-City, Aichi-Pref. Tel: (0566) 22-2211	Friend RA4-1	2.5	jointed arm (V)	4	electric	servo CP	619H 670V	0.05	–	–	assembly, handling	
	Friend RA4-2	6	jointed arm (V)	4	electric	servo CP	923H 1000V	0.05	–	–	assembly, handling	
	Friend RA6-2	6	jointed arm (V)	6	electric	servo CP	870H[23] 1680V	0.05	–	–	handling, arc welding	[23]in both directions
	Friend RC2-2	6	jointed arm (S)	3	electric	servo CP	803H	0.05	–	–	assembly	
	Friend RC4-2	6	jointed arm (S)	4	electric	servo CP	803H 125V (T)	0.05	–	–	assembly	
Yaskawa Co. Ltd., Robot and Material Handling Div. 16-9 Uchikanda 2-chome, Chiyado-ku, Tokyo Tel: (03) 256-7271	L3	3	jointed arm	5	electric dc	servo PTP/CP	919H 1215V	0.1	1.0	1000 steps	**arc welding,** assembly, sealant application	
	L10	10	jointed arm	5	electric dc	servo PTP	1186H 1719V	0.2	0.8H 1.1V	1000 steps	**arc welding,** sealant application, handling	
	S-30	30	jointed arm (H)	3	electric dc	servo CP	1000H 1250H 759V	0.3	1.3	1000 steps	welding, plasma cutting	

Japanese industrial robot manufacturers and overseas distributors

Japan	*UK*	*USA*	*France*	*Germany*	*Other*
Aida Engineering Ltd., Automatic Machine Dept., 2-10 Ohyamacho, Sagamihara, Kanagawa-ken 229 (Tel: 0427 (72) 5231)				**Chori GmbH,** *Graf Adolfstrasse 59,* *4 Dusseldorf*	**Chori Co. Ltd.,** *Jernbanegade 17,* *Copenhagen,* *4 DK-1608 Denmark*
Daido Steel Co. Ltd., 7-13 Nishi Shinbashi 1-chome, Minato-ku, Tokyo (Tel: 03/501 5261)		**Daido Steel,** *Suite 342 East,* *Pan American Building,* *200 Park Avenue,* *New York, NY 10017*			
Daikin Industries Ltd., Robot Systems Division, 700-1 Hitotsuya, Settsu City, Osaka (Tel: 06/349 7361)					
Daini Seikosha Co. Ltd., 31-1, 6-chome, Kameido, Koto-ku, Tokyo 136 (Tel: (03) 682 1111)	**Airstead Industrial Systems Ltd.,** *New England House,* *New England Street,* *Brighton BN1 4GH* *(Tel: 0273 689793)*	**Seiko Instruments USA Inc.,** *2990 W. Lomita Blvd.,* *Torrance, CA 90505* *(Tel: (213) 530-8777)*			
Dainichi Kiko Ltd., Kosai Ind. Complex, Kosaimachi, Nakakoma-gun, Yamanashi (Tel: 05528 (2) 5581)	**Dainichi-Sykes Robotics,** *Walton Summit Centre,* *Bamber Bridge, Preston,* *Lancs PR5 8AE* *(Tel: 0772 322444)*	**GCA/PAR Systems,** *3460 Lexington Avenue N,* *St Paul, MN 55112* *(Tel: 612 484-7261)*			
Fanuc Ltd., Engineering Administration Dept., 5-1 Asahigaoka 3-chome, Hino City, Tokyo 191 (Tel: 0425/841111)	**600 Fanuc Robotics Ltd.,** *Hythe Station Road,* *Colchester,* *Essex* *(Tel: 0206 868848)*	**GMF Robotics Corp.,** *5600 New King Street,* *Troy, MI 48098* *(Tel: 313 641-4222)*	**Halbroun Freres,** *4 Rue Jean Moulin,* *94130 Nogent-sur-Marne* *(Tel: (1) 873-69-45)*	**Siemens AG,** *Postfach 103,* *D 8000 Munich 1*	
Fuji Electric Co. Ltd., Hibiya Park Building, 8-1 Yurakucho 1-chome, Choyoda-ku, Tokyo (Tel: 03-201 1971)	**Haynes & Fordham,** *Unit 4,* *Moorfield Industrial Estate,* *Yeadon, Leeds LS19 7BM* *(Tel: 0532 507090)*				

Japan	UK	USA	France	Germany	Other
Fujitsu Ltd., Industrial Engineering Division, 1015 Kamiodanaka, Nakahara-ku, Kawasaki-shi, Kanagawa-ken (Tel: (044) 777-1111)					
Harmo Japan, 7621-10 Fujizuka, Nisha-minoco, Ina-City, Nagano P-ec. (Tel: 399 45)	**Harmo Robots Ltd.,** *Unit 5, Babbage Road,* *Cavendish Trading Estate,* *Stevenage, Herts.* *(Tel: 0438 728444)*				
Hikawa Industry Co. Ltd., Kinomoto 49-2, Miyoshicho, Nishi Kamo-gun, Aichi (Tel: 05613 (4) 1611)					
Hirata Industrial Machines, 5-4 Myotaiji-machi, Kumamoto 860, (Tel: 0963-44-2266)	**Airstead Industrial Systems Ltd.,** *New England House,* *New England Street,* *Brighton BN1 4GH* *(Tel: 0273 689793)*	**Hirata Corp. of America,** *Indianapolis* *(Tel: 317 846 8859)*			
Hitachi Ltd., Industrial Components & Equipment Division 6 Kanda Surugadai, 4-chome, Chiyoda-ku, Tokyo (Tel: 03 (258) 1111)	**GEC Robot Systems Ltd.,** *Boughton Road,* *Rugby CV21 1BD* *(Tel: 0788 2144)* **Lansing Industrial Robots** *Kingsclere Road,* *Basingstoke,* *Hants.* *(Tel: 0256 3131)*	**Hitachi America,** *6 Pearl Court,* *Allendale, NJ 07401* *(Tel: 201 825 8000)* **General Electric,** *Automation Systems,* *1285 Boston Avenue,* *Bridgeport, CN 06602* *(Tel: 203 382 2876)* **Automatix Inc.,** *1000 Tech Park Drive,* *Billerica, MA 01821* *(Tel: 617 667 7900)*		**Hitachi Europe GmbH,** *Jaegerhofstr 32,* *D-4000 Dusseldorf*	
Ishikawajima-Harima Heavy Industries Co. Ltd., F. Project Group, Control Systems Division, 2-9-7 Yaesu, Chuo-ku, Tokyo (Tel: 03 277-4161)					
Kawasaki Heavy Industries Ltd., Hydraulic Machinery Division, 4-1 Hammatsu-cho 2-chome, Minato-ku, Tokyo (Tel: 03/435 6853)	**Unimation (Europe) Ltd.,** *Unit C, Stafford Park 18,* *Telford, Salop* *(Tel: 0952 618931)*	**Unimation Inc.,** *Shelter Rock Lane,* *Danbury, CT 06810* *(Tel: 203 744 1800)*			

Japan	UK	USA	France	Germany	Other
Kayaba Industry Co. Ltd., Engineering Administration Dept., 4-1 Hammatsu-cho 2-chome, Minato-ku, Tokyo (Tel: 03/435 3511)					
Kobe Steel Ltd., Machinery & Engineering Division, 1-8-2 Marunouchi 1-chome, Chiyoda-ku, Tokyo (Tel: 03/218 7013)					
Komatsu Ltd., 3-6 Akasaka 2-chome, Minato-ku, Tokyo (Tel: 03/584 7111)					
Marol Company Ltd., 1.34 2-chome, Ohashi-cho, Nagata-ku, Kobe (Tel: (078) 611 2151)	**Telehoist Ltd.,** *Manor Road,* *Cheltenham* *Glos. GL51 9SH* *(Tel: 0242 21355)*		**CGMS,** *98 rue d'Ambert, BP 1825,* *45008 Orleans Cedex* *(Tel: (38) 86 2514)*		
Matsushita Industrial Equipment Co. Ltd., 1-1, 3-chome, Inazu-cho, Toyonaka-city, Osaka (Tel: 06 862 1121)		**The DeVilbiss Co.,** 300 Phillips Ave., P.O. Box 913, Toledo, OH 43692, (Tel: (419) 470 2169)			
Matsushita Electric Industrial Co. Ltd., Manufacturing Equipment Division, 2-7 Matsuba-cho, Kadoma-shi, Osaka 571 (Tel: 06 901 1171)					
Meidensha Electric Manufacturing Co. Ltd., Mech-Electronics Division, 1-17, 2-chome Ohsaki, Shinagawa-ku, Tokyo 141 (Tel: 492/1111)					
Mitsubishi Electric Corp., Engineering Dept., 2-3 Marunouchi 2-chome, Chiyoda-ku, Tokyo (Tel: 03 (218) 2164)	**Fairey Automation Ltd.,** *Techno Trading Estate,* *Bramble Road,* *Swindon SN2 6HB* *(Tel: 0793 615111)*				
Mitsubishi Heavy Industries Co. Ltd., Precision and Machinery Division, 5-1 Marunouchi 2-chome, Chiyoda-ku, Tokyo (Tel: 03/212 3111)	**Mitsubishi Corp.,** *Bow Bells House,* *Broad Street,* *London EC4M 9BQ* *(Tel: 01-248 8321)*	**Mitsubishi Corp.,** *277 Park Ave.,* *New York, NY 10017* *(Tel: 212 826 2188)*			**Voest Alpine AG,** *Postfach 2,* *A 4010 Linz,* *Austria* *(Tel: 0732 5851)*

Japan	UK	USA	France	Germany	Other
Motoda Electronics Co. Ltd., 32-9 Kamikitazawa 4-chome, Setagaya-ku, Tokyo (Tel: 03/303 8491)			**CGMS,** *98 rue d'Ambert BP 1825,* *45008 Orleans Cedex,* *(Tel: (38) 86 25 14)*		
Nachi-Fujikoshi Corporation, Machine Tool Division, World Trade Centre, 4-1 Hammatsucho, 2-chome, Minato-ku, Tokyo (Tel: 03/435 5031)	**Nachi (UK) Ltd.,** *70 Cardigan Street,* *Gosta Green, Birmingham* *(Tel: 021 359 5011)*	**Nachi American Inc.,** *223 Veterans Bld.,* *Carlstradt, NJ 07072* *(Tel: 0201 935 8620)*			
Nippon Electric Co. Ltd., Production Facilities Development Division, 1-17 Shibaura 2-chome, Minato-ku, Tokyo (Tel: 03/451 5131)		**NEC America,** *277 Park Avenue,* *New York, NY 10017* *(Tel: 212 758 1666)*			
Nippon Robot Machine Co. Ltd., 73 Yohge Nihongi-cho, Anjoh City, Aichi Pref. (Tel: 0566/74 1101)					
Nitto Seiko Co. Ltd., Assembly Machine Division, 20 Inokura, Ayabe-city, Kyoto 623 (Tel: 0773 42-3111)		**VSI Automation Assembly Inc.,** *165 Park St.,* *Troy, MI 48084*			
Okamura Corporation, Uragocho 5-chome, Yokosuka City, Kanagawa Pref. (Tel: 0468/65 8201)		**C Itoh America Inc.,** *21415 Civic Center Drive,* *Southfield, MI 48076* *(Tel: 313 352 6250)*			
Orii Corporation, 22-12 Haginaka 3-chome, Ohta-ku, Tokyo					
Osaka Transformer Co. Ltd., 1-11 Tagawa 2-chome, Yodogawa-ku, Osaka (Tel: 06/301 1212)	**Norman Butters Co. Ltd.,** *P.O. Box 100,* *Coventry CV5 6HW* *(Tel: 0203 71121)*				
Pentel Co. Ltd., Production Department, 1-8 Yoshi-cho 4-chome, Sohka City, Saitama Pref. 340 (Tel: 0489 22 1111)	**Sale Tilney Technology plc,** *Weybridge,* *Surrey KT15 2RH* *(Tel: 0932 48311)*	**Pentel of America,** *2715 Columbia St.,* *Torrance, CA 90503*			

Japan	*UK*	*USA*	*France*	*Germany*	*Other*
Sanki Engineering Co. Ltd., Industrial Plant and Machinery Division, Sales No. 1 Department, Sanshin Building, 4-1 Yurakucho 1-chome Chiyoda-ku, Tokyo (Tel: 03/502 6111)					
Sankyo Seiki Manufacturing Co. Ltd., Machine Tools Division, 1-17-1 Shinbashi 1-chome, Minato-ku, Tokyo 105 (Tel: 03/591 6886/03 508 1156)	**IBM UK,** *P.O. Box 41,* *Portsmouth PO6 3AU*	**IBM,** *1000 NW 51st Street,* *Baco Raton, FL 33432* *(Tel: 305 998 2000)*	**CGMS,** *98 rue d'Ambert BP 1825,* *45008 Orleans Cedex* *(Tel: (38) 86 25 14)*		
Shinko Electric Co. Ltd., 2-12.2 Nihonbashi, Chuo-ku, Tokyo (Tel: 03 274 1111)					
Shinmeiwa Industry Co. Ltd., Machinery and Plant Division, 1-1 Shinmeiwa-cho, Takarazuka City, Hyogo Pref. (Tel: 0798/52 1234		**Nichimen Colne,** *6 North Michigan Ave.,* *Chicago, IL 60602* *(Tel: 312 346 9339)*	**Commercy,** *55200 Commercy* *(Tel: (29) 01 01 04)*		
Shoku Corporation, Engineering Department, 1010 Minorudai, Matsudo City, Chiba Pref. (Tel: 0473/64 1211)				**VFW Fokker GmbH,** *Werk Haykemkamp,* *2870 Delmenhorst* *(Tel: 04221 131)*	
Star Seiki Co. Ltd., 252 Kawachiya Shinden, Komaki City, (Tel: 0568/75 5211)	**Haynes & Fordham,** *Unit 4,* *Moorfield Industrial Estate,* *Yeadon, Leeds LS19 7BM* *(Tel: 0532 507090)*	**Sterltech,** P.O. Box 23421, Milwaukee, WI 53223 (Tel: 414 354 0493)		**DME Deutschland,** *D 7106 Neuenstadt* *(Tel: 07139 511)*	
Taiyo Ltd., 1-1 Kita-Eguchicho, 1-chome, Higashi- Yodogawa-ku, Osaka (Tel: 06 340-1111)	**Haynes & Fordham,** *Unit 4,* *Moorfield Industrial Estate,* *Yeadon, Leeds LS19 7BM* *(Tel: 0532 507090)*		**CGMS,** *98 rue d'Ambert BP 1825,* *45008 Orleans Cedex* *(Tel: (38) 86 25 14)*		
Tokico Ltd., Equipment Export Department, Hitachi Kamakurabashi Building, 1-1-14 Uchikanda, Chiyoda-ku, Tokyo 101 (Tel: 03 258-0431)		**Tokico America Inc.,** *388 W. Lomita Blvd./Suite E,* *Torrance, CA 90505*			

Japan	*UK*	*USA*	*France*	*Germany*	*Other*
Tokyo Keiki Co. Ltd., 2-16 Minami Kamata, Ohta-ku, Tokyo (Tel: 03/732 2111)					
Toshiba Seiki Co. Ltd., Sales Division, 14-33 Higashi Kashiwagaya 5-chome, Ebina City, Kanagawa Pref. (Tel: 0462/31 8111)	**Evershed Robotics Ltd.,** *Bridge Road,* *Chertsey,* *Surrey KT16 8LJ* *(Tel: 09328 61181)*		**CGMS,** *98 rue d'Ambert BP 1825,* *45008 Orleans Cedex* *(Tel: (38) 86 25 14)*	**Datasem GmbH,** *Richthausener strasse 2,* *8501 Winkelhaid* *(Tel: 09187 3222)*	
Toyoda Machine Works Ltd., 1 Asahi-cho, 1-chome, Kariya City, Aichi Pref. (Tel: 0566/22 2211)					
Yaskawa & Co. Ltd., Robot Material Handling Division, 16-9 Uchikanda 2-chome, Chiyoda-ku, Tokyo (Tel: 03/256 7271)	**Torsteknik UK Ltd.,** *1 Swan Industrial Estate,* *Banbury, Oxon.* *(Tel: 0295 61245)* **Hazmac (Handling) Ltd.,** *Norreys Drive,* *Cox Green,* *Maidenhead, Berks SL6 4BY* *(Tel: 0628 34645)*	**Machine Intelligence Corp.,** *330 Potero Ave.,* *Sunnyvale, CA 94086* *(Tel: 408 737 7960)* **Yaskawa Electric America Inc.,** *305 Era Drive,* *Northbrook, IL 60062* *(Tel: 312 564 0770)* **Nordson Corp.,** *555 Jackson Street,* *Amherst, OH 44001* *(Tel: 216 988 9411)*	**SCEMI,** *61 rue de Funas,* *38300 Bourgoin-Jaillieu* *(Tel:(74) 93 20 04)*	**Messer Greisheim GmbH,** *Industriegase,* *Homberger Strasse 12,* *4000 Dusseldorf* *(Tel: (0211) 4303-242)*	**Torsteknik,** *Box 130,* *S 38500 Torsa,* *Sweden*

Japanese industrial robot manufacturers by process application

	Spot welding	Arc welding	Grinding/fettling	Deburring	Plasma/flame cutting	Surface coating	Sealant/glue laying	Assembly	Machine loading	Forging	Palletising	Handling
Aida Engineering Ltd.									●			
Daido Steel Co. Ltd.									●	●	●	●
Daikin Industries Ltd.		●							●			
Dainichi Kiko Ltd.	●	●	●		●		●	●	●		●	●
Daini Seikosha Co. Ltd.								●				
Fanuc Ltd.	●	●		●	●		●	●	●			●
Fujitsu Ltd.								●				
Harmo Japan									●			
Hikawa Industry Co. Ltd.		●					●					●
Hirata Industrial Machines								●				
Hitachi Ltd.		●				●	●	●				●
Ishikawajima-Harima Heavy Industries Co. Ltd.		●										
Kawasaki Heavy Industries Ltd.	●	●			●	●	●	●	●	●		●
Kayaba Industry Co. Ltd.									●		●	●
Kobe Steel Ltd.		●				●						
Komatsu Ltd.			●							●		
Marol Company Ltd.											●	●
Matsushita Industrial Equipment Co. Ltd.		●										
Matsushita Electric Industrial Co. Ltd.								●				●
Meidensha Electric Mfg. Co. Ltd.												●
Mitsubishi Electric Corp.	●	●						●				
Mitsubishi Heavy Industries Co. Ltd.	●					●						
Motoda Electronics Co. Ltd.												●
Nachi-Fujikoshi Corporation	●	●				●	●	●				●
Nippon Electric Co. Ltd.								●				
Nippon Robot Machine Co. Ltd.									●			
Nitto Seiko Co. Ltd.								●				
Okamura Corporation											●	●
Orii Corporation									●			
Osaka Transformer Co. Ltd.		●										
Pentel Co. Ltd.								●				
Sanki Engineering Co. Ltd.								●				●
Sankyo Seiki Mfg. Co. Ltd.								●				
Shinko Electric Co. Ltd.									●		●	●
Shinmeiwa Industry Co. Ltd.		●										
Shoku Corporation									●			
Star Seiki Co. Ltd.									●			
Taiyo Ltd.							●					●
Tokico Ltd.						●						
Tokyo Keiki Co. Ltd.											●	●
Toshiba Seiki Co. Ltd.	●							●	●			●
Toyoda Machine Works Ltd.		●						●				●
Yaskawa & Co. Ltd.		●			●		●	●				●

Source of material

All articles in this book were originally published by IFS (Publications) Ltd. in the quarterly journals: *The Industrial Robot*, *Assembly Automation*, *Sensor Review* and *The FMS Magazine*.

The Industrial Robot

Decade of Robotics

Special Tenth Anniversary issue of *The Industrial Robot*

Assembly Automation

The FMS Magazine

Sensor Review